7th International Conference on Mechatronics and Manufacturing (ICMM 2016)

MATEC Web of Conferences Volume 45 (2016)

Singapore
13 - 14 January 2016

Editors:

Hao Gong
Jae Hoon Lee
Yan Biao

ISBN: 978-1-5108-2211-5

Printed from e-media with permission by:

Curran Associates, Inc.
57 Morehouse Lane
Red Hook, NY 12571

Some format issues inherent in the e-media version may also appear in this print version.

Printed by Curran Associates, Inc. (2016)

For additional information, please contact EDP Sciences – Web of Conferences
at the address below.

EDP Sciences – Web of Conferences
17, Avenue du Hoggar
Parc d'Activité de Courtabœuf
BP 112
F-91944 Les Ulis Cedex A
France

Phone: +33 (0) 1 69 18 75 75
Fax: +33 (0) 1 69 28 84 91

contact-edps@webofconferences.org

Additional copies of this publication are available from:

Curran Associates, Inc.
57 Morehouse Lane
Red Hook, NY 12571 USA
Phone: 845-758-0400
Fax: 845-758-2633
Email: curran@proceedings.com
Web: www.proceedings.com

TABLE OF CONTENTS

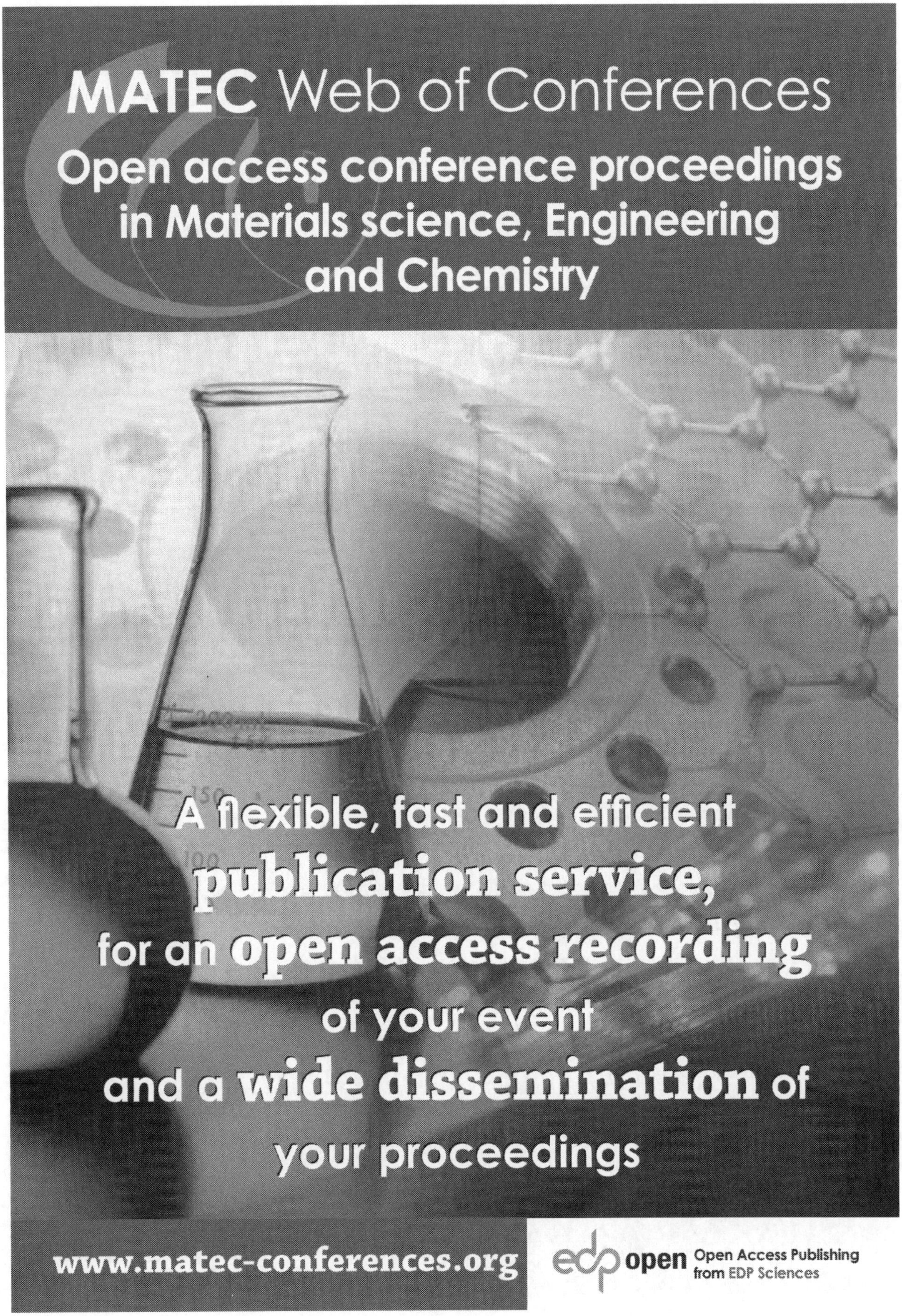
MATEC Web of Conferences
Open access conference proceedings in Materials science, Engineering and Chemistry
A flexible, fast and efficient publication service, for an open access recording of your event and a wide dissemination of your proceedings
www.matec-conferences.org
edp open Open Access Publishing from EDP Sciences

Aims & scope

MATEC Web of Conferences is an open-access journal devoted to the publishing of conference proceedings dealing with Materials science, Engineering and Chemistry.

The journal aims to make conference proceedings visible, citable and widely usable by the scientific community. It offers outstanding services for the publication and dissemination of contents in order to ensure long-term archiving and maximum exposure of scientific events.

Coverage includes all engineering disciplines: civil, naval, mechanical, chemical, and electrical engineering as well as nanotechnology and metrology. The journal also deals with the physical-chemical characterization of materials, their implementation, their resistance in their environment, etc. Others subdisciplines of chemistry, such as analytical chemistry, petrochemistry, organic chemistry, etc., and even pharmacology, are also welcome.

Why publish with MATEC Web of Conferences?

- **Open access:** conference papers available worldwide.
- **Fast publication:** less than 3 months.
- **Total flexibility:** no page restrictions, free unlimited use of colour in the online version, any kind of scientific document welcomed (traditional articles, posters, abstracts and slideshows).
- **High visibility:** all documents indexed in main databases and easily found online.
- **Fair author rights:** CC-BY Creative Commons license.
- **Range of additional publishing options:** printed proceedings, CD-ROMs, copy-editing, etc.

Contact: Isaline Augusto
matec@webofconferences.org

ISSN (electronic edition): 2261-236X

www.matec-conferences.org

edp open Open Access Publishing from EDP Sciences

Introduction

The organizing committee warmly welcomed our distinguished delegates and guests to the 2016 7th International Conference on Mechatronics and Manufacturing (ICMM 2016) held during January 13-14, 2016 in Singapore.

ICMM conference is initiated by American Society for Research, assisted by many other universities and institutes every year. This is the annual conference and 2016 is the seventh year of ICMM conference. The conference committee chairs are Prof Hao Gong, Prof Yan Biao, Prof Jae Hoon Lee and Prof. Vadim Rashitovich Gasiyarov.

The conference was organized to gather members of our international community scientists so that researchers from around the world could present their cutting edge work, expanding our community's knowledge and insight into the significant challenges currently being addressed in that research. The conference program committee was quite diverse and truly international, with membership from the Americas, Europe, Asia, Africa and Oceania.

This proceedings volume contains fully refereed papers presented at the ICMM 2016. The main conference themes and tracks are Mechatronics and Manufacturing. The main goal is to provide international scientific forums for exchange of new ideas in a number of fields that interact in-depth through discussions with their peers from around the world. The conference solicited and gathered technical research submissions related to all aspects of major conference themes and tracks. All of the submitted papers in these proceedings have been peer reviewed by the reviewers drawn from the scientific committee, external reviewers and editorial board depending on the subject matter of the paper. Reviewing and initial selection were undertaken electronically. After the rigorous peer-review process, the submitted papers were selected on the basis of originality, significance, and clarity for the purpose of the conference. The selected papers and additional late-breaking contributions to be presented as lectures made for an exciting technical program. The high quality of the program – which was guaranteed by the presence of an unparalleled number of internationally recognized top experts – can be assessed when reading the contents of the program. The conference was therefore a unique event, where attendees were able to appreciate the latest results in their field of expertise, and to acquire additional knowledge in other fields.

We are grateful to all those who have contributed to the success of ICMM 2016. We would like to thank the program chairs, organization staff, and the members of the program committees for their work. We hope that all participants and other interested readers benefit scientifically from these proceedings and also find it stimulating in the process.

Finally, we hope that you had a unique, rewarding, and enjoyable week at ICMM 2016 in Singapore.

With our warmest regards,

Prof Hao Gong, National University of Singapore, Singapore

Prof Yan Biao, Tongji University, China

Assoc. Prof. Jae Hoon Lee, Ehime University, Japan

Assoc. Prof. Vadim Rashitovich Gasiyarov, South Ural State University , Russia

Committee Members

Conference Chairs

Prof. Hao Gong, National University of Singapore, Singapore

Prof. Vadim R. Gasiyarov, South Ural State University, Russia

Assoc. Prof. Jae Hoon Lee, Ehime University, Japan

Program Chairs

Prof Yan Biao, Tongji University, China

Prof. Zahari Taha, Universiti Malaysia Pahang, Malaysia

Technical Committees

Prof. Asif Iqbal, Department of Mechanical & Aerospace Engineering, Air University, Pakistan

Asst. Prof. Ramin Ghasemiasl, Islamic Azad University, Iran

Dr. Shantha Gamini Jayasinghe, Australian Maritime College, Australia

Assoc. Prof. Ranganath M. S, Delhi Technological University, India

Prof. Ho-Sung Lee, Korea Aerospace Research Institute, Korea

Dr. Andrew Randles, Institute of Microelectronics, Singapore

Dr. YANG Weifeng, Institute of Materials Research and Engineering, Singapore

Dr. ZHANG Hongbao, Institute of Microelectronics, Singapore

Dr. Nguyen-Hieu HOANG, SINTEF, Norway

Asst. Prof. Dr. V. ANANDAKRISHNAN, National Institute of Technology, India

Statement of Peer review

In submitting conference proceedings to MATEC Web of Conferences, I certify to the Publisher that I adhere to the **Policy on Publishing Integrity** of the journal in order to safeguard good scientific practice in publishing.

1. All articles have been subjected to peer review administered by the proceedings editors.
2. Reviews have been conducted by expert referees, who have been requested to provide unbiased and constructive comments aimed, whenever possible, at improving the work.
3. Proceedings editors have taken all reasonable steps to ensure the quality of the materials they publish and their decision to accept or reject a paper for publication has been based only on the merits of the work and the relevance to the journal.

Title, date and place of the conference

2016 7th International Conference on Mechatronics and Manufacturing (ICMM 2016)
January 13-14, 2016 / Singapore

Proceedings editor(s):

Prof. Hao Gong
Assoc. Prof. Jae Hoon Lee
Prof. Yan Biao

Date and editor's signature

MATEC Web of Conferences 45, 01001 (2016)
DOI: 10.1051/matecconf/20164501001

Acoustic Velocity and Attenuation in Magnetorhelogical fluids based on an effective density fluid model

Min Shen[1,2] , Qibai Huang [1]

[1] Huazhong University of Science and Technology in Wuhan,China
[2] Wuhan Textile University in Wuhan,China

Abstract. Magnetrohelogical fluids (MRFs) represent a class of smart materials whose rheological properties change in response to the magnetic field, which resulting in the drastic change of the acoustic impedance. This paper presents an acoustic propagation model that approximates a fluid-saturated porous medium as a fluid with a bulk modulus and effective density (EDFM) to study the acoustic propagation in the MRF materials under magnetic field. The effective density fluid model derived from the Biot's theory. Some minor changes to the theory had to be applied, modeling both fluid-like and solid-like state of the MRF material. The attenuation and velocity variation of the MRF are numerical calculated. The calculated results show that for the MRF material the attenuation and velocity predicted with this effective density fluid model are close agreement with the previous predictions by Biot's theory. We demonstrate that for the MRF material acoustic prediction the effective density fluid model is an accurate alternative to full Biot's theory and is much simpler to implement.

1 Introduction

Magnetorheological fluids (MRFs) are materials whose properties change when an external electro-magnetic field is applied. They are considered as "smart materials" because their physical characteristics can be adapted to different conditions. Hence, MRFs material can be used in design of active sound barriers to control noise transmission loss and for modification of noise absorption characteristics of sound absorbing materials [1].

The MRFs consist of micron size magnetically permeable particles suspended in a carrying liquid such as the mineral or silicon oil. From the point of view of the inner structure, the particles are scattered randomly in the liquid carrier in the absence of magnetic fields shown in Figure.1(a). When a magnetic field is applied to the fluid, the particles magnetized and chain-like structures that align in the direction of the applied field as depicted in Figure.1(b). In the presence of shear strain, the particles aligned as columnar microstructure, in turn dramatically change its acoustic impedance control magnetic field. However, if the magnetic field is removed, the suspension turns to a Newtonian fluid and the transition between these two phases is highly reversible, which provides a unique feature of electric- or magnetic-field controllability of acoustic impedance of MRF. This feature has inspired the design of new active sound barriers to control noise transmission loss or as acoustic metamaterials with negative effective dynamic bulk modulus [2-3].

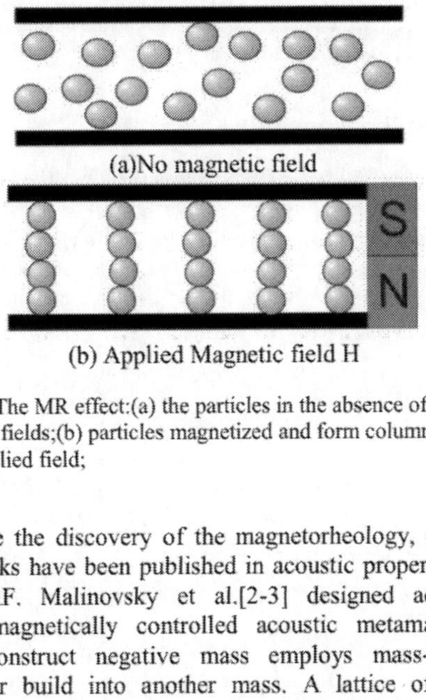

Figure.1 The MR effect:(a) the particles in the absence of magnetic fields;(b) particles magnetized and form columns when applied field;

Since the discovery of the magnetorheology, only a few works have been published in acoustic properties of the MRF. Malinovsky et al.[2-3] designed actively electro-magnetically controlled acoustic metamaterial. They construct negative mass employs mass-spring oscillator build into another mass. A lattice of such "mass-in-mass" elements could be highly attenuated for acoustic/vibration waves. An one dimensional chain model discussed based on discrete elements with effective mass density and elastic modulus. In their study, the theoretical model of acoustic wave propagation in MRF is based on one dimensional wave equation in the elastic medium.

Howarth et al[4] conducted a series of acoustic experiments at the National High Magnetic Field Laboratory in Tallahassee, Florida. The acoustic sound speed of MR fluids was measured as functions of applied magnetic field strength, normal and orthogonal field orientations, and acoustic frequency. Their presentation discussed measurement methodology and preliminary results about the MRF material.

In previous work, Shen and Huang [5] present the acoustic model of MRF materials based on the Biot-Stoll model[6-8]. The Biot's model is an useful extension of the equivalent fluids description, which takes into account the stiffness of the frame and demonstrates that two compressional waves and one shear wave can propagate in porous materials. For the case where a strong coupling exists between the fluid and the frame, the two compressional wave exhibit very different properties, and are identified as the slow wave and the fast wave. The Biot-Stoll model is a comprehensive model. It is necessary to derive an approximate equation for simplify practical application acoustic wave propagation in MRF suspensions.

This paper presents an effective density fluid model that studies the acoustic propagation in the MRF material under magnetic field to simple the Biot's model. The effective density fluid model (EDFM) approximates a fluid-saturated porous medium as a fluid with a bulk modulus and effective density derived from Biot's theory.

2 Acoustic Model

2.1 The Biot's model

When an acousitc wave propagates in a porous, elastic,solid medium saturated by a visous fluid. The Biot's model considers that two compressionl waves and one shear wave can propagate in a porous medium simultaneously[6-8]. Let u is displacement of the frame, U is displacement of the pore fluid relative to the frame.

$$u=\nabla\phi_s+\nabla\times\psi_s \tag{1}$$

$$\beta(u-U)=\nabla\phi_f+\nabla\times\psi_f \tag{2}$$

where β is the porosity. Biot's equations for the scalar potensials are then given as (using a plane wave solution $\exp[i(kx-\omega t)]$)

$$(-k^2H)\Phi_s+k^2C\Phi_f=(-\omega^2)\rho\Phi_s+\rho\omega^2\Phi_f \tag{3}$$

$$(-k^2C)\Phi_s+k^2M\Phi_f=(-\omega^2)\rho_f\Phi_s+\omega^2\alpha\rho_f\Phi_f/\beta + i\omega F\eta\Phi_f/\kappa \tag{4}$$

in Eqs.(3) and (4), $k=\omega/c$ is the acoustic wave number, f is the frequency, c is the speed of sound, α is the structure constant or tortusotity, η is the pore fluid viscosity, κ is the permeability.

$$\rho=\beta\rho_f+(1-\beta)\rho_s \tag{5}$$

where ρ is the toal mass density , ρ_f is the pore fluid mass density and ρ_s is the particle mass density.

The material coefficients are

$$H=[(K_r-K_b)^2/(D-K_b)]+K_b+(4\mu/3) \tag{6}$$

$$C=(K_r-K_b)K_r/(D-K_b), \tag{7}$$

$$M=K_r^2/(D-K_b), \tag{8}$$

$$D=K_r[1+\beta(K_r/K_r-1)] \tag{9}$$

We note that K_r is the bulk modulus of individual particles and K_f is th bulk moulus of the base fluid.The parameter F representsthe deviation from Poiseuille flow as frequency increases. Biot derived an expression for F that in the present notation is

$$F(\xi)=\frac{\xi T(\xi)/4}{1-2iT(\xi)/\xi} \tag{10}$$

with

$$T(\xi)=\frac{(-\sqrt{i})J_1(\xi\sqrt{i})}{J_0(\xi\sqrt{i})} \tag{11}$$

where J_1 and J_0 are cylindrical Bessel functions and

$$\xi=a\sqrt{\frac{\omega\rho_f}{\eta}} \tag{12}$$

The parameter a is the pore size parameter. Johnson et al[9]. Examined the problem of a pressure gradient applied across a porous sample and found that for a porous media consisting of nonintersecting canted tubes the relationship for a

$$a=\sqrt{\frac{8\alpha\kappa}{\beta}} \tag{13}$$

Using Eqs.(3) and (3),a fourth-order polynomial for the complex wave number k can be found. The soltion gives two different wave numbers corresponding to the Biot fast and slow waves. Both these waves show significant dispersion.

2.2 The effetive density fluid model

Taking the K_b=μ=0 limit of Biot theory is motivated by the fact that for magnetroheological fluids the frame and shear moduli are much lower than the other moduli[10-12].

Using Eqs.(6-8), it is easy to shown that, when K_b and μ are set to zero.

$$H=C=M=(\frac{1-\beta}{K_r}+\frac{\beta}{K_f})^{-1} \tag{14}$$

The result states that the compressibility is a linear function of the concentration of the particles in a suspension as assumed by Urick in

$$(-k^2H)\ \phi_s + k^2H\phi_f = (-\omega^2)\rho\phi_s + \rho_f\omega^2\phi_f \quad (15)$$

$$(-k^2H)\ \phi_s + k^2H\phi_f = (-\omega^2)\rho_f\phi_s + \frac{\omega^2\alpha\rho_f}{\beta}\phi_f + \frac{i\omega^2F\eta}{\kappa}\phi_f \quad (16)$$

Using Eqs.(15) and (16), a second-order equation for the wave number k can be found. This solution and the relation from Eq.(3-4) that $k=\omega/c$ can be used to find that

$$k^2 = \frac{\omega^2\rho_{\text{eff}}(\omega)}{H} \quad (17)$$

where

$$\rho_{eff(\omega)} = \rho_f\left(\frac{\alpha(1-\beta)\rho_s + \beta(\alpha-1)\rho_f + \frac{i\beta\rho F\eta}{\rho_f\omega\kappa}}{\beta(1-\beta)\rho_s + (\alpha-2\beta+\beta^2)\rho_f + \frac{i\beta F\eta}{\omega\kappa}}\right) \quad (18)$$

is a complex effective density.

3 Numerical Results and Discussion

To verify the correctness of this model, the longitudinal wave velocity and attenuation coefficient of the MRF are calculated. Using the equations in the preceding section, the longitudinal wave velocity and attenuation coefficient are calculated for MRF-132DG material. Table 1 lists the parameters of the model for the MRF investigated which is typical of MRF-132DG manufactured by Lord Corporation. All parameters of MRF in braces are either directly supplied by the manufacturer or have been calculated[5]. The calculations for the linear equations are numerically performed. For the MRF description, the fluid bulk modulus can be assumed purely elastic, because in our case the base fluid is low-viscosity silicone oil.

Figure 2 shows the real and imaginary parts of the effective density as a function of frequency from 10Hz to 100KHz.The physical reason for the transition between these two values can be understood from Biot theory[6-7]. At low frequencies the pore size parameter is much smaller than the viscous skin depth and there is no relative motion between the grains and the water. At high frequencies the skin depth is much smaller that the pore size parameter (typical pore radius) and it is the tortuosity determines the relative motion of the fluid and solid.

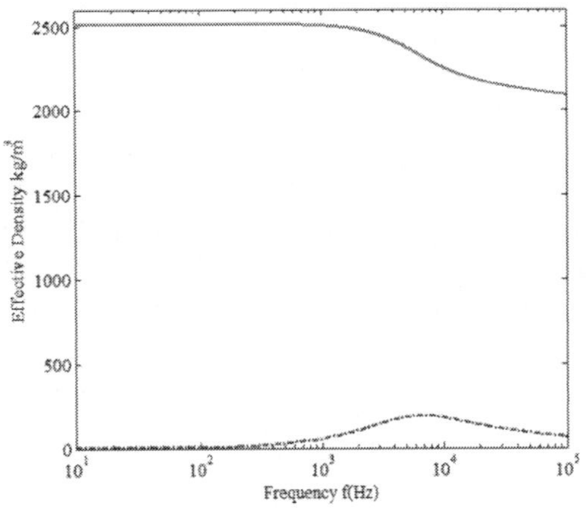

Figure.2 The real (solid curve) and imaginary (dot-dashed) parts of the effective density given in Eq.(12) for the MRFs in Table 1.

Table.1 The EDFM model Parameters of MRF-132DG

Parameters	Symbol(Unit)	MRF
Solid density	$\rho_r(\mathrm{kg}/m^3)$	3160
Fluid density	$\rho_f(\mathrm{kg}/m^3)$	1000
Grain bulk modulus	$K_r'(N/m^2)$	1.26×10^{11}
Volumetric Specitic loss	δ_{kr}	0.4
Fluid bulk modulus	$K_f(N/m^2)$	9.8×10^{8}
Grain diameter	$d_m(\mathrm{m})$	1□10×10^{-6}
Fluid viscosity	$\eta(Ns/m^2)$	0.075□10^{6}
Frame shear modulus	$G_b'(N/m^2)$	0□8×10^{6}
Shear specific loss	δ_{Gb}	0□0.5
Frame bulk modulus	$K_b'(N/m^2)$	0□4×10^{8}
Volumetric specific loss	δ_{Kb}	0□0.5
Porosity	β	0.1□0.5
Permeability	$B_0(m^2)$	4×10^{-12}

Figure 3 and 4 shows the sound speed and attenuation as a function of frequency for both Biot model (solid lines) were used to derive the dispersion equation and the EDFM (dot dashed line-Eq.(18)). Both models produce similar dispersion in sound speed. However, the EDFM predict speeds that are about 10m/s slower at the lowest frequencies and about 5m/s slower at the highest frequencies. The average sound speed error is approximately 0.5% over the frequency range plotted. The bulk and shear moduli of the frame obviously increase in sound speed slightly. It is demonstrated that the sound speeds predicted with this effective density fluid model are very close to the predictions of Biot theory and much simpler than the full Biot model.

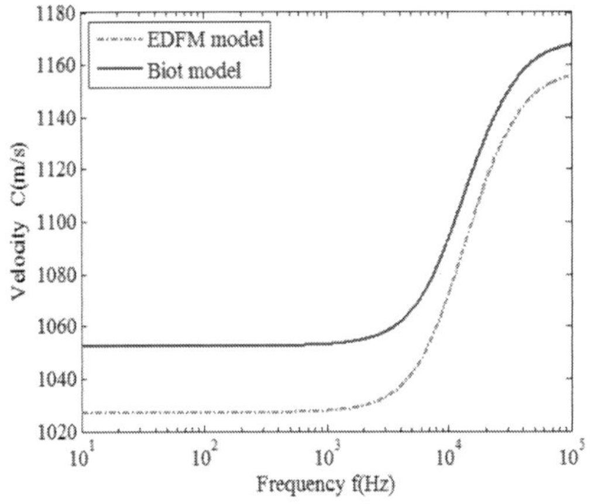

Figure.3 The sound speed as a function of frequency calculated using the Biot model (solid curve) and the effective density fluid model(dashed cruve) for the MRFs in Table 1.

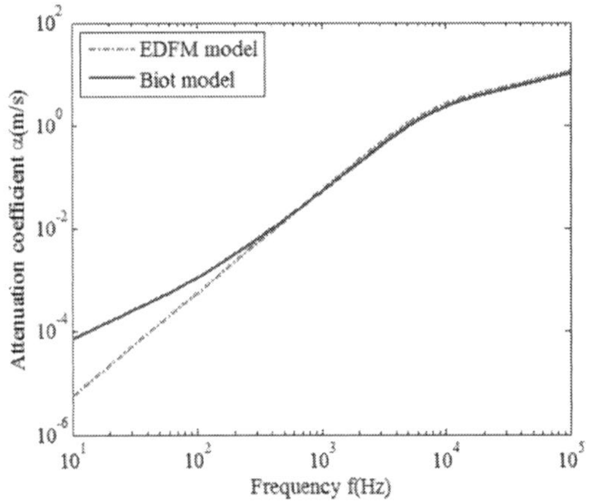

Figure.4 Attenuation as a funciton of frequency calculated using the Biot model(solid) curve and the effective density fluid model (dashed curve) for the MRFs in Table.1

Figure 4 shows that the attenuation predicted by the EDFM is slightly lower at the lowest frequencies plotted. This discrepancy increases as the frequency is reduced below the range plotted due to internal frictional effects of the frame. The attenuation due to frame moduli varies linearly with frequency. Thus, at low enough frequencies the frame attenuation will begin to become important.

The continuous line was obtained by averaging just the density values, taking into account that for a dilute suspension one can neglect the effects of suspended particles on the elastic properties of the composite medium.

Figure 5 shows the three-dimensional representations of longitudinal wave velocity in the frequency of 1Hz to 1MHz, and the viscosity range of 0.087 to 10^2 when placed in a weak magnetic field. The viscosity of MRF trends to saturation above 10^3 . It is seen from Fig.5 that the longitudinal wave velocity increases dependent on the field strength, but when the viscosity above the 60, the wave velocity is nearly kept 1750m/s and indepent on frequency.

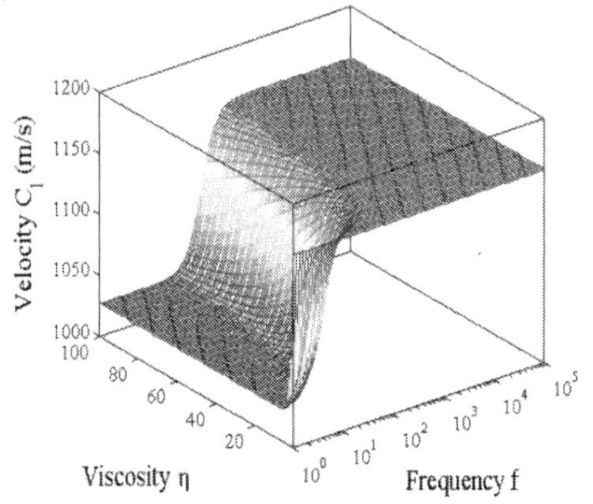

Figure.5 Three-dimensional representation of the longitudinal wave velocity as functions of viscosity and frequency for MRF

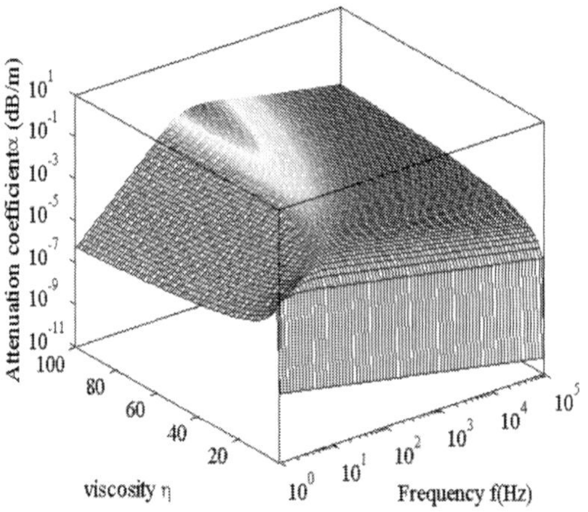

Figure.6 Three-dimensional representation of the attenuation as functions of viscosity and frequency for MRF

Figure 6 shows the three-dimensional representations of attenuation coefficient variation changes with frequency and viscosity. It is seen from Fig.6 that the attenuation coefficient increases as the viscosity increases above the frequency 10^2Hz. However, there is a sharp jump on the curve below the frequency 10^2Hz, the attenuation coefficient is very complex dependent on both frequency and field strength. Hence, the MRF has better active acoustic absorption properties with the applied magnetic field in the mid-high frequency range. We should improve the dynamic acoustic absorption performance in the low frequency range by appropriately select other geometric parameters of the MRF.

Figure 7 shows the three-dimensional representations of sound speed variation changes with frequency and porosity. It is seen from Fig.7 that the sound speed increases as the porosity decreases from the frequency

10Hz to 100k Hz. The velocity increases occurring over a broad band of frequency. It is well known that in a colloidal suspension, only one longitudinal mode can exist when an elastic excitation propagates in it if the wavelength is much larger than the characteristic size of the suspended particles.

In calculating the sound speed, we assume that in the first mode, the elastic perturbation propagates through channels of almost clear fluid. Provided that the perturbation wavelength is much larger than the characteristic size of the iron particles.

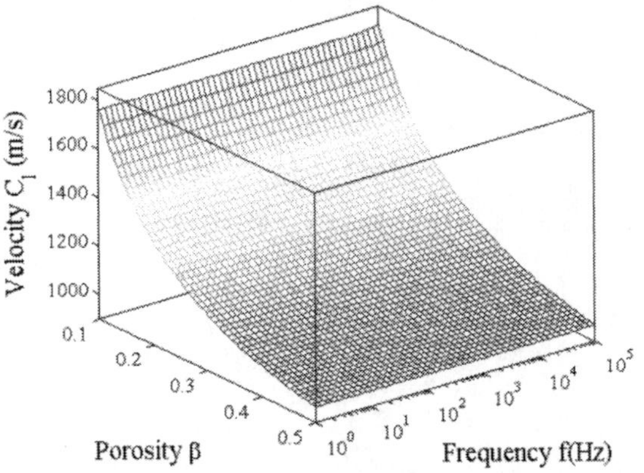

Figure.7 Three-dimensional representation of the sound speed as functions of porosity and frequency for MRF

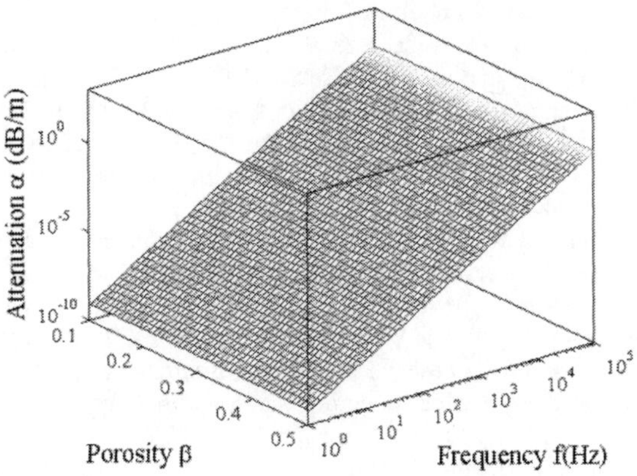

Figure.8 Three-dimensional representation of the attenuaiton coefficient as functions of porosity and frequency for MRF

Figure8 shows the three-dimensional representations of attenuation coefficient variation changes with frequency and porosity. It is seen from Fig.8 that the attenuation coefficient is almost increasing as the porosity increases. We can assume that the effect of iron spheres in this dilute dispersion is only of inertial nature,i.e, the elastic properties of the fluid will not be affected by the presence of a few non-interacting particles. Then, neglecting dissipation effects, the only average which must be considered in this calculation.

ACKNOWLEDGEMENTS

The present research was supported by the China National Science Foundation (No.51175195 , No.51505344). The magneto-rheological investigations were performed at State Key Laboratory of Digital Manufacturing Equipment and Technology, Huazhong University of Science and Technology,Wuhan,China.

References

1. M.Szary, The analytical model of rheological fluid for vibration and noise control, Proceedings of Meetings on Acoustics, Montreal, Canada, 19(2013):1-8.
2. S.Malinnovsky and D.M.Donskoy, Electro-magnetically controlled acoustic metamaterials with adaptive properties, Journal of the Acoustical Society of America, 132(2012):2866-2872.
3. D.M.Donskoy, V.S.Malinovsky, Broadband acoustic metamaterials with electro-magnetically controlled properties, Proceedings of Meetings on Acoustics, Montreal,Canada, 19(2013):1-8.
4. T.R.Howarth,F.Fratantonio,J.E.Boisvert,A.Bruno,C. L.Scandrett,W.M.Wynn and P.S.Davis, Acoustic behavior of magnetorheological fluids in magnetic fields, Journal of the acoustical society of America, 13(2011):2359.
5. M.Shen and Q.B.Huang, Acoustic properties of magnetorheological fluids under magnetic fields, Applied Mechanical and Materials,721(2015):818-823.
6. M.A.Biot, Theory of propagation of elastic waves in a fluid-saturated porous solid, Part I:Low frequency range, Journal of the Acoustical Society of America, 28,(1956):168-178.
7. M.A.Biot, Theory of propagation of elastic waves in a fluid-saturated porous solid, Part II:Higher frequency range, Journal of the Acoustical Society of America,28(1956):179-191.
8. R.D.Stoll, Acousitc waves in ocean sediments, Geophysics, 42,4,(1977):715.
9. D.L.Johnson, J.Koplik and R.Dashen, Theory of dynamic permeability and tortuosity in fluid-saturated porous media, Journal of Fluid Mechanics, **176**(1087):379-402.
10. K.L.Williams, An effective density fluid model for acoustic propagation in sediments derived from Biot theory, Journal of the Acoustical Society of America, 110,5(2001),2276-2280.
11. K.L.Williams and D.R.Jackson, Bistatic bottom scattering:Modle, experimens, and model/data comparison, Journal of the Acoustical Society of America, 103,1,(1998):169-181.
12. K.L Willams, D.R Jackson, E.I Thorsos,Tang and J, S.G Schock, Comparison of sound speed and attenuation measured in a sandy sediment predictions based on the Biot theory of porous media,IEEE Journal Ocean Engineering, 27,3(2002): 413-428.

MATEC Web of Conferences 45, 01002 (2016)
DOI: 10.1051/matecconf/20164501002

Investigation of Microstructure in Solid State Welded Al-Cu-Li alloy

Kookil No[1], Ye Rim Lee[1+], Kyung-Ju Min[2] and Ho-Sung Lee[1,2]

[1]*University of Science & Technology, Republic of Korea*
+currently at Korea Institute of Materials Science, Republic of Korea
[2]*Korea Aerospace Research Institute, Republic of Korea*

Abstract. Al-Li alloys have been extensively used in aerospace vehicle structure since the presence of lithium increases the modulus and reduce the density of the alloy. Especially the third generation Al-Cu-Li alloy shows enhanced fracture toughness at cryogenic temperatures so that the alloy has been used on the fuel tank of space launchers, like Super Lightweight External Tank of the Space Shuttle. Since the commercial size of the plate cannot accommodate the large tank size of the launcher, joining several pieces is required. However, lithium is highly reactive and its compounds can decompose with heat from conventional fusion welding and form different types of gases which result in formation of defects. In this study, the microstructure change is investigated after solid state welding process to join the Al-Cu-Li sheets with optical and transmission electron microscopic analysis of precipitates.

1. Introduction

Aluminum lithium alloy has been used for aerospace structure because its higher modulus and lower density comparing to conventional aerospace aluminum alloy [1]. The alloy also shows superior mechanical properties in higher specific strength, enhanced resistance to high cycle fatigue, and high fracture toughness at cryogenic temperatures [2]. One of the well-known application of aluminum lithium alloys is for space launch vehicle. By using aluminum lithium alloy and friction stir welding to super lightweight external tank for space shuttle, total dry weight is reduced by about 11% [3, 4]. It is known that fusion welding of Al-Li alloys is difficult due to the reactive property of element Li during melting and causes porosity, cracking and low joint efficiency [5, 6]. Solid state welding generally involves heating the metal with a suitable amount of pressure so that homogeneous and complete microstructural joining is possible with no melting of the parts to be joined. It means the process offers advantages over fusion welding which exhibit solidification problems (e.g. porosity, hot cracking, segregation etc).

Solid state diffusion welding is an attractive manufacturing method for aerospace applications where mechanical properties in the bond area and a sound metallurgical bond are important [7]. Diffusion welding is such a process in which two matched surfaces are held together at a temperature range between 0.5 of the absolute melting temperature of the materials and the room temperature under a low pressure without causing a macroscopic plastic deformation in the materials.. Diffusion welding consisted of initial creep, which is followed by vacancy diffusion to allow for complete void closure and homogeneous welding at the interface [8]. This process of void closure is the opposite to the cavity nucleation process of the final stage of creep rupture. Cavitation model for nucleation and growth of cavity requires the critical size of stable cavity based on surface energy and applied stress [9]. Therefore, thermo-dynamically there is a critical size of void for enclosing and complete welding. The whole mechanism is governed by surface diffusion, volume diffusion, grain boundary diffusion, and power-law creep. The welding is formed from atomic migration across an interface in a solid state, and there is no metallurgical discontinuity at the interface and hence mechanical properties and microstructure at the bonded region are not different from those of the base metal. This is different from a brazing process in which a foreign metal with a lower melting point is used to weld together.

In friction welding, rotating welding tool generates frictional heat and causes material at the interface to deform and with a profiled pin is plunged into the joining areas between two pieces of plate material [10]. By keeping the tool rotating and moving it along the seam to be joined, heat from friction between the welding tool and the workpiece causes the plate to soften without melting. The softened material is transferred to the trailing edge of the tool pin as the tool travels along the weld line. The workpiece is forged by the intimate contact of the tool shoulder and the pin profile. After cooling, a solid phase bond is created between the pieces of metal that have been stirred and mixed together. This process requires low energy input and joins materials without filler metals or shielding gas. In addition, in most cases the localized heat during friction welding results in a refined microstructure, which can provide improved strength relative to the parent material.

The present work is to investigate the microstructure change including phase and microstructure evolution of diffusion welded and friction stir welded Al-Cu-Li alloy. Microstructures present formation and evolution in representative precipitates including θ (Al_2Cu), δ (Al_3Zr), and T (Al_xLi_yCu) series.

2. Experimental

The Al-Li alloy was obtained from Constellium Global Aerospace, France. The contents of the alloy was examined by inductively coupled plasma optical emission spectrometer (ICP-OES, 720, Agilent, CA, USA).

Table 1. Composition of the Al-Li alloy (wt/o)

Ag	Al	Cr	Cu	Fe	Li	Mg
0.19	94.56	0.0004	3.76	0.05	1.00	0.28
Mn	**Ni**	**Pb**	**Si**	**Ti**	**Zn**	**Zr**
0.02	0.0015	0.0003	0.02	0.01	0.001	0.11

For diffusion welding, two sheets in thickness of 5mm were prepared and the surface of sheets was carefully prepared by special procedure [4] since the surface condition was vitally important for complete bonding. The surfaces to be bonded were machined and grinded with 600 grit sand paper before chemical cleaning. Ultrasonic rinsing in a high-purity solvent, and in distilled water followed. The rinsed surfaces were then air dried with clean filtered air. Prior to diffusion welding, the specimens were immediately cleaned again with a high-purity solvent and dried. Then, two sheets are sequentially stacked in the tool according to the present embodiment. After stacking, the tool is sealed and vacuumed and then, the tool is heated to 500 ℃. Hydrostatic gas pressure of 50MPa is applied for 1, 2, and 3 hours. Solid state welding system is shown in Figure 1

Two plates were butt-joined by friction stir welding and the welding direction was parallel to rolling direction with coupons of 300mm in length and 100mm in width. One plate where the direction of rotation is the same as that of welding is called the advancing side (AS), with the other plate designated as being the retreating side (RS). The range of welding parameter is 600 RPM for rotation speed and 300 mm/min for travel speed. After welding, all of the specimens were inspected by x-ray radiography and one with defects are discarded. For optical microstructure, the specimens were etched with Graff-Sargent solution followed by Keller's solution [11]. The etched surfaces were observed by optical microscopy (Metallux 3, Leica Microsystems). The details were examined by transmission electron diffraction microscopy (TEM; Titan G2 Double Cs corrected TEM, Philips/FEI) with an acceleration voltage of 200~300 kV, and composition was determined by energy-dispersive spectroscopy (EDS), electron energy loss spectroscopy (EELS) and scanning TEM-high angle annular dark field (STEM-HAADF) in the TEM. Samples for the TEM observation were prepared on Molybdenum-grid by a focused ion beam (FIB)-SEM hybrid system equipped with a Ga+ focused ion beam (FIB) column (FIB; SMI3050SE, SII Nanotechnology Inc.)

(a) (b)

Figure 1. Photographs of solid state joining apparatus: (a) diffusion bonding press (b) friction stir welding machine

3. Results and Discussion

Figure 2 shows metallographic microstructure with identification of some of precipitates. Precipitates are found at the grain boundaries, as shown in Figure 2(b). The precipitates are easy to nucleate at the gain boundaries because the high dislocation density is suitable for the diffusion of the solute elements at the grain boundaries. The precipitates such as θ'', α', and δ' phases, formed at relatively low temperature, are known as a spherical shape. Furthermore, as θ phase is grown from GP zone, θ'' and θ' phases, the size of the θ series is varied from 10 nm to 5 μm. In contrast, the high-temperature precipitates including T_1 and T_2 are reported as a pin- and platelet-like shape. It is known that T_1 phase has a plate-shaped morphology and is the primary phase responsible for the strengthening effect in this alloy [12]. The nucleation of this phase is based on the presence of secondary alloying elements such as Mg and Ag and T1 phase usually coexists with the θ' (Al_2Cu) phase. Zr contributes to form coherent spherical β '(Al_3Zr) precipitates which are very stable at high temperatures.

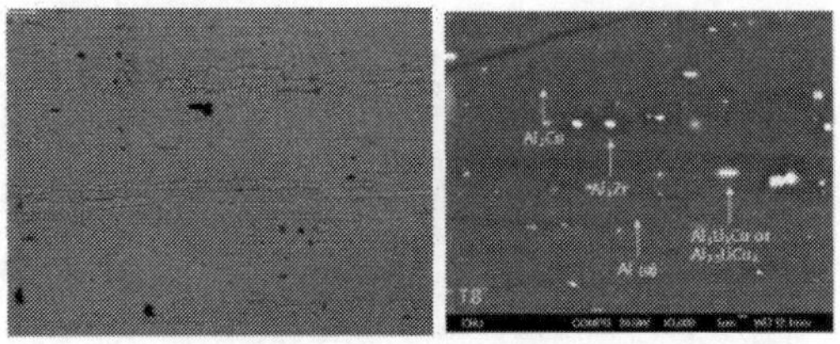

(a) (b)

Figure 2. (a) Optical microscopy structure and (b) TEM photograph with phase identification

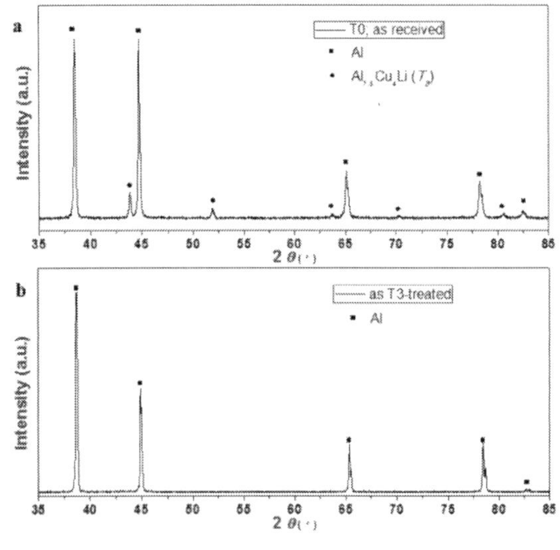

Figure 3. XRD pattern of this alloy

Figure 3 reveals XRD patterns of AA2195 samples processed by different. The pattern of AA2195-T0 in Figure 3(a) reveals two phases such as Al and $Al_{7.5}Cu_4Li$. The phase of $Al_{7.5}Cu_4Li$ was produced by a natural aging process. Changes in the phases were detected after T3 solution treatment. As shown in Figure 3(b), AA2195-T3 reveals only Al peaks. This result indicated that all contents in AA2195-T0 were successfully solution-treated and supersaturated into the Al phase. Thus, the peaks for the Al phase were shifted to higher angle compared to ICDD (International Center for Diffraction Data) [13]. This shift indicates that the Al atoms were substituted by other atoms like Mg, Ag and Cu. Furthermore, the Al lattice was shrunken and distorted by the atoms replaced.

The microstructure of diffusion welded interface is shown in Figure 4. After 2 hours of bonding, the interface is still clearly visible, but the grain growth after recrystallization is apparent for the longer time as shown in Figure 4(c). Therefore, it was difficult to obtain an appropriate welding condition at this temperature. It is recommended that the diffusion welding temperature would be lower or welding time shorter for this alloy.

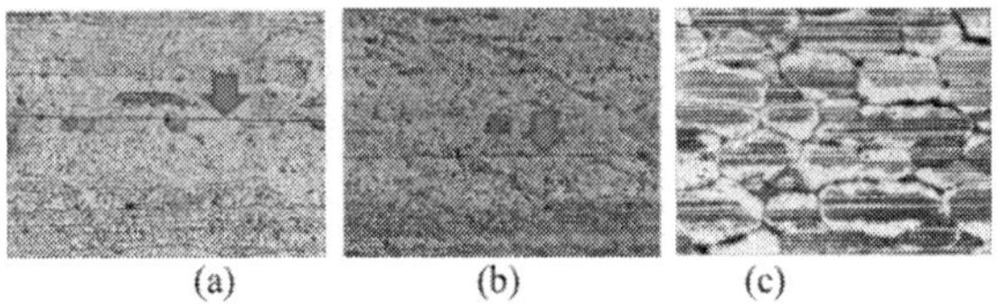

Figure 4. Micrographs (X200) of diffusion welded interface welded for (1) one hour, (2) two hours, and (3) 3 hours under 50MPa at 500℃

The cross sectional view of friction stir welded joint is shown in Figure 5 with an example of onion rings, which is the prominent feature during the FSW. The distinguished lines in Figure 5 is called onion rings. In this region, it shows that the materials are well mixed along with rotating threaded pin. The rotating tool produce frictional heat and extrude a cylindrical shaped material around to the retreating side of the joint. Therefore, the formation of onion rings was caused by the rotation and advancing effects.

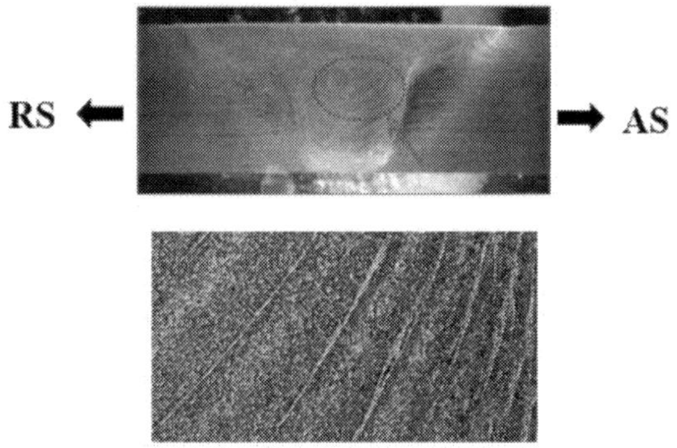

Figure 5. Cross sectional view of friction stir welded joint and a micrograph of onion rings

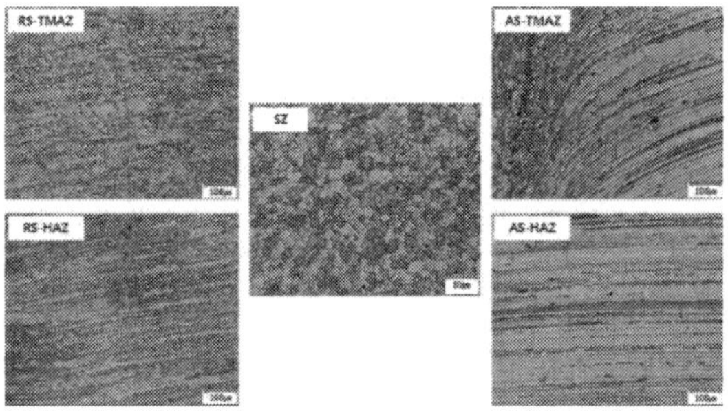

Figure 6. Microstructure of friction stir welded AA2195 at 600rpm-360mm/min condition [14]

The grain structure of friction stir welded joint is shown in Figure 6 and five distinct regions are stir zone (SZ), heat-affected zone (HAZ) and thermo-mechanical affected zone (TMAZ), which are obviously distinguished by their grain characteristics and distribution of precipitates for retreating side and advancing side. Stir zone (SZ) is the region, which has experienced temperatures and deformation to become fully recrystallized. This area had immediate interaction with the pin tool. Thermomechanically affected zone (TMAZ) is the region that has experienced plastic deformation in addition to heat from the process. This area has not undergone recrystallization and grains are deformed, elongated and rotated due to the strain to which they were subjected. Heat affected zone (HAZ) is the region without deformation, but has experienced only temperature increase during welding affecting the microstructure and/or mechanical properties. The microstructure is similar to that of parent metal and the grains are slightly overgrown as a result of the exposure to welding heat. The material located at the advancing side is mainly influenced by the friction from shoulder and pin, while at the retreating side, the dominating effects are extrusion and backfill from the advancing side and the shoulder. Figure 6 (AS-TMAZ) has shown the boundary between TMAZ and nugget in the advancing side and dynamic recrystallization has taken placed in this region.

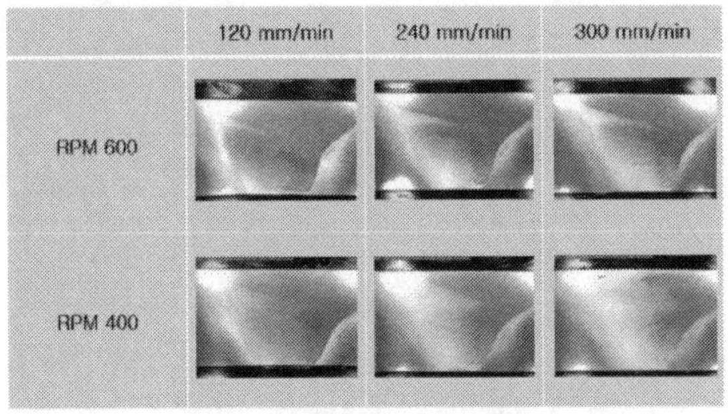

Figure 7. Macrostructure of friction stir welded joints

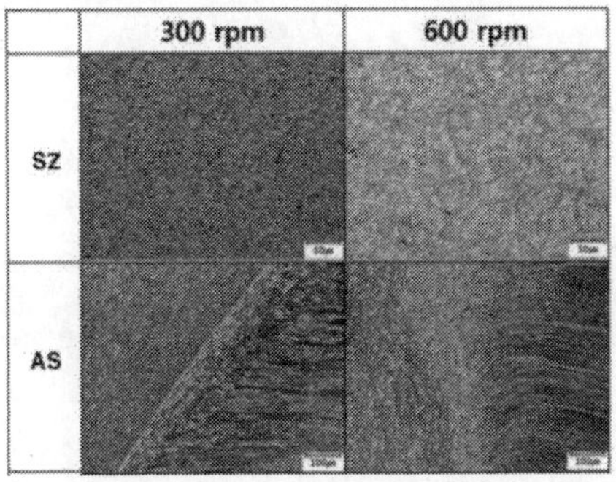

Figure 8. Effect of rotating speed on microstructure welded at 120 mm/min

Figure 7 shows typical macrostructure of friction stir welded joints at different process parameters. In macrostructure, it is hard to notice any distinguishable change on microstructure except at higher traveling speed, the orange ring marks become clearly visual. Effect of rotating speed is shown in Figure 8 for the specimen welded at 120mm/min. At higher rotating speed, the distinction between HAZ and TMAX becomes disappear and this is due to the temperature increase at this region.

The grain size at high rotating speed (600RPM) is larger than that at 300RPM due to the heat generated by the plastic deformation.

Material in the stirred zone experience sufficient deformation and heat input which cause the complete dynamic recrystallization. However in case of insufficient or excessive rotational speed combined with too low downward force, wormhole defect was created as shown in Figure 9. Some of other weld defects observed in this study is summarized in Figure 10. The wormhole or tunnel defect is a kind of volumetric defect that may be discrete or continuous. Tunnel defects may occur at high or low tool advance per tool revolution. Defect location is dependent on the alloy and the welding parameters but is normally on the advancing side of the weld [15]. The occurrence of wormhole defects is more difficult to control in materials with high strength and low thermal conductivity. It is known that aluminum alloys are particularly prone to crown-side overheating defect due to local melting with production of brittle solidification products. The defect is not necessarily detectable by conventional non destructuve inspection. Recent development of automated phased array ultrasonics is known to be effective and under implementation [16].

Figure 9. Detection of wormhole defect of friction stir welded at 400rpm-360mm/min condition

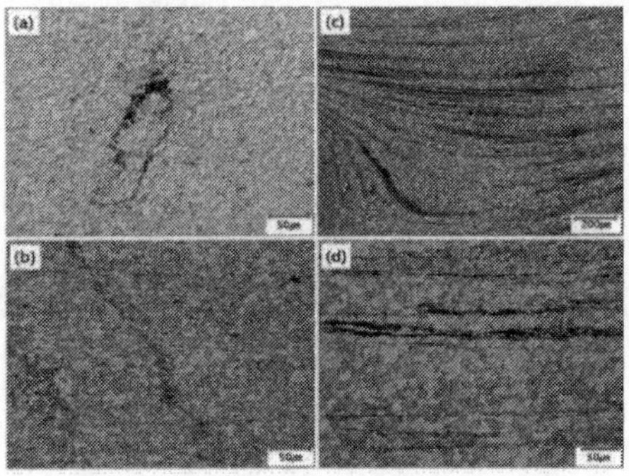

Figure 10. Examples of various weld defects : (a), (b)void (c) oxides alignment (d) cross direction voids

4. Conclusion

In this work, two different types of solid state welding methods are performed to join Al-Cu-Li alloy sheets. An attempt to diffusion welding results in grain growth after complete welding so that it needs further study to obtain optimum welding condition. For friction stir welding, a very good microstructural quality has been achieved and will be validated with mechanical testing.

References

1. A. R. C. Westwood, Material Science Technology, **6**, 958-961 (1990).
2. R. J. H. Wanhill, Internal Journal of Fatigue, **16** 3-20 (1994).
3. G. Adams, P. Pareti, J. Thompson, and K. Lawless, AEROMAT 2001, June 11-14, 2001, Los Ang3les, CA., U.S.A.
4. Information on http://www.nasa.gov/pdf/3752main_ET Overview_Wanda_print.pdf
5. NASA FS-2005-04-025-MSFC, April 2005.

6. J. H. Vickers, J. Fikes, J. R. Jackson, T. F. Johnson, J. K. Sutter, and R. A. Martin, "NASA Composite Cryotank Technology Demonstration: Overview", SAMPE 2012 Conference, Baltimore, MD, 2012.
7. H. S. Lee, in : Welding and Joining of Aerospace Materials, edited by M.C. Chaturvedi, Woodhead Publishing, Philadelphia, PA, USA (2012), Ch. 10 320-344
8. B. P. Kashyap and A. K. Mukherjee: Res. Mechanica, 17 (1986), pp. 293-355
9. A. H. Chokshi, A. K. Mukherjee, and T.G. Langdon: Mater. Sci. Eng., Vol. RI0(1993), pp. 237-74.
10. R. S. Mishra, Scripta Materialia, 58 pp. 325-326 (2008)
11. Standard Guide for Preparation of Metal Surfaces for Adhesive Bonding, ASTM D 2651-01, 2001.
12. P.S. Chen and B.N. Bhat, NASA/TM 211548, Alabama, 2002.
13. Powder Diffraction File, International Center for Diffraction Data, Newtown, PA. U.S.A.
14. Y. R. Lee, *Effect of process* parameters on friction stir welds of AA2195, M.S. Thesis, August 2015, University of Science & Technology at Korea.
15. R. S. Mishra, U.S. patent (6,923,362) on "Integral channels in metal components and fabrication thereof," August 2, 2005.
16. https://www.niar.wichita.edu/niarworkshops/Portals/0/WSU_Burford.pdf

MATEC Web of Conferences **45**, 01003 (2016)
DOI: 10.1051/matecconf/20164501003

Effects of Friction Stir Welding Speed on AA2195 alloy

Ho-Sung Lee[1,2], Ye Rim Lee[2+] and Kyung Ju Min[1]

[1]*Korea Aerospace Research Institute, Republic of Korea*
[2]*University of Science & Technology, Republic of Korea*
+currently at Korea Institute of Materials Science, Republic of Korea

Abstract. The application of friction stir welding (FSW) to aerospace has grown rapidly due to the high efficiency and environmental friendly nature of the process. FSW is achieved by plastic flow of frictionally heated material in solid state and offers many advantages of avoiding hot cracking and limiting component distortion. Recently low density, high modulus and high strength AA2195 are used as substitute for conventional aluminum alloys since the weight saving is critical in aerospace applications. One of the problems for this alloy is weld metal porosity formation leading to hot cracking. Combination of FSW and AA2195 provides synergy effect to improve mechanical properties and weight saving of aerospace structure such as cryogenic fuel tanks for launch systems. The objective of this paper is to investigate the effect of friction stir welding speed on mechanical and microstructural properties of AA2195. The friction stir welded materials were joined with four different tool rotation speeds (350~800 rpm) and five welding speeds (120~360 mm/min), which are the two prime welding parameters in this process.

1. Introduction

Friction stir welding (FSW) is a solid-state joining process, in which no solidification microstructures are produced so to eliminating the brittle phases common in fusion welding of high strength aerospace aluminum alloys [1, 2]. In FSW, a rotating tool is pressed against the surface of two plates. The side of the weld for which the rotating tool moves in the same direction as the traversing direction, is commonly known as the advancing side (AS) and the other side, where tool rotation opposes the traversing direction, is known as the retreating side (RS). Frictional heat generated by rotation and traversing of the tool due to the high normal pressure and shearing action of the shoulder along the joint line causes a softened zone of material without melting. This softened material cannot escape outside as it is under constrained extrusion by the tool shoulder. As the tool travels along the joint line, material is moved around the tool probe between the retreating side of the tool and the surrounding non-deformed material. The mixed and extruded material is deposited to form a mechanically mixed joint in solid state behind the tool. The main process parameters are tool rotation rate and travelling speed along the welding direction. The rotation of the tool results in heating and mixing of the material around the pin. Therefore, higher tool rotation rates generate higher temperature because of higher friction heating and results in more severe mixing of material [3]. The travelling motion of the tool moves the mixed material from the front to the back of the pin. The details of the tool geometry are also important and are usually restricted from publication.

AA2195 is the third generation aluminum-lithium alloy and shows high specific strength (strength to density ratio). Lithium is the most effective addition to aluminum to reduce density and exhibits desirable precipitation hardening characteristics. Furthermore, the elastic modulus of the material is increased by each addition of Li. It also attain very high strengths with or without prior cold work, a necessity to attain the high strength tempers comparing to most aerospace aluminum alloys. Since this alloy is susceptible to hot cracking during fusion welding, a solid state welding, such as FSW, is adequate.

The purpose of the current study was to study the effect of friction stir welding speed on mechanical and microstructural properties of AA2195. Since FSW is both a deformation and a thermal process, temperature distribution during FSW is measured. The microstructure and mechanical properties of weldments were investigated for different welding conditions.

2. Experimental

The material used for this study is AA2195 which is obtained from Constellium Global Aerospace, France. The welds of this alloy were made on a three-axis milling machine equipped with a special fixture for friction stir welding. Two plates of 300mm in length and 100mm in width were joined by friction stir welding, in which one is for the advancing side (AS), and the other for the

Article available at http://www.matec-conferences.org or http://dx.doi.org/10.1051/matecconf/20164501003

retreating side (RS). The welding direction was parallel to rolling direction. The range of welding parameter is four rotation speeds (350, 400, 600 and 800 rpm) and five travel speeds (120, 180, 240, 300, and 360 mm/min). During welding, temperature distribution of welding area was monitored with K-type thermocouples with Kyowa UCAM-550A. Ten K-type thermocouples were set on the surface of the plates on a line 100 mm distant from the weld-start point to measure the temperature distribution. There are 3 thermocouples on the advancing side located at 100mm from the start of welding and each 3 thermocouples are attached on advancing side and retreating side located at another 100mm. After welding, all of the specimens were inspected by x-ray radiography and one with defects are discarded. Tensile specimens with 200 mm long and 12.8 mm wide gauge part were cut along the welded joints and tested at room temperature using a 100 KN MTS testing machine under displacement control. Metallographic specimens were prepared and JEM-2100F FE-TEM was used for electron microscopic study.

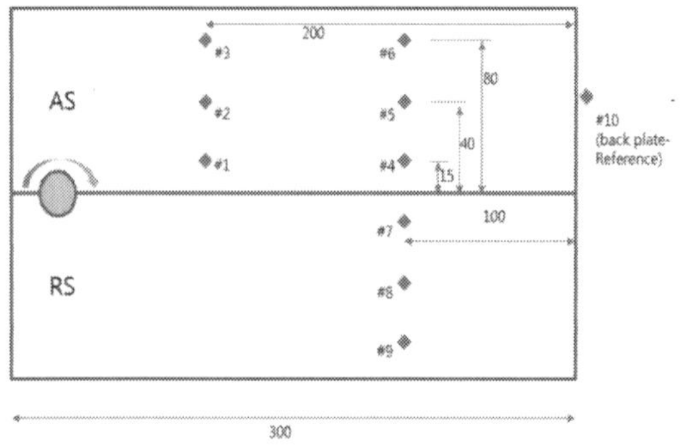

Figure 1. Location map of thermocouples

3. Results and Discussion

Figure 2 presents a typical grain microstructure of the joint welded at 400 rpm with travel speed of 240mm/min showing a well recrystallized structure [4]. From upper part to the bottom part of the joint, the grain size tends to gradually decrease and this is because the heat dissipates through the back plate and results in fine recrystallized grains. The grain size in the advancing side is larger than that in the retreating side due to the heat input difference generated by the plastic deformation. Temperature measurement confirms the maximum temperature on advancing side is higher than that on retreating side as reported in literature [5].

Results of temperature measurement during FSW process are shown in Figure 3. General trend of temperature change is that higher tool rotation rate or lower traveling speed produce higher temperature. At the measurement from the specimen welded in 400 rpm and 240 mm/min, the temperatures at #1 and #4 increased to over 205℃ and the maximum temperature at #7 was 175℃. It is notable that even though Tang [6] did not observe any temperature variation between the advancing and retreating side, recent study by Commin [5, 7] showed clear difference of temperature distribution between the advancing and retreating side, which supports the results of this study.

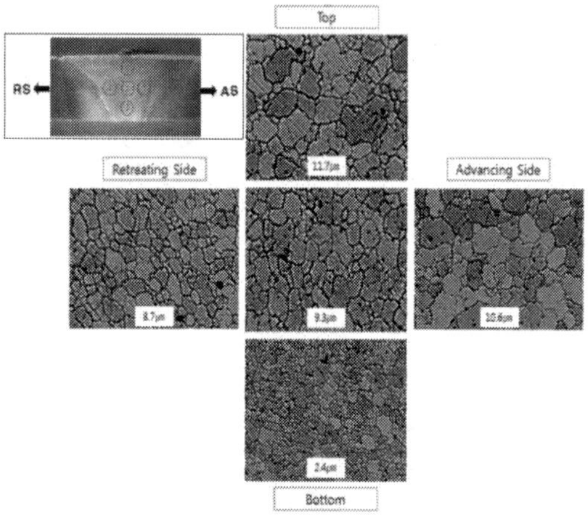

Figure 2. Grain structure of stir zone of typical joint welded at 400 rpm with travel speed of 240mm/min [4]

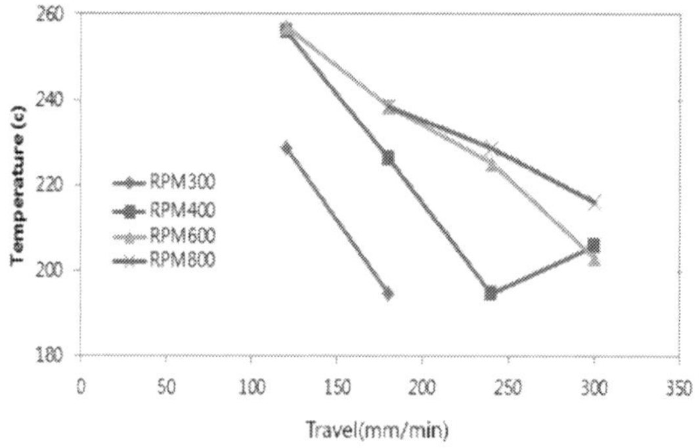

(a) Maximum temperature measured with different welding conditions

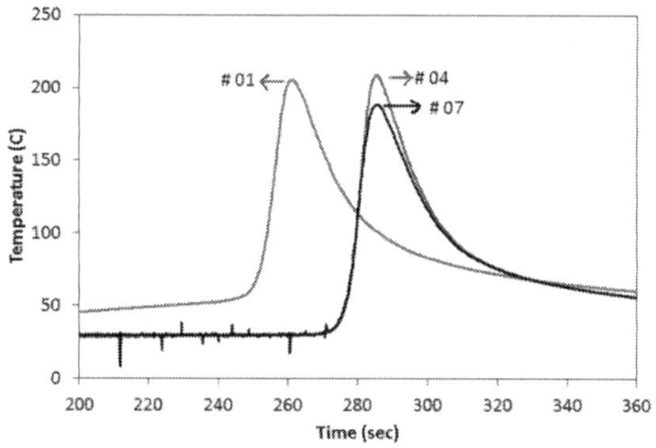

(b) Temperature profle at different location measured at 400 RPM and 240 mm/min conditions

Figure 3. Results of temperature measurement during friction stir welding process

The experimental tensile properties of welded AA 2195 alloy joints are presented in Figure 4. Welded specimens from rotation rate of 350 rpm are not presented here since the radiographic inspection detects relatively large amount of voids and therefore invalidated. The unwelded parent metal shows a tensile strength and elongation of 552 MPa and 6%, respectively. It is shown that the highest strength was obtained from the joint welded with rotation rate of 400 rpm and traveling speed of 240 mm/min. The strength of the specimens welded at 400 rpm are relatively insensitive to the traveling speed, while as that of the specimens welded at 800 rpm decreases as the traveling speed increases. After inspecting the failure mode, it is shown that the specimens welded at 800 rpm failed on advancing side of the joint and the specimen welded at 400 rpm mostly failed on retreating side. The specimens failed on the advancing side generally show lower strength and this may be explained by the temperature difference between the two regions as mentioned earlier.

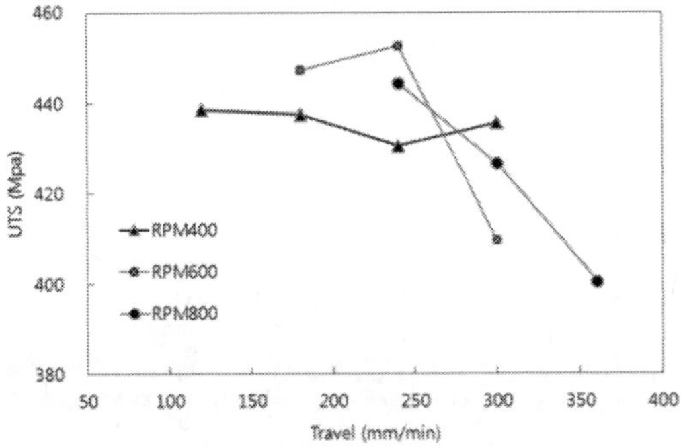

(a) Effect of welding travel speed on tensile strength

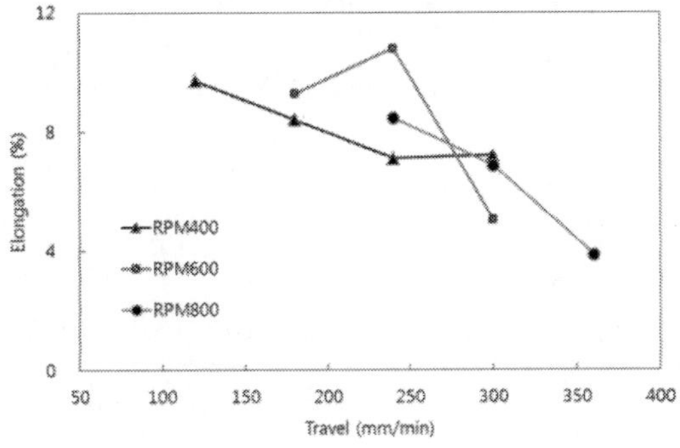

(b) Effect of welding travel speed on elongation of FSW joints

Figure 4 Effect of welding travel speed on mechanical properties of FSW joints

In Figure 5 is shown a TEM bright field image sampling microstructure at retreating side and advancing side for the joint welded with rotation rate of 400 rpm and traveling speed of 240 mm/min. In retreating side, smaller precipitates are aligned in the elongated grain boundaries and the size of precipitates in advancing side is larger than that in retreating side. The results of TEM microstructural study will be published elsewhere [8]. There are still some controversy in microstructure study of stir zone. In the previous study of Li [9], no T_1 was observed, but T_B was observed in the stir zone. This was supported by Oertelt [10]. Howerer, Schneider, et. al. presented coarse T_1, θ', and T_B precipitates were observed in the stir zone [11]. Shukla studied in detail the evolution of precipitates in friction stir welds and observed different levels of T_1, θ' dissolution in the heat affected zone and thermo-mechanical affected zone [12]. Further TEM study is on-going to investigate the effect of traveling speed on the precipitation.

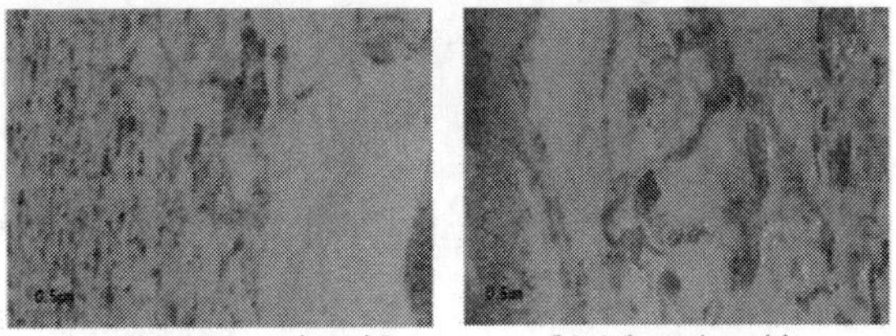

(a) Retreating side (b) Advancing side

Figure 5. TEM photographs after friction stir welding [4]

4. Conclusion

The effect of friction stir welding speed on mechanical and microstructural properties of AA2195 is investigated. The materials were joined with four tool rotation speeds and five welding speeds. It is shown that the recrystallization has resulted in very fine grain in the region adjacent to the back plate. It is confirmed that the grain size depends on the amount of the heat input which is a function of rotating rate and traveling speed. Difference of temperature distribution between the advancing and retreating side is reported to support the Commin's work [5, 7]. The effect of welding speed is sensitive to the mechanical properties of the specimens welded at 800 rpm, which failed on advancing side of the joint. It is shown that the highest strength was obtained from the joint welded with rotation rate of 600 rpm and traveling speed of 240 mm/min.

References

1. S. Shah, J. Wagner, and H. Bable: Thermal Exposure Effects on Al-Li Alloy Plate Products. Presented at the 14th Annual Advanced Aerospace Materials and Processes Conference, Dayton, OH, June 10, 2003.
2. D. E. Glass, Airframe Technology Development for Next Generation Launch Vehicles, IAC-04-V.5.09.
3. R. S. Mishra, M. W. Mahoney, S. X. McFadden, N. A. Mara, and A. K. Mukherjee, "High strain rate superplasticity in a friction stir processed 7075 al alloy," Scripta Materialia, **42**, 163 (2000)
4. Y. R. Lee, *Effect of process parameters on friction stir welds of AA2195*, M.S. Thesis, August 2015, University of Science & Technology at Korea.
5. L. Commin, M. Dumont, J. Masse, and L. Barrallier, Acta Materialia, **57** 326-334 (2009)
6. W. Tang, X. Guo, J. C. McClure, and L. E. Murr, J. Mater. Process Manufac. Sci, **7** 163 (1998)

7. M. Maeda, L. Huijie, H. Fujii, and T. Shibayanagi, Welding in the World, **19** 69-75 (2013)
8. J. H. Kim and H. S. Lee, submitted to J. Alloys and Compounds.
9. Z. X. Li, W.J. Arbegast, P.J. Hartley, E.I. Meletis, in: Proceedings of the Fifth International Conference on Trends in Welding Research, Pine Mountain, Georgia, June 1998, ASM International, Materials Park, Ohio, 568.
10. G. Oertelt, S. S. Babu, S. A. David, E. A. Kenik, Weld. J. **80** 71 (2001)
11. J. A. Schneider, A. C. Nunes Jr., P.S. Chen, G. Steele, J. Mater. Sci. **40** 4341 (2005)
12. A. K. Shukla and W. A. Baeslack, Sci. Technol. Weld. Join. **14** 376 (2009)

MATEC Web of Conferences 45, 01004 (2016)
DOI: 10.1051/matecconf/20164501004

Formulation of Forming Load in V-Bending

Yuki Koumura, Zhigang Wang

Dept. of Mechanical Engineering, Gifu University, 1-1 Yanagido, Gifu, 501-1193, Japan

Abstract. A novel method is described to calculate the forming load in V-bending by a press brake. The data of forming load are collected by FEM analysis. With an increase of the punch stroke in V-bending, the forming load increases gradually after the elastic limit, and then decreases after showing the maximum value. The proposal formulation to trace the variations in the forming load curve includes the calculating method of the load of the elastic limit, the maximum load in air bending and the variations of the forming load before/after the bending stroke of the maximum load. The calculated precision is confirmed by comparing with the measured load-stroke curves in V-bending with a press brake.

1 Introduction

V-bending by a press brake is a common method for small lot production of sheet metal parts [1]. The most important issue in V-bending is to decide the punch stroke to obtain the product angle precisely [2]. However, the proper punch stroke is difficult to predict accurately because it depends sensitively on mechanical properties and the thickness of the sheet metal and the width of the die [3]. Although some methods to identify mechanical properties of a sheet being bent from real time measurements have been developed, the performance is still not satisfactory for a high end machine [4-7]. The mainstream is the adjustment system with laser sensors [8], but the method is expensive because a single press uses many tools at the same time to finish a part.

In previous papers [9-11], the proper punch stroke was formulated based on geometric modeling and the numerical results of FEM analysis. The feature of the formulation is based on two new concepts; bend factor and inflection point. The bend factor is the length of bending round part of the product and depends on the mechanical properties of the sheet and the ratio of the width of die to the sheet thickness. The inflection point is the point that the bending angle under punch loading equals to the angle of the die groove. A series of experiments under various conditions showed that the control system based on the formulation is practical even for a high end machine. However, the estimation accuracy of the bending load is still to be improved.

Forming load in V-bending causes a deflection deformation of the flame in the press brake and results in a change of the product angle along the length direction. Thus estimation accuracy of the bending load is very important to make the proper crown in the machine bed to cancel the deflection deformation of the flame.

The aim of the present study is to improve the estimation accuracy of the bending load. **Fig.1** illustrates the relationship between the bending load and stroke in V-bending. The bending load increases gradually after the elastic limit W_p, and then decreases after showing the maximum value W_a. In order to trace the curve, the formulation is done as follows.

(1) Calculation method of point ① where bending load shows the maximum

(2) Calculation method of point ②, the elastic limit

(3) Formulation of the curve ③ between the elastic limit point and the maximum load point

(4) Formulation of the curve ④ after the maximum load point

It should be noted that the formulation in the present paper does not include the inflection point that appears at the end of the bending process when the angle of product under bending load equals the die angle [10].

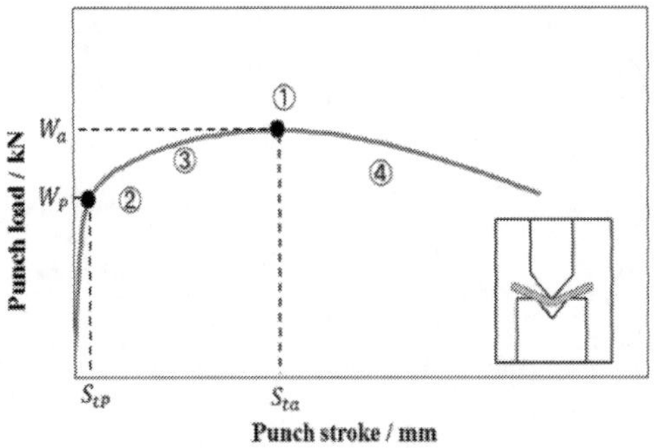

Figure 1. Relationship between punch load and punch stroke

2 Formulation of bending load

2.1 Maximum load in air bending W_a

The equation (1) shows the formula for calculating the maximum load in air bending that is used widely in bending industries. In the present study, to improve the calculation accuracy, a correction factor k_m that depends mainly on the sheet material is introduced and given in **Table 1**.

$$W_{ac} = \frac{k_g \sigma_B w t^2}{1000V} \tag{1}$$

$$W_a = k_m W_{ac} \tag{2}$$

Where σ_B is tensile strength of the sheet material, w is the bending length, t is the sheet thickness, V is the die width. As shown in the equation (3), k_g is a factor that depends on the combination of the sheet thickness and the die width.

$$k_g = -0.055\frac{V}{t} + 1.753 \tag{3}$$

Table 1 Value of k_m

Sheet material	SPCC	A5052	SUS304
k_m	1.17	0.88	1.09

Fig.2 shows the FEM analysis results of the punch stroke S_{ta} of the maximum load. S_{ta} changes linearly with V/t. Thus S_{ta} is formulated by the equation (4). As shown in **Fig.3** and **Fig.4**, a_1 and b_1 vary simply with the sheet thickness t.

$$S_{ta} = a_1(t)\frac{V}{t} + b_1(t) \tag{4}$$

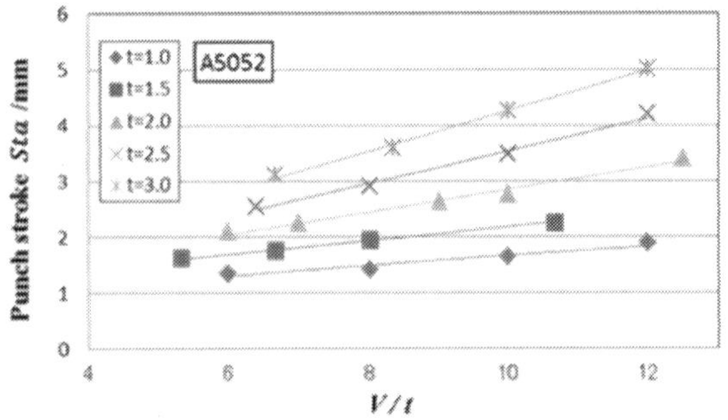

Figure 2. Relationship between punch stroke S_{ta} and V/t

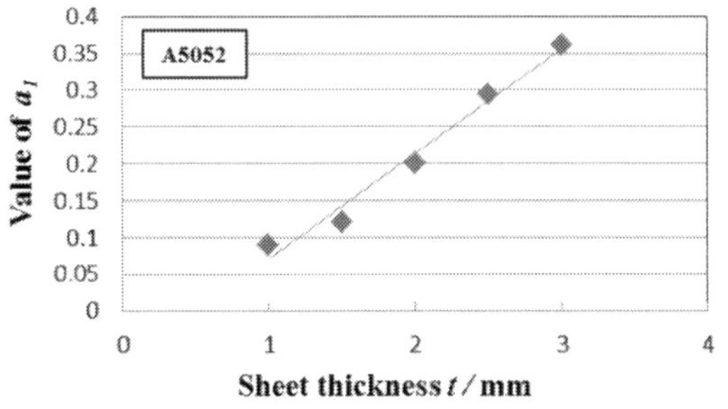

Figure 3. Relationship between value of a_1 and sheet thickness

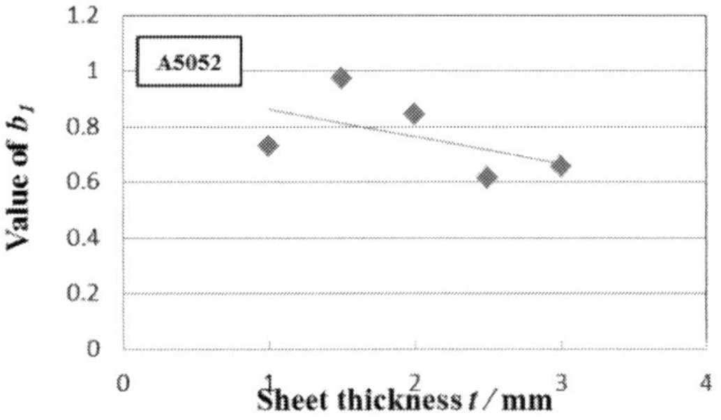

Figure 4. Relationship between value of b_1 and sheet thickness

2.2 Elastic limit load W_p

Fig.5 shows FEM analysis results of W_p for different V/t. Based on the results, W_p is expressed by the equation (5) As shown in **Fig.6** and **Fig.7**, a_2 and b_2 vary simply with the sheet thickness t.

$$W_p = a_2(t)\frac{V}{t} + b_2(t) \tag{5}$$

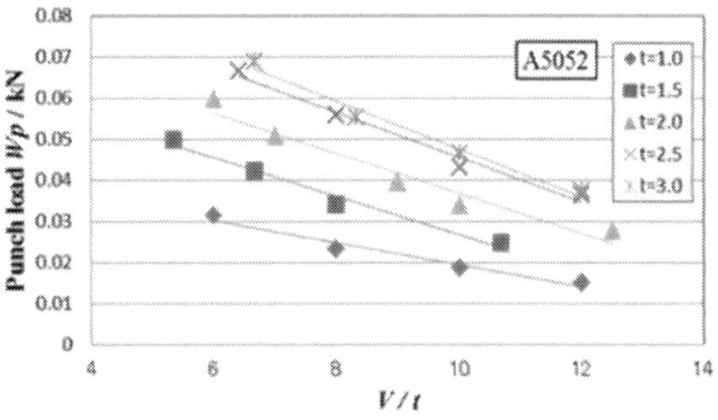

Figure 5. Relation between punch load W_p and V/t

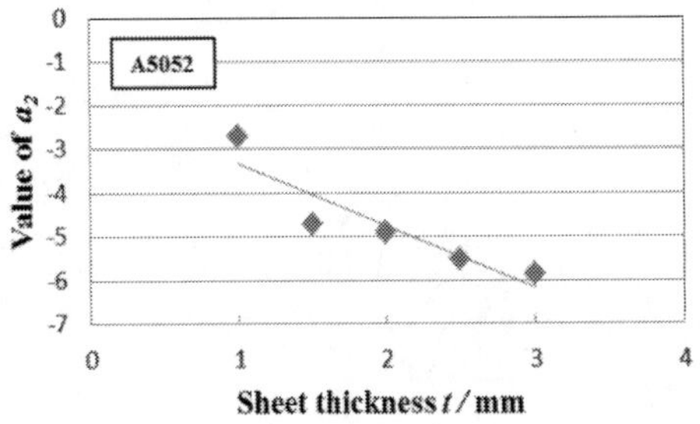

Figure 6. Relationship between value of a_2 and sheet thickness

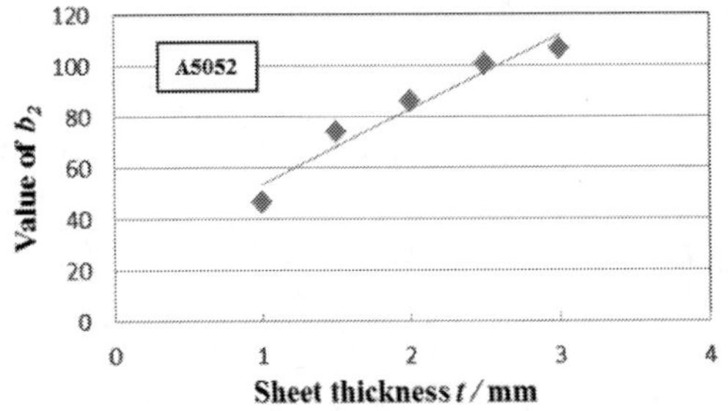

Figure 7. Relationship between value of b_2 and sheet thickness

Fig.8 shows FEM analysis results of S_{tp} for different V/t. Based on the results, S_{tp} is expressed by the equation (6). As shown in **Fig.9** and **Fig.10**, c_2 and d_2 vary simply with the sheet thickness t.

$$S_{tp} = c_2(t)\frac{V}{t} + d_2(t) \qquad (6)$$

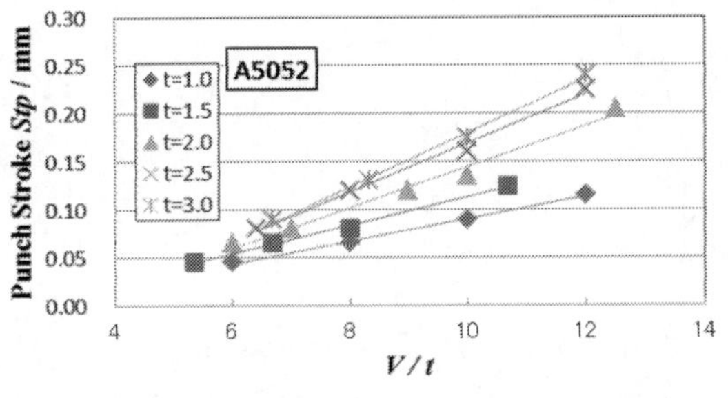

Figure 8. Relationship between punch stroke S_{tp} and V/t

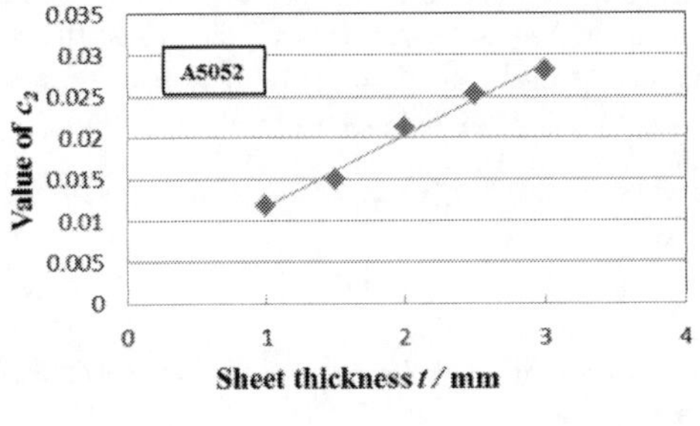

Figure 9. Relationship between value of c_2 and sheet thickness

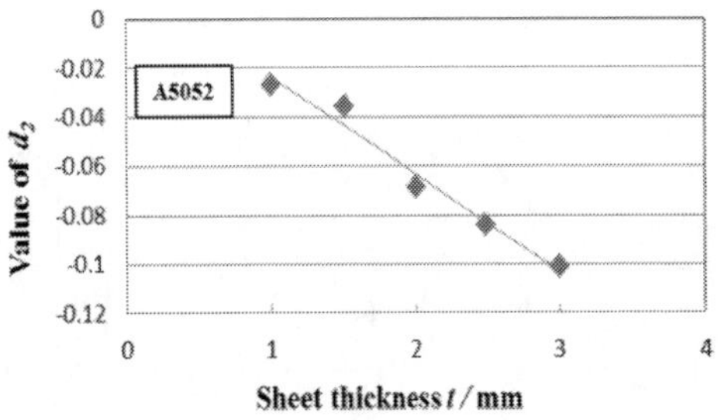

Figure 10. Relationship between value of d_2 and sheet thickness

2.3 Bending load W_3 between the elastic limit and the maximum load point

The formulation of relationship between the bending load W_3 and the punch stroke is difficult, the formulation of relationship between the angle of product and the punch stroke is considered. As reported in the previous paper [9], the stroke S_t for the angle of product θ can be calculated by the equation (7).

$$S_t = \left(t + R_d + \frac{A}{\pi - \left(\frac{\pi}{180} \times \theta\right)} \right)\left(1 - \frac{1}{sin\frac{\theta}{2}} \right) + \left\{\frac{V}{2} + R_d \tan\left(45 - \frac{\emptyset}{4}\right)\right\} cot\frac{\theta}{2} + S_{tp} \qquad (7)$$

Where R_d is the radius of the die shoulder, A is the bend factor, $\emptyset$ is the angle of die groove.

Fig.11 shows FEM analysis results of W_3/W_a as a function of (180-θ)/180. Thus W_3 can be expressed by the equation (8). As shown in **Fig.12** and **Fig.13**, a_3 and b_3 vary simply with V/t.

$$W_3 = a_3\left(\frac{V}{t}\right)\left(\frac{180-\theta}{180}\right)^{b_3\left(\frac{V}{t}\right)} W_a \qquad (8)$$

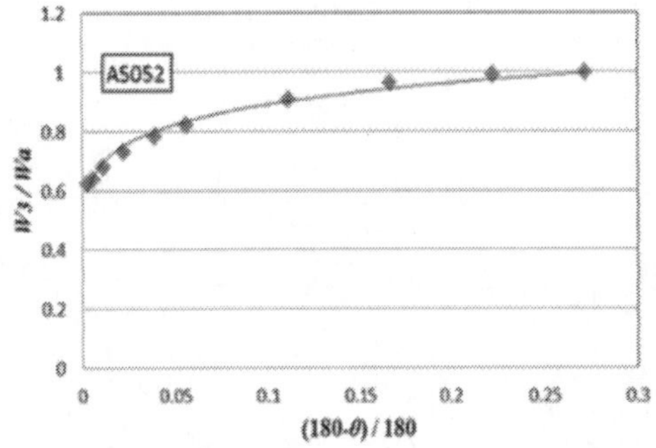

Figure 11. Relationship between W_3/W_a and $(180 - \theta)/180$

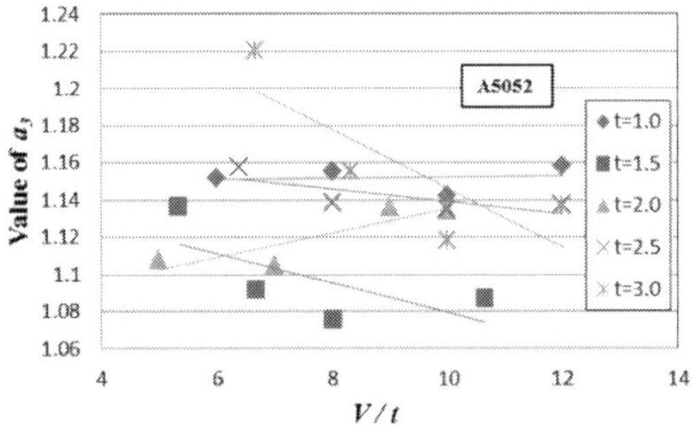

Figure 12. Relationship between value of a_3 and V/t

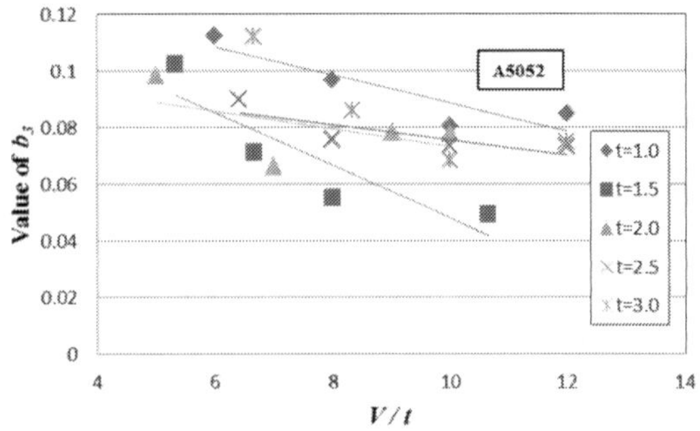

Figure 13. Relationship between value of b_3 and V/t

2.4 Bending load W_4 after the maximum load point

Fig.14 shows FEM analysis results of W_4/W_a. Therefore W_4 is expressed by the equation (9). As shown in **Figs.15 -17**, a_4, b_4 and c_4 vary simply with V/t.

$$W_4 = W_a\left\{a_4\left(\frac{V}{t}\right)\left(\frac{180-\theta}{180}\right)^2 + b_4\left(\frac{V}{t}\right)\left(\frac{180-\theta}{180}\right) + c_4\left(\frac{V}{t}\right)\right\} \tag{9}$$

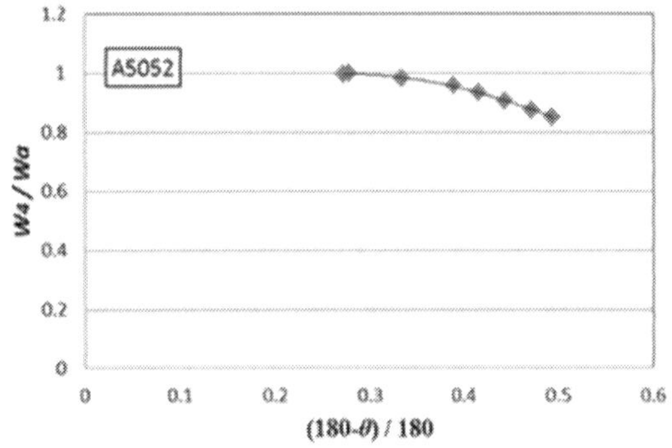

Figure 14. Relationship between W_4/W_a and (180-θ)/180

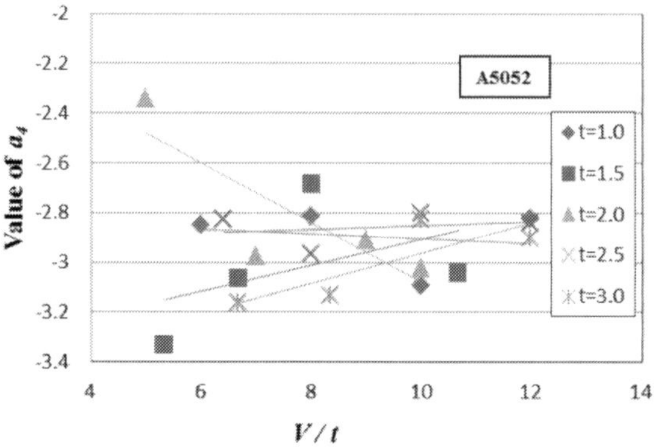

Figure 15. Relationship between value of a_4 and V/t

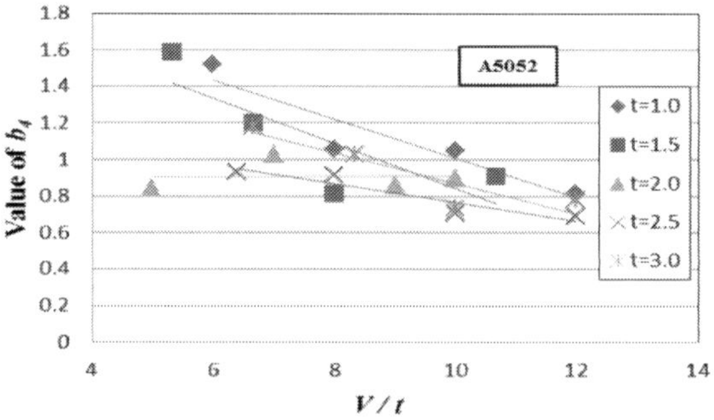

Figure 16. Relationship between value of b_4 and V/t

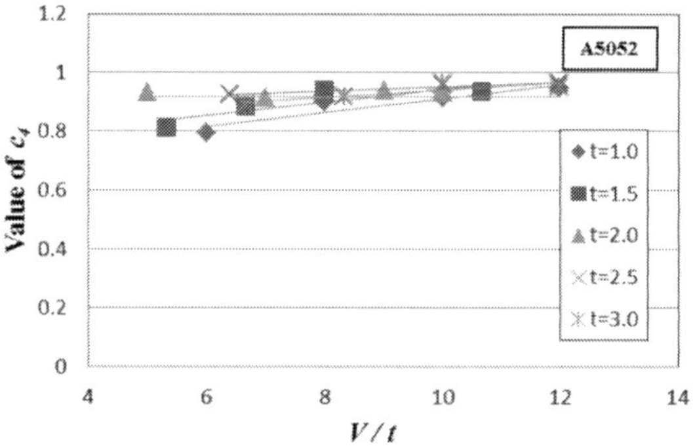

Figure 17. Relationship between value of c_4 and V/t

2.5 Effect of friction on bending load

The above mentioned data of FEM analysis is obtained by a commercial code DEFORM-2D. In the FEM simulation, friction coefficient is assumed zero to ensure the simulation stability and shorten the calculation time. The increase in the bending load due to friction is taken into account by using the equation (10).

$$\Delta W = \frac{\mu W}{\mu + \cot\beta} \tag{10}$$

$$\beta = 90° - \frac{\theta}{2} \tag{11}$$

Where μ is the friction coefficient between the sheet and the die.

3 Verification experiments

To confirm the validity of the proposed formulation, a series of experiments are carried out by using three types of dies. **Table 2** shows the experimental conditions. The punch stroke is measured with a linear gauge; the punch load is measured by a load cell. **Figs.18-20** shows the comparison results. The calculated curve traces precisely the experimental curve in all cases.

Table 2 Experimental conditions

Material	Die width *V* [mm]	Sheet thickness *t* [mm]	Bending width *w* [mm]
SPCC	8	1.53	60.09
	10	1.53	60.62
	12	1.52	60.26

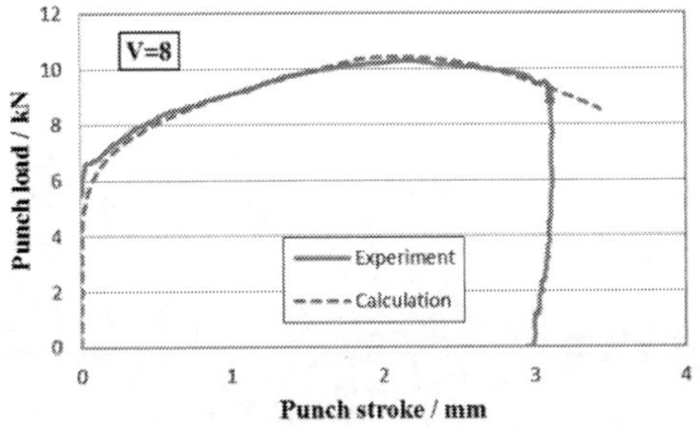

Figure 18. Comparison between measured and calculated load (V=8)

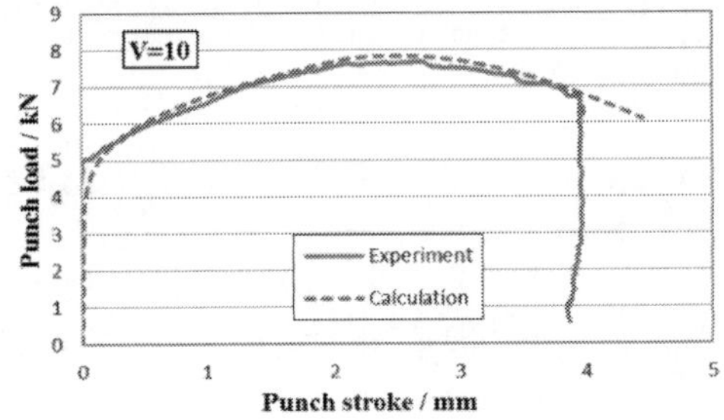

Figure 19. Comparison between measured and calculated load (V=10)

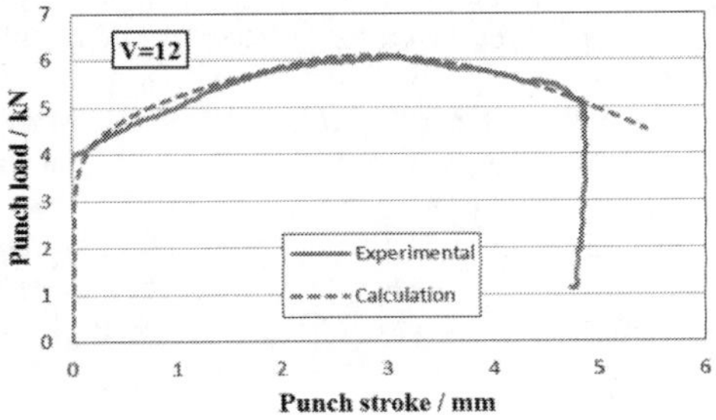

Figure 20. Comparison between measured and calculated load (V=12)

4 Conclusions

The forming load in V-bending is formulated based on the numerical results by FEM analysis. The proposal formulation includes the calculating method of the load of the elastic limit, the maximum load in air bending and the variations of the forming load before/after the bending stroke of the maximum load. The calculated precision is confirmed by comparing with the measured load-stroke curves in V-bending with a press brake.

References

1. Http://www.wila.nl/
2. Http://www.jp.trumpf.com/ja.html
3. D.C. Gossard and B.T. Allison, *The proceedings of the 8th NAMRC*, 252-256 (1980).
4. K.A. Stelson and D.C. Gossard, *Trans. of the ASME, Journal of Engineering for Industry*, **104**, 389-393(1982).
5. K.A. Stelson, *Trans. of the ASME, Journal of Engineering for Industry*, **105**, 45-53 (1983).
6. S. Kim and K.A. Stelson, *Trans. of the ASME, Journal of Engineering for Industry*, **110**, 218-222 (1988).
7. A. Chandra, *Trans. of the ASME, Journal of Engineering for Industry*, **109**, 265-273 (1987).
8. Http://www.amada.co.jp/
9. Z.G. Wang, S. Ikeido, and N. Kanada, *The proceedings of the 2011 Japanese spring conference for the technology of plasticity*, 299-300 (2011).
10. Z.G. Wang, S. Ikeido, and N. Kanada, *The proceedings of the 2011 Japanese spring conference for the technology of plasticity*, 301-302 (2011).
11. Z.G. Wang, S. Ikeido, and N. Kanada, *The proceedings of the 2011 Japanese spring conference for the technology of plasticity*, 303-304 (2011).

MATEC Web of Conferences **45**, 02001 (2016)
DOI: 10.1051/matecconf/20164502001

Modelling and Simulation of Three Level Inverters for Main Drive of the Plate Mill Rolling Stand

Vadim R. Gasiyarov[1], Andrey A. Radionov[1] and Alexander S. Maklakov[1]

[1]*South Ural State University, 454080, 76 Lenin Avenue, Chelyabinsk, Russia*

Abstract. This paper addresses a modelling and a simulation of three level inverters for main drive of the plate mill rolling stand. The main purpose of the article is to develop of a mathematical description of the three level voltage source inverter based on a space vector pulse width modulation and to simulate a consumption current by the main electric drive based on 18-pulse rectification circuit. A twelve-megawatt synchronous motor as a load has been used. The modelling results were simulated at the Matlab/Simulink program. It can allow us to estimate an electromagnetic compatibility of the main electric drive based on transients of the consumption current. The research results can be used to assess the electromagnetic compatibility of powerful three level natural point clamped converters with different types of the space vector pulse width modulation on a load or a grid.

1 Introduction

Nowadays medium voltage reversible electric drives based on synchronous motors and power converters, which consist of active front end rectifiers (AFEs) and voltage source inverters (VSIs), are widely used in electrotechnical systems of rolling stands. Such consumers are able to create a bidirectional power flow at a low negative effect on the grid. Usually medium voltage drives have a high rated power, so a modelling and a researching of these drives are more effective to increase an energy-efficiency of real industrial objects. Power quality and the energy efficiency are very important factors to keep the production costs [1].

Very often the medium voltage AFEs and VSIs have a neutral point clamped (NPC) stricture. A parallel connecting of the AFEs or the VSIs is commonly used to create more than six-pulse rectifier circuits. For example, several transformers with a shift angle of secondary voltages on 20°, 0° and -20° for 18-pulse circuit and 15°, 0°, -15° and 30° for 24-pulse circuit can be applied [2].

A structural feature of powerful AFEs and VSIs is three level NPC topology. It allows one to improve an electromagnetic capability with the grid or a load at a low swathing frequency of semiconductors modules. Total harmonic distortional (THD) of output current or voltage for three level NPC converters is a small magnitude due to the increasing of output voltage levels. High voltage insulate gate bipolar transistors (HV IGBTs), integrated gate-commutated thyristors (IGCTs) and injection enhanced gate transistors (IEGTs) are generally used as the semiconductors modules for AFEs and VSIs [4].

A control of the modules is done by the pulse width modulation (PWM). The most common PWM technology is a space vector (SVPWM) and a selective harmonics elimination (SHE PWM). In AFEs operate based on the SHEPWM and in VSIs are used the SVPWM [5, 6].

2 Problem definition

The main purpose of this paper is to develop a mathematical model of a power circuit of the electric drive of the plate mill rolling stand based on three level NPC inverters with SVPWM. Additional purpose is to evaluate the electromagnetic compatibility of consumption current by the synchronous motor.

3 Development of a mathematical description

3.1 Typical features of the main drive of the plate mill rolling stand

Generally, a main electric drive of plate mill rolling stands is system, in which synchronous motor of each mill roll are connected to the grid through several medium voltage three level NPC converters. The NPC converters consist of AFEs and VSIs. The maximal power of each converter can achieve more than 20 MVA. 18-pulse or 24-pulse rectification circuits are typically used to improve the quality of direct voltage and to reduce the harmonic level of the consumption current. The inductor with high voltage short circuit are applied at the input of motors [7-11]. The characteristic power circuit for main electric drive of plate mill rolling stand is shown in Figure 1.

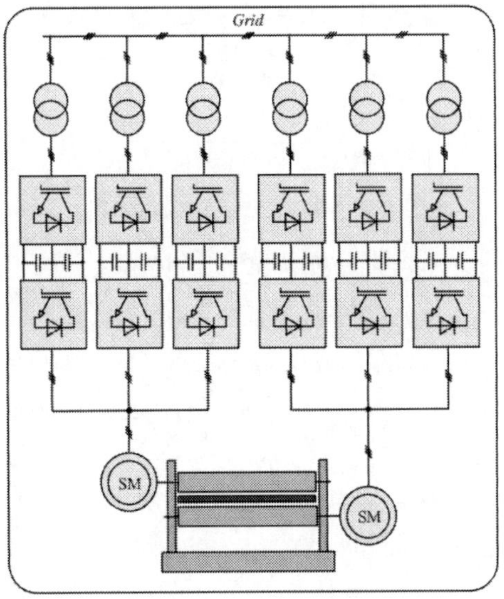

Figure 1. Characteristic power circuit for main electric drive of plate mill rolling stand

3.2 Modelling of a three-level inverter

There are a lot of control methods of three-level inverters, but at present the SVPWM technology is widely used. The exponential increase of digital signal processors (DSPs) has allowed one to develop the complex algorithms SVPWM for multilevel NPC inverters. It is possible to reduce switching losses using low switching frequency of semiconductor modules [12-14]. The three-level VSI based on the NPC structure is shown in Figure 2.

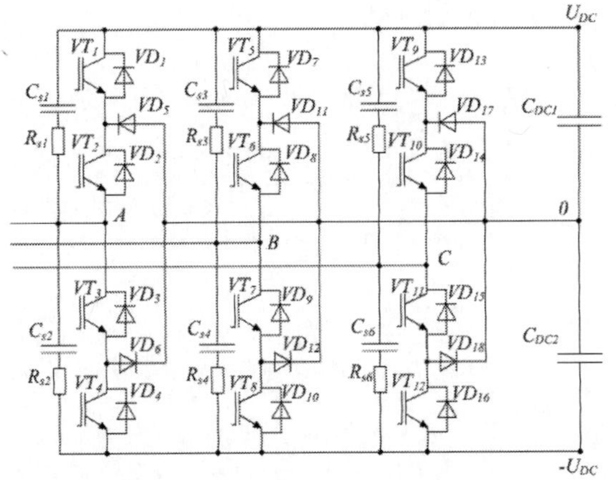

Figure 2. Three level VSI based on NPC structure

The three level VSI in Figure 1 contains twelve HV-IGBT or IEGT transistors ($VT_1 \ldots VT_{12}$) and eighteen diodes ($VD_1 \ldots VD_{18}$). Two equivalent capacitors C_{DC1} and C_{DC2} create a neutral point 0. Snubbers consist of the resistors ($R_{S1} \ldots R_{S6}$) and the capacitors ($C_{S1} \ldots C_{S6}$), which are parallel connected the semiconductor modules [15].

An output phase voltage of a VSI has magnitudes $+U_{dc}/2$, 0, and $-U_{dc}/2$. If the symmetric phase system is considered, all curves of the phase voltages have an odd harmonic spectrum.

According to Kirchhoff's laws, the system equations have been given and written based on snubbers current i_{S1} and i_{S2}, and current i_1, i_2, i_0 flowing through the semiconductor modules for all operating conditions:

$$\begin{cases} i_{S1} = \dfrac{u_{S2} - u_{S1} + u_{DC1} + u_{DC2} + R_S \cdot i_A}{2 \cdot R_S} - \\ \quad - \dfrac{u_{S1} + u_{S1} + u_{DC1} + u_{DC2} + R_S \cdot i_A}{2 \cdot R_S + 4 \cdot R_{on}} \\ i_{S2} = \dfrac{u_{S1} - u_{S2} + u_{DC1} + u_{DC2} + 2 \cdot R_{on} \cdot i_A}{2 \cdot R_S + 4 \cdot R_{on}} - \\ \quad - \dfrac{u_{S1} + u_{S1} - u_{DC1} - u_{DC2}}{2 \cdot R_S} \\ i_1 = \dfrac{u_{S1} + u_{S2} - u_{DC1} - u_{DC2} + R_S \cdot i_A}{R_S + 2 \cdot R_{on}} \end{cases};$$

$$\begin{cases} i_{S1} = \dfrac{u_{S2} - u_{S1} + u_{DC1} + u_{DC2} + 2 \cdot R_{on} \cdot i_A}{2 \cdot R_S - 4 \cdot R_{on}} - \\ \quad - \dfrac{u_{S1} + u_{S1} - u_{DC1} - u_{DC2}}{2 \cdot R_S} \\ i_{S2} = \dfrac{R_{on}(u_{S1} + u_{S2} - u_{DC1} - u_{DC2}) - R_S(u_{S2} + R_{on} \cdot i_A)}{R_S^2 + 2 \cdot R_S \cdot R_{on}} \\ i_2 = \dfrac{u_{S2} - u_{S1} + u_{DC1} + u_{DC2} + R_S \cdot i_A}{R_S - 2 \cdot R_{on}} \end{cases};$$

$$\begin{cases} i_{S1} = \dfrac{u_{S2} + u_{DC1} - u_{S1} - u_{DC2} - 2 \cdot R_{on} \cdot i_A}{2 \cdot R_S + 4 \cdot R_{on}} - \\ \quad - \dfrac{u_{S1} + u_{S1} - u_{DC1} - u_{DC2}}{2 \cdot R_S} \\ i_{S2} = \dfrac{R_{on} \cdot (u_{DC1} - u_{DC2} - u_{S1} - u_{S2})}{R_S^2 + 2 \cdot R_S \cdot R_{on}} \\ \quad + \dfrac{R_S \cdot (u_{DC2} - u_{S2} + R_{on} \cdot i_A)}{R_S^2 + 2 \cdot R_S \cdot R_{on}} \\ i_0 = \dfrac{u_{S2} - u_{S1} + u_{DC1} - u_{DC2} + R_S \cdot i_A}{R_S + 2 \cdot R_{on}} \end{cases};$$

$$\begin{cases} i_{S1} = -\dfrac{1}{2} \cdot \left(i_A + \dfrac{u_{S1} + u_{S2} - u_{DC1} - u_{DC2}}{2 \cdot R_S} \right) \\ i_{S2} = \dfrac{1}{2} \cdot \left(i_A - \dfrac{u_{S1} + u_{S2} - u_{DC1} - u_{DC2}}{2 \cdot R_S} \right) \end{cases},$$

where $R_S = R_{S1} = R_{S2}$ – active resistance of the first and the second snubbers; $R_{on} = R_{onVT1} + R_{onVT2} = R_{onVD1} + R_{onVD2}$ – total active resistance of the first and the second transistors or diodes; i_{S1} and i_{S2} – snubber currents; i_1 – current flowing through the first and the second transistors or diodes; u_{DC1}, u_{DC2} – direct current (DC) voltages of the equivalent capacitors C_{DC1} and C_{DC2}; u_{S1} and u_{S2} – snubber voltages; i – load current of the phase A.

3.3 Modelling of SVPWM

The output three-phase voltages of the VSI has the following system equation:

$$\begin{cases} U_{A0} = \sum_{n=1,3,\ldots\infty}^{\infty} U_n \cdot sin(n \cdot \omega \cdot t) \\ U_{B0} = \sum_{n=1,3,\ldots\infty}^{\infty} U_n \cdot sin\left(n \cdot \omega \cdot t - \frac{2 \cdot \pi}{3}\right) \\ U_{C0} = \sum_{n=1,3,\ldots\infty}^{\infty} U_n \cdot sin\left(n \cdot \omega \cdot t + \frac{2 \cdot \pi}{3}\right) \\ U_n = m \cdot \frac{2 \cdot U_{dc}}{\pi \cdot n} \cdot \left[\sum_{j=1}^{n} (-1)^{j+1} \cdot cos(n \cdot \alpha_i)\right] \end{cases} \quad (1)$$

where α_i – angle switching ($0 < \alpha_i < 180°$); n – harmonics number ($1,3,\ldots,\infty$); ω – angular frequency of synchronous motor; m – modulation index; U_{A0}, U_{B0}, U_{C0} – output phase voltage of the VSI; t – time.

If first harmonic of the system (1) is considered, the space vector theory can be used to write the following system equation [16]:

$$\begin{cases} U_{A0} = U_1 \cdot sin(\omega \cdot t) \\ U_{B0} = U_1 \cdot sin\left(\omega \cdot t - \frac{2 \cdot \pi}{3}\right) \\ U_{C0} = U_1 \cdot sin\left(\omega \cdot t + \frac{2 \cdot \pi}{3}\right) \end{cases} \quad (2)$$

Based on the system equation (2) we can use the space vector theory and apply the following expression:

$$\overline{U} = \frac{2}{3} \cdot \left(U_{A0} \cdot e^{j \cdot 0} + U_{B0} \cdot e^{j \cdot \frac{2 \cdot \pi}{3}} + U_{C0} \cdot e^{j \cdot \frac{4 \cdot \pi}{3}}\right) \quad (3)$$

The equation (3) allows us to determine a reference space vector for an each switching state of the three-level VSI [17]. In Figure 3, the space vector diagram of three level NPC inverters is presented.

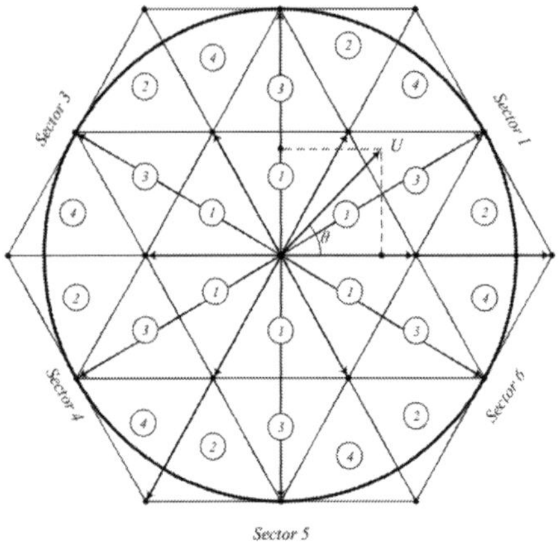

Figure 3. Space vector diagram of three level NPC inverters

The space vector diagram in Figure 3 is a symmetric hexagon with shift on 60° for the each sectors. The angle θ determines the sector number based on the following conditions:

$0° \le \theta < 60°$ – Sector 1;

$60° \le \theta < 120°$ – Sector 2;

$120° \le \theta < 180°$ – Sector 3;

$180° \le \theta < 240°$ – Sector 4;

$240° \le \theta < 300°$ – Sector 5;

$300° \le \theta < 360°$ – Sector 6.

The times calculation T_1, T_2, T_3 for each sector and region are as the following equations [18, 19]:

for region 1:

$$T_1 = T_S - T_2 - T_3 = T_S \cdot \left(1 - 3\frac{U}{U_{dc}}\cos(\theta) - \sqrt{3}\frac{U}{U_{dc}}\sin(\theta)\right)$$

$$T_2 = T_S \cdot \left(3\frac{U}{U_{dc}}\cos(\theta) - \sqrt{3}\frac{U}{U_{dc}}\sin(\theta)\right)$$

$$T_3 = T_S \cdot 2\sqrt{3}\frac{U}{U_{dc}}\sin(\theta)$$

for region 2:

$$T_1 = T_S - T_2 - T_3 = T_S \cdot \left(2 - 3\frac{U}{U_{dc}}\cos(\theta) - \sqrt{3}\frac{U}{U_{dc}}\sin(\theta)\right)$$

$$T_2 = T_S \cdot \left(3\frac{U}{U_{dc}}\cos(\theta) - \sqrt{3}\frac{U}{U_{dc}}\sin(\theta) - 1\right)$$

$$T_3 = T_S \cdot 2\sqrt{3}\frac{U}{U_{dc}}\sin(\theta)$$

for region 3:

$$T_1 = T_S - T_2 - T_3 = T_S \cdot \left(3\frac{U}{U_{dc}}\cos(\theta) + \sqrt{3}\frac{U}{U_{dc}}\sin(\theta) - 1\right)$$

$$T_2 = T_S \cdot \left(1 - 3\frac{U}{U_{dc}}\cos(\theta) + \sqrt{3}\frac{U}{U_{dc}}\sin(\theta)\right)$$

$$T_3 = T_S \cdot \left(1 - 2\sqrt{3}\frac{U}{U_{dc}}\sin(\theta)\right)$$

for region 4:

$$T_1 = T_S - T_2 - T_3 = T_S \cdot \left(2 - 3\frac{U}{U_{dc}}\cos(\theta) - \sqrt{3}\frac{U}{U_{dc}}\sin(\theta)\right)$$

$$T_2 = T_S \cdot \left(3\frac{U}{U_{dc}}\cos(\theta) - \sqrt{3}\frac{U}{U_{dc}}\sin(\theta)\right)$$

$$T_3 = T_S \cdot \left(2\sqrt{3}\frac{U}{U_{dc}}\sin(\theta) - 1\right)$$

The example of space vector formation in the Sector 1 and the region 1 is shown in Figure 4.

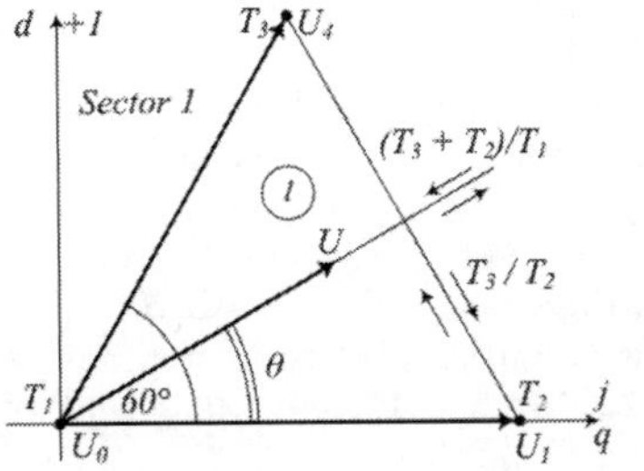

Figure 4. Space vector formation

A ratio of the times T_1, T_2, and T_3 regulate the space vector depth and the angle by using the following expression:

$$\frac{T_3 + T_2}{T_1}$$ – the space vector depth;

$$\frac{T_3}{T_2}$$ – the space vector angle.

4 Simulation of the main drive of the plate mill rolling stand

The simulation model of the main drive of the plate mill rolling stand based on the three level NPC inverters with SVPWM in Matlab/Simulink was developed. The block diagram of this model is shown in Figure 5.

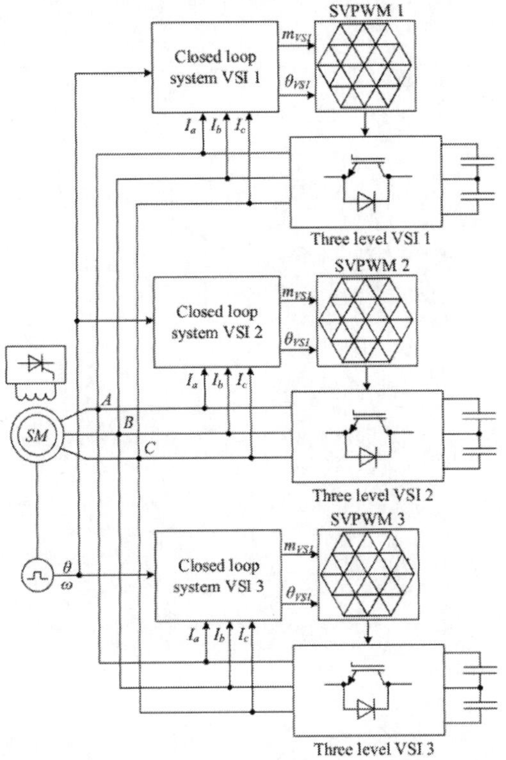

Figure 5. Simulation model of the main drive of the plate mill rolling

A closed loop system of three level NPC inverter has been developed on the basis of a classical structure orientated according to the direction of angular position of synchronous motor, which in the articles [18, 20, 21] has been considered. Synthesis of regulators was developed using the step-by-step correction principle. The modeling results of a consumption phase current by the synchronous motor and the output phase-to-phase voltage of VSI are shown in Figure 6 and Figure 7.

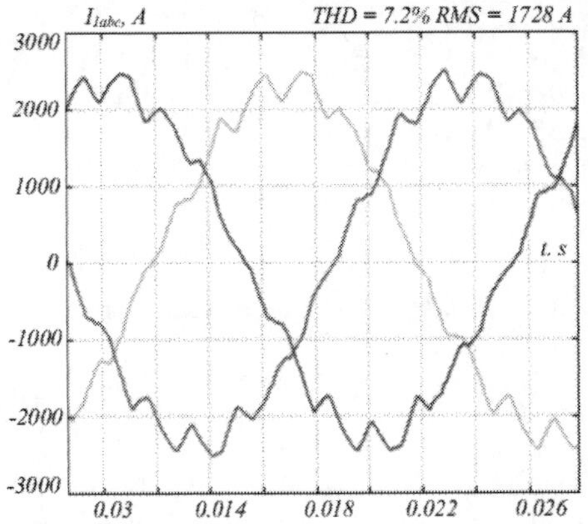

Figure 6. Consumption phase current by the synchronous motor

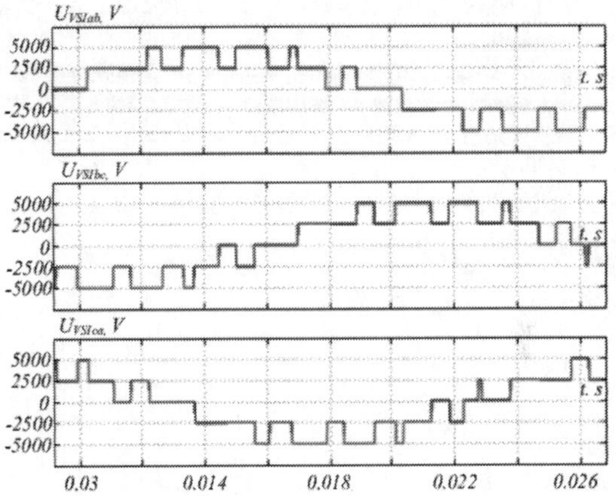

Figure 7. Output phase-to-phase voltage of VSI

5 Conclusion

The mathematical description and the simulation of three level VSI based on NPC structure for main drive of the plate mill rolling stand have been made in the article. The mathematical description of the three level VSI based on the SVPWM and to simulate the consumption current by the main electric drive based on 18-pulse rectification circuit have been developed. The modelling results were simulated at the Matlab/Simulink program. This model can be used to estimate the electromagnetic compatibility of the main electric drive based on transients of the consumption current. The research results can be employed to assess the electromagnetic compatibility of the powerful three level NPC converters with different types of the SVPWM on the load or the grid.

References

1. Leopoldo G. Franquelo, Jose Rodríguez, Jose I. Leon, Samir Kouro, Ramon Portillo and Maria A.M. Prats, IEEE industrial electronics magazine, **2**, 28-39 (2008)
2. J. Rodriguez, J.-S. Lai, F.Z. Peng, IEEE Trans. Ind. Electron., **49**, 4, 724-738 (2002)
3. J. Rodriguez, S. Bernet, B. Wu, J. Pontt, and S. Kouro, IEEE Trans. Ind. Electron., **54**, 6, 2930-2945 (2007)
4. A. Nabae, I. Takahashi, and H. Akagi, IEEE Trans. Ind. Applicat., **1A-17**, 5, 518-523 (1981)
5. ABD Almula G. M. Gebreel, *Simulation and implementation of two-level and three-level inverters by matlab and RT-LAB*, 136 (2011)
6. Mr. Anupam Mishra, Mr. Sunil Panda, Mr. B. Srinivas, *Control of voltage source inverters using PWM/SVPWM for adjustable speed drive applications*, **43** (2009)
7. A.S. Maklakov, A.A. Radionov, *Actual Problems of Electronics Instrument Engineering (APEIE), 2014 12th International Conference*, 770-774 (2014)
8. A.S. Maklakov, A.A. Radionov, Russian Internet Journal of Industrial Engineering, **2**, 40-47 (2013)
9. B. Wu, Y. Lang, N. Zargari, S. Kouro, *Power Conversion and Control of Wind Energy Systems*, 453 (2011)
10. B. K. Bose, *Power Electronics and Motor Drives: Advances and Trends*, (2006)
11. B. Wu, *High-Power Converters and AC Drives*, (2006)
12. Dong-Myung Lee, Jin-Woo Jung, and Sang-Shin Kwak, Journal of Power Electronics, **11**, 5, 688-696 (2011)
13. Haibing Hu, Wenxi Yao, and Zhengyu Lu, Trans. on Power Electr., **22**, 6, 2234-2244 (2007)
14. Remus Teodorescu, Marco Liserre, Pedro Rodrıguez. *Grid converters for photovoltaic and wind power systems*, 398 (2011)
15. T.R. Khramshin, D.S. Krubtsov, G.P. Kornilov, Russian Internet Journal of Electrical Engineering, **1**, 1, 3-7 (2014)
16. L. Malesani, L. Rossetto, P. Tomasin, P. Tenti, IEEE Trans. on Ind. App., **31**, 2, 287-292 (1995)
17. V. Blasko, V. Kaura. IEEE Trans. Ind. Electron., **12**, 1, 116–123 (1997)
18. A.S. Maklakov, Russian Internet Journal of Industrial Engineering, **3**, 16-25 (2014)
19. A.S. Maklakov, A.A. Radionov, *Proceedings of the 2015 IEEE North West Russia Section Young Researchers in Electrical and Electronic Engineering Conference, ElConRusNW 2015*, 236-240 (2015)
20. A.A. Radionov, A.S. Maklakov, E.A. Karyakina, *Proceedings 2015 International Siberian Conference on Control and Communications, SIBCON 2015*, (2015).
21. A.A. Radionov, A.S. Maklakov, V.R. Gasiyarov, *Proceedings of 2014 International Conference on Mechanical Engineering, Automation and Control Systems, MEACS 2014*, (2014).

MATEC Web of Conferences **45**, 02002 (2016)
DOI: 10.1051/matecconf/20164502002

An Improved VI-CFAR Detector Based on GOS

Likai Zhao, Sen Li and Guangzhao Hu

School of Information and Electronics, Beijing Institute of Technology, Beijing 100081, P. R. China

Abstract. In combination with the advantages of CA-CFAR, GO-CFAR and SO-CFAR algorithm, the VI-CFAR has strong adaptability both in homogeneous and non-homogeneous environment. However, if the interfering targets are present in both the halves of the reference sliding windows, the use of the window with the smallest mean is affected by them and therefore results in a performance degradation. In order to overcome the shortcoming, an improved VI-CFAR detector based on GOS (IVI-CFAR) is proposed in this paper. We introduce the IVI-CFAR detector and make performance simulation and analysis in homogenous and non-homogenous environment. In the homogeneous environment, the IVI-CFAR detector has some CFAR loss relative to the CA-CFAR detector. In the clutter edge environment, the IVI-CFAR detector keeps the good performance of the VI-CFAR detector. In multiple interfering targets environment, the IVI-CFAR detector performs robustly, which is similar to the OS-CFAR detector. In addition, the IVI-CFAR detector shortens the sample sorting time of the OS-CFARR detector.

1 Introduction

In radar and sonar signal detection, in order to get a constant false alarm (CFAR) performance, the actual average power of interfering background will be estimated by the reference cells near the test cell to adaptively set the detection threshold. For the different method of background power estimation, the method of CFAR is also different. Conventional CFAR detectors include average cell CFAR (CA–CFAR[1]) detector and order statistics CFAR (OS-CFAR[2]) detector. Under the homogeneous background, the CA-CFAR detector can get close to the optimal detection performance, but in the non-homogeneous environment, the detection performance of CA-CFAR seriously declines. When the actual number of the interfering targets is less than the predetermined number, the OS-CFAR detector has good detection performance, and in the homogeneous environment the CFAR loss can also be acceptable. However, in the clutter edge environment, the OS-CFAR detector has a high false alarm peak, then the false alarm control ability is poor. In order to overcome the disadvantages of the CFAR detectors, the VI-CFAR detector was proposed by Smith and Varshney [3, 4]. In this detector, the data in the reference sliding window is used to compute two statistics VI and MR. Based on these statistics, two tests are performed in order to select the algorithms (CA-CFAR, GO-CFAR [5] and SO-CFAR [6]) to be used for the estimation of the clutter power in the test cell [7]. So the detector combines the advantages of the CA-CFAR, GO-CFAR and SO-CFAR algorithms. Under the homogeneous background, the VI-CFAR detector has a small CFAR loss relative to the CA-CFAR detector. Under the non-homogeneous background, the VI-CFAR detector also has strong robustness. However, when the interfering targets appear in both the leading and lagging sliding window, the possibility of the VI-CFAR detector selecting the SO-CFAR algorithm increases, this will lead to the serious deterioration of the performance of the VI-CFAR detector. Based on the OS-CA CFAR algorithm [8, 9], an improved VI-CFAR (IVI-CFAR) detector is introduced .Through the Monte-Carlo simulation trials, this paper simulates the IVI-CFAR detectors detection and analyses the detection performance.

This paper is organized as follows. Section 2 introduces the principle of the VI-CFAR detector. In Section 3, an IVI-CFAR detector is proposed and the principle of the IVI-CFAR detector is introduced in detail. Simulation results and discussions are given in Section 4. And we make conclusions in Section 5.

2 The Principle VI-CFAR Detector

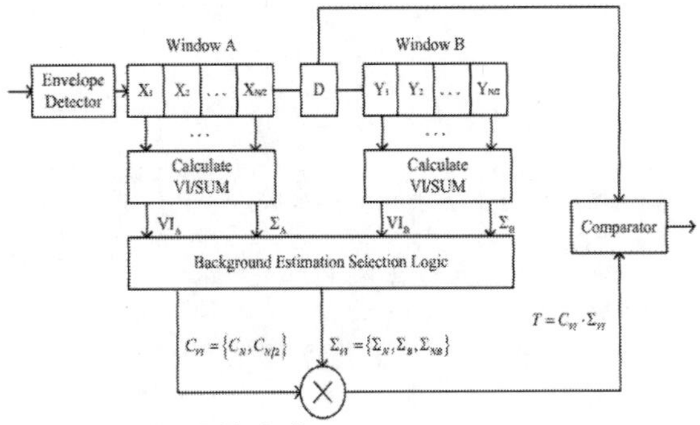

Figure 1. VI-CFAR block diagram.

Figure 1 is the block diagram of the VI-CFAR detector, where D is the test cell, the length of the leading reference sliding window A or the lagging reference sliding window B is N/2. Through VI and MR, the VI-CFAR detector adaptively selects the reference sliding window to estimate the background clutter power. Adaptive detection threshold T is equal to the product of the threshold factor C_{VI} and the background clutter power estimation. According to the comparison result of the test cell with the detection threshold T, the VI-CFAR detector judges whether the target exists or not. Adaptive decision criterion is: if D>T, the result is H_1 and the target exists in the test cell; if D≤T, the result is H_0, and the target does not exist in the test cell.

This article assumes that the signal is independent, identically distributed (IID), zero mean, Gaussian random process. Consequently, the envelope amplitude at the output of the square-law detector is an exponentially distributed random variable. The samples in the reference sliding window are independent of each other and of the sample in the test cell. When a target is present in the test cell, the use of guard cells (not shown in Figure 1) between the test cell and the reference sliding window prevents target energy from corrupting the reference sliding window.

The VI-CFAR detector utilizes the VI as well as the MR to select the subset of reference cells used for background clutter estimation. The test statistics are defined by formula (1) and (2) respectively. In order to reduce the computational, VI can be simplified as VI*, of which the defined formula is shown by formula (3). In formula (1)-(3), $\hat{\sigma}^2$ is the estimated population variance, and $\hat{\mu}$ is the estimated population mean. X_i is the samples of the leading or lagging half of the reference sliding window. $\overline{X}$ is the arithmetic mean of X_i in a half reference sliding window, and n equals N/2, N is the size of the whole reference sliding window.

$$VI = 1+\frac{\hat{\sigma}^2}{\hat{\mu}^2} = 1+\frac{1}{n-1}\cdot\sum_{i=1}^{n}(X_i-\overline{X})^2\Big/(\overline{X})^2 \tag{1}$$

$$MR = \overline{X}_A/\overline{X}_B = \sum_{i\in A}X_A\Big/\sum_{i\in B}X_B \tag{2}$$

$$VI^* = 1+\frac{\hat{\sigma}^2}{\hat{\mu}^2} = 1+\frac{1}{n}\cdot\sum_{1}^{n}(X_i-\overline{X})^2/(\overline{X})^2 = n\cdot\sum_{i=1}^{n}(X_i)^2\Big/\left(\sum_{i=1}^{n}X_i\right)^2 \tag{3}$$

By the following two hypothesis tests formula (4) and formula (5), the VI-CFAR detector determines whether the background is homogeneous or not, and adaptively selects the reference sliding window to adopt corresponding CFAR algorithm

$$\begin{cases} VI \le K_{VI} \Rightarrow \text{Homogeneous clutter background} \\ VI > K_{VI} \Rightarrow \text{Non-homogeneous clutter background} \end{cases} \tag{4}$$

$$\begin{cases} K_{MR}^{-1} \le MR \le K_{MR} \Rightarrow \text{Same Mean} \\ MR > K_{MR} \text{ or } MR < K_{MR}^{-1} \Rightarrow \text{Different Mean} \end{cases} \tag{5}$$

K_{VI} and K_{MR} are VI and MR decision thresholds respectively. The selections of them have relation to predetermined false-alarm probability and confidence levels. The corresponding error probability is shown in equation (6) and equation (7) respectively.

$$\alpha_0 = P(VI > K_{VI}\,|\text{Homogeneous background}) \tag{6}$$

$$\beta_0 = 1-P(K_{MR}^{-1} \le MR \le K_{MR}\,|\text{Same Mean}) \tag{7}$$

By the above equations (6)-(7), the improvement of threshold K_{VI} and K_{MR} can reduce the error probability respectively. But at the same time, the judgment sensitivity of non-homogeneous background also can be reduced.

3 An Improved VI-CFAR Detector Based on GOS

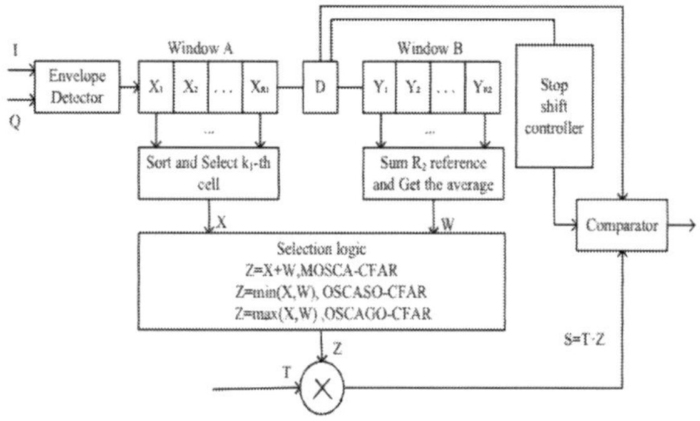

Figure 2. OS-CA CFAR block diagram.

According to the Figure 2, the OS-CA CFAR detector block diagram can be seen. The common characteristic of the OS-CA CFAR detector is: the leading reference sliding window makes local estimation in OS method, the lagging reference sliding window makes local estimate in CA method. At the same time, stop shift control logic constitutes the automatic screening technology with a reference cell shift register. When a target is declared in the test cell D, the lagging reference sliding window stops shift to eliminate the sample in the test cell.

According to the outcomes of the VI hypothesis test and MR hypothesis test, the IVI-CFAR detector adaptively chooses corresponding CFAR detection algorithm. When the clutter edge is present, the IVI-CFAR detector adopts the OSCAGO-CFAR and the order value k_1 of the OSCAGO-CFAR is set to a large value. When the interfering targets exist in the leading reference sliding window or the lagging reference sliding window, the IVI-CFAR detector also uses the OSCAGO-CFAR, but the order value k_2 of the OSCAGO-CFAR is set to a small value. When the interfering targets exist in both the halves of the reference sliding window, the IVI-CFAR detector adopts the OSCASO-CFAR.

The specific adaptive threshold production rules and the corresponding CFAR algorithm selections of the IVI-CFAR detector are shown in **Table 1**. C_N, C_{OSCAGO} and C_{OSCASO} are the threshold multiplier factors, which correspond to different CFAR method. Σ_{AB}, Σ_A and Σ_B are the sums of the whole and the half reference sliding respectively. $X_A(k_1)$, $X_A(k_2)$ and $X_A(k_3)$ are the local estimation of the leading reference sliding window in OS method respectively. $X_B(k_2)$ is the local estimation of the lagging reference sliding window in OS method.

Table 1. IVI-CFAR Adaptive Threshold.

Leading Window Variable?	Lagging Window Variable?	Different mean?	Adaptive Threshold	CFAR Algorithm
No	No	No	$C_N \cdot \Sigma_{AB}$	CA
No	No	Yes	$C_{OSCAGO}(k_1)\cdot OSCAGO_{\max\left(X_A(k_1),\frac{1}{n}\Sigma_B\right)}(k_1)$	OSCAGO
Yes	No	-	$C_{OSCAGO}(k_2)\cdot OSCAGO_{\max\left(X_A(k_2),\frac{1}{n}\Sigma_B\right)}(k_2)$	OSCAGO
No	Yes	-	$C_{OSCAGO}(k_2)\cdot OSCAGO_{\max\left(X_B(k_2),\frac{1}{n}\Sigma_A\right)}(k_2)$	OSCAGO
Yes	Yes	-	$C_{OSCASO}(k_3)\cdot OSCASO_{\min\left(X_A(k_3),\frac{1}{n}\Sigma_B\right)}(k_3)$	OSCASO

Under the homogeneous background, the threshold multiplier factor C_N is determined based on CA-CFAR with N cells as shown in formula (8) [10]:

$$C_N = P_{fa}^{-1/N} - 1 \tag{8}$$

When the interfering targets are present in a single half of the reference sliding window or the clutter edge is around the test cell, the calculation of C_{OSCAGO} is based on OSCAGO-CFAR as shown in formula (9):

$$\begin{aligned} P_{fa}^{\text{OSCAGO}} = & \frac{R_1!\Gamma(R_1 - \mathrm{k} + C_{OSCAGO} + 1)}{\Gamma(R_1 + C_{OSCAGO} + 1)} \\ & - \frac{1}{R_2}\sum_{i=0}^{R_2-1}\sum_{j=0}^{k-1}\binom{k-1}{j}(-1)^j \\ & \times\left(\frac{R_2}{R+1-k+j+C_{OSCAGO}}\right)^{i+1} \\ & + \sum_{i=k}^{R_1}\binom{R_1}{i}\sum_{j=0}^{i}\binom{i}{j}(-1)^i \\ & \times\left(\frac{R_2}{R-i+j+C_{OSCAGO}}\right)^{R_2} \end{aligned} \tag{9}$$

In the presence of interfering targets in both the halves of the reference sliding window, the value of C_{OSCAGO} is based on OSCASO-CFAR and obtained by formula (10):

$$\begin{aligned} P_{fa}^{\text{OSCASO}} = & \frac{1}{\left(1+{}^{C_{OSCASO}}\!/_{R_2}\right)^{R_2}} + k\binom{R_1}{k}\frac{1}{R_2}\sum_{i=0}^{R_2-1}\sum_{j=0}^{k-1}\binom{k-1}{j}(-1)^j \\ & \times\left(\frac{R_2}{R+1-k+j+C_{OSCASO}}\right)^{i+1} \\ & - \sum_{i=k}^{R_1}\binom{R_1}{i}\sum_{j=0}^{i}\binom{i}{j}(-1)^i \\ & \times\left(\frac{R_2}{R-i+j+C_{OSCASO}}\right)^{R_2} \end{aligned} \tag{10}$$

4 Simulation results and discussions

To illustrate the detection performance of the IVI-CFAR detector in homogeneous and non-homogeneous environments characterized by the presence of interfering targets and the presence of clutter edge, we adopt 100000 Monte-Carlo simulation trials. The whole reference sliding window size N is 36, the P_{fa} is fixed to 10^{-4}, and the main target(in test cell) and interfering targets are both Swerlling Ⅱ. The confidence level of VI hypothesis test α_0 is fixed to 3.3×10^{-4}, the confidence level of MR hypothesis test β_0 is set as 0.08 and corresponding K_{VI} and K_{MR} are equal to 4.76 and 1.806 respectively[11].

4.1. Performance in Homogeneous Environment

Figure 3 shows the probability of detection (P_d) of the IVI-CFAR, CA-CFAR, GO-CFAR, SO-CFAR and VI-CFAR in the homogeneous environment. The symbol "OS(30)" denotes that the order value of the OS-CFAR is 30. The symbol "IVI(15,6,10)" denotes that the order value of the OSCAGO-CFAR is 15 in clutter edge environment and 6 when interfering targets exist in a single half of the reference sliding window. In addition, the order value of the OSCASO-CFAR is 10 when the interfering targets are present in both halves of the reference sliding window. From Figure 3,we can see that the IVI-CFAR detector has a little CFAR loss relative to the CA-CFAR and outperforms the GO-CFAR and VI-CFAR detectors. When P_d=0.5, the CFAR loss of the IVI-CFAR relative to the CA-CFAR is about 0.1 dB. The CFAR loss for the VI-CFAR relative to the IVI-CFAR is about 0.15dB.

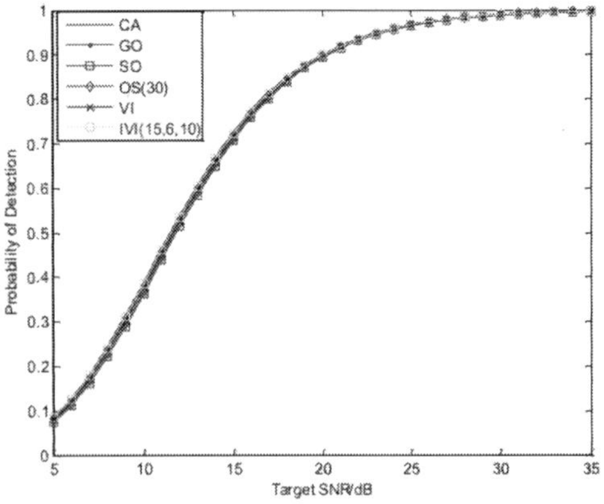

Figure 3. P_d comparison of the CFAR detectors in homogeneous environment.

4.2. Performance in Multi-target Environment

It is assumed that interfering-to-noise ratio (INR) is equal to the signal-to-noise ratio (SNR) in the multi-target environment in this paper[12]. In order to analyse the different effect of interfering targets location in the reference sliding window on the detection performance of the VI and IVI-CFAR, this paper analyses in two different conditions [12], i.e. the interfering targets only appear in the leading reference sliding window A and the interfering targets exist in the both halves of reference sliding window at the same time.

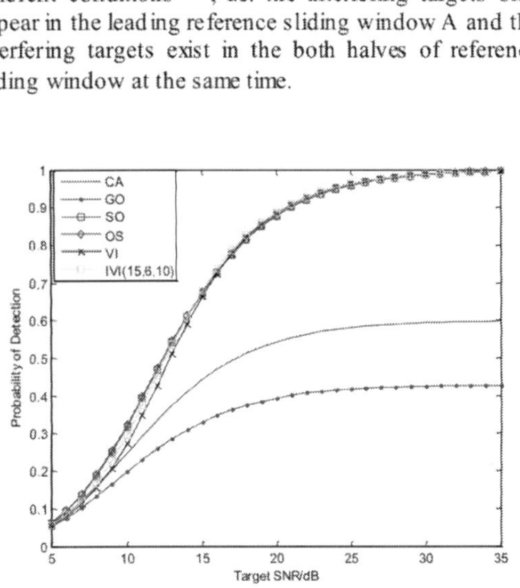

Figure 4. P_d comparison of the CFAR detectors when two interfering targets in window A.

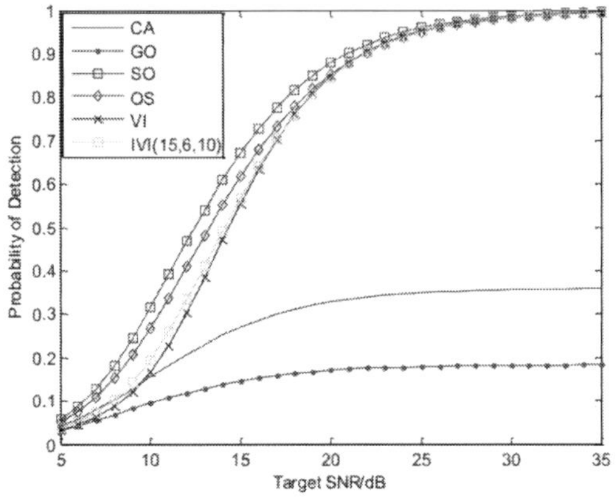

Figure 5. P_d comparison of the CFAR detectors when four interfering targets in window A.

Figure 4 and **Figure 5** show the P_d of the CA-CFAR, GO-CFAR, SO-CFAR, OS-CFAR, VI-CFAR and IVI-CFAR in the case of two interfering targets in the leading reference sliding window and four interfering targets in the lagging reference sliding window respectively. As can be seen, the IVI-CFAR exhibits a low CFAR loss relative to the SO-CFAR and outperforms the VI-CFAR and VI-CFAR. With the increase of the value of SNR, the P_d of IVI-CFAR and VI-CFAR approach that of OS-CFAR. In Figure 4,when P_d=0.5, the CFAR loss of the IVI-CFAR relative to the SO-CFAR is about 0.15 dB. The CFAR loss of the VI-CFAR relative to the IVI-CFAR is about 0.3dB.

Figure 6 shows the P_d of the CA-CFAR, GO-CFAR, SO-CFAR, OS-CFAR, VI-CFAR and IVI-CFAR in the case of two interfering targets in both the halves of the reference sliding window. It can be seen from Figure 6 that the thresholds of both the VI-CFAR and SO-CFAR are overestimated, so the detection performance degerade quickly while the OS-CFAR and IVI-CFAR perform well.

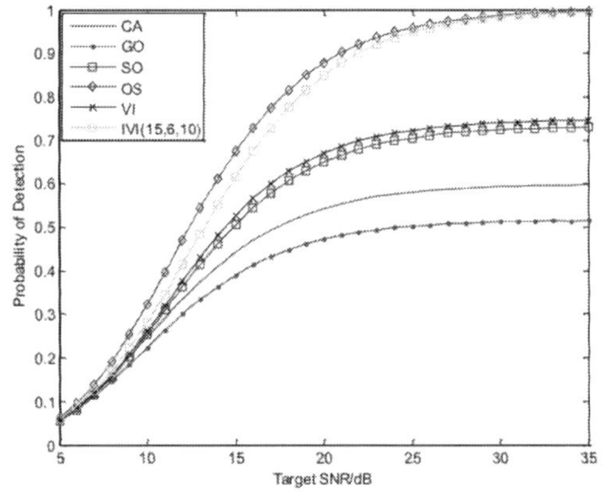

Figure 6. P_d comparison of the CFAR detectors when one interfering target in each half window.

4.3. Performance in Clutter Edge Environment

It is assumed that clutter envelope obeys Rayleigh distribution, and clutter-to-noise ratio is CNR. In Monte-Carle simulation trails, the clutter edge progressed from left to right(window A to window B). The P_{fa} of the CA,GO,SO,OS,VI and IVI-CFAR in clutter edge environment is shown in **Figure 7** where the CNR is equal to 10dB. In this case, it can be seen that the false-alarm regulation properties of the IVI-CFAR and VI-CFAR are almost consistent. The false-alarm regulation properties of the CA-CFAR and OS-CFAR are poor relatively and that of SO-CFAR is the worst.

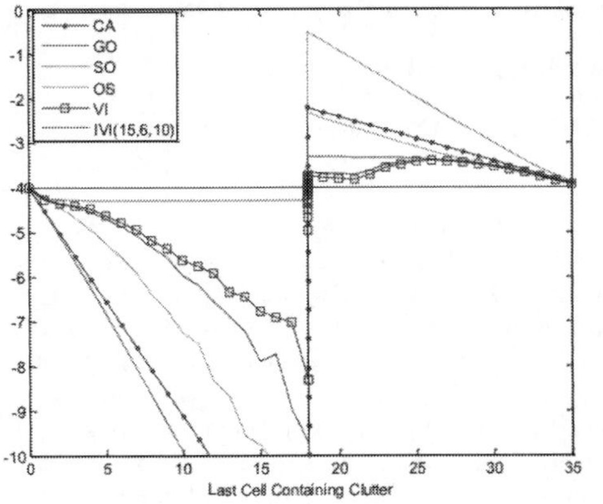

Figure 7. P_{FA} comparison of the CFAR detectors in clutter edge (CNR=10dB) environment

5 Conclusions

In this paper, we have presented an improved version of VI-CFAR, called IVI-CFAR detector and have analyzed the performance of the proposed IVI-CFAR detector in a homogeneous and non-homogeneous environments. In the homogeneous environment, the detection performance of the IVI-CFAR detector has some CFAR loss relative to the CA-CFAR detector and outperforms the GO-CFAR and VI-CFAR detectors. In the multiple targets situations, the IVI-CFAR detection is more robust than the VI-CFAR detector when the interfering targets are present in both the halves of the reference sliding window. In clutter edge environment, the performance of the IVI-CFAR detector is almost consistent with that of the VI-CFAR. In addition, the false-alarm regualtion property of the IVI-CFAR is even better than that of the GO-CFAR.

Acknowledgment

This research was supported in part by the National Natural Science Foundation of China under Grant No. 61421001, 61331021, 61172176 and 61490691

References

1. HM. Finn, RS.Johson, RCA Review.**29**, 50(1968)
2. H. Rohling, IEEE Transaction on AES, **19**,13 (1983)
3. ME.Smith,PK.Varshney.IEEEINTLRADAR,6 (1997)
4. ME.Smith,PK.Varshney. IEEE Transaction on AES,**36**,11 (2000)
5. V.G. Hansen, IEEEINTLRADAR,8 (1973)
6. G.V. Trunk., IEEE Transaction on AES,**14**,6 (1978)
7. K. Cheikh, F. Soltani, ICSIPA2011,4 (2011)
8. H. You, G. Jian, P. Yning, L. Dajin, CIC, 3 (1996)
9. Y. He, J. Guan, IEEE, 3 (1995)
10. K.Siddiq,IEEE, 4(2006)
11. A.Cong, Y.Xu, Q.Yao. IEEE, **9**,4(2011)
12. W.Hu, Y.Wang, S.Wang, IEEE, 4(2006)

MATEC Web of Conferences 45, 02003 (2016)
DOI: 10.1051/matecconf/20164502003

A Finite Element Modeling Study on the Fingertip Deformation under Pressure Stimulation

Huiling Chen[1], Liguo Shuai[1], Lei zhang[2], Mei Miao[3]

[1]*School of Mechanical Engineering, Southeast University, Nanjing 211189, China*
[2]*School of Mechanical Engineering, Nantong University, Nantong 226019, China*
[3]*Institute of Mathematics and Computer, Jiangsu Second Normal University, Nanjing 210013, China*

Abstract. Pressure stimulus causes skin deformation and tactile sensation on the fingertip. Both theoretical approach and experimental technique may be used to investigate the relationship between the deformation and the sensation. Building an appropriate skin model is the most important step for further theoretical and experimental analysis. In this paper, a two dimensional (2D) fingertip biomechanical model employing finite element (FE) method is proposed based on the physiological structure of skin. With biomechanical and electrophysiological simulations, the predicted distributions of the strain energy density (SED) and the stress/strain are obtained. The relation between the predicted biomechanical responses of the subcutaneous tissue and the discharged rate reported in the literatures are investigated. The results show that the soft tissues of fingertips are very sensitive to the external stimulus, and the spatial distribution characteristics of SED within soft tissues can explain the evoked charging rate of mechanoreceptors effectively. The simulation data of the proposed FE model is highly consistent with the verified data.

1 Introduction

The touch feeling on skin mainly includes pressure, vibration and temperature perception. When someone touches an object with his finger, his skin will be deformed under force, and the inside receptors may also have a subtle deformation, which will trigger a tactile related electrical impulse. This signal is transferred via the peripheral nerve system and finally transmitted to the brain to tell the subject that there is a touch action on the fingertip. The nerve impulse frequency is associated with both the mechanical environment such as stress/strain and the SED around the receptors [1]. The investigation in biomechanics of tactile perception may help us to understand the mechanism of tactile sensation.

Two categories of fingertip models have been proposed on the tactile discrimination of one or two-point tactile stimuli. One category is the structural model, which refers to linear elastic half-space model and the multilayer FE model [2]. The other is the physical model constructed based on the quasi-viscoelastic constitutive equation [3]. [4] developed an elastic half-space model based on standard continuum mechanics theory. It was assumed that for small deformations the mechanical properties of skin were adequately approximated by a homogeneous, elastic, and isotropic medium. However, the model, without considering the multi-layers of skin, could not describe the mechanics property of fingertip exactly under loading. [5] built a 'waterbed' model on the basis of [4], however, it did not take the skin thickness into account. Based on the waterbed model, when skin was forced by a line load, the surface deformation did not agree well with the real skin. [6] constructed a finger model considering the material heterogeneity and geometric structure, and used the finite deflection theory to adapt the large deformation under loading. This model could simulate the surface deflection profiles of finger under load, but it was only suitable for the applications of the given load. An FE model based on anatomical structure of fingertip may potentially overcome those limitations mentioned above. [7] developed an FE model to analyze the influence of the complexity and the fine structure of finger skin (such as epidermal ridges and papillate) on the deformation of the soft tissue inside the fingertip. [8] built an FE model to study the injury of soft tissue and bone in the fingertip arising from the operating work. The results demonstrated that it was feasible for the FE model to simulate the response of soft tissues to loading. In most of the FE models, the skin and subcutaneous tissue layer were regarded as being linearly elastic, and the analyzed data got along well with the experimental data under the condition of small deformation. However, the skin layers and the subcutaneous tissues are nonlinear and time-dependent [9, 10, 11]. The response to the mechanical stimulation should also be nonlinear and time-dependent correspondingly. It is therefore inadequate to consider the linear condition only. It is well known skin is a multilayer structure including epidermis, dermis, subcutaneous tissue, bone and nail. There are different types of mechanoreceptors in each layer. In literatures of [2,3,12],

Article available at http://www.matec-conferences.org or http://dx.doi.org/10.1051/matecconf/20164502003

they showed that the epidermis was of hyper-elastic behavior and the dermal tissues were of linearly viscoelastic behavior when skin experienced large deformation. [13] tested that the mechanical behavior of the skin and the subcutaneous tissue can be simulated using the nonlinear biphasic constitutive model. [11] proposed a large deformation biomechanical model for pressure ulcers. [14] defined the fingertip skin layers as hyper-elastic and viscoelastic and considered subcutaneous tissue as a porous elastic material, and developed a structural model based on the definition of different tissue materials. The validity of the model was tested with analyzing the skin biomechanics under a line load. However, when solving complex contact problems, it was prone to be of non-convergence and low efficiency.

2 The Fingertip Finite Element Model

In this study, we take into account the following aspects:1) The epidermis is considered to be linear elastic under a small deflection for its thinness. The Young's moduli of the epidermis, the bone and nail are assumed, according to the published experimental data [15], to be 0.072MPa, 17.0 GPa and 170.0 MPa, respectively; while Poisson's ratio is assumed to be 0.30 for them. 2) For the deformation parameters of dermis and subcutaneous tissue, the stress vector are the sum of elastic stress vector and viscous stress vector, and the parameters of the soft tissue material model are based on [4,11,16].

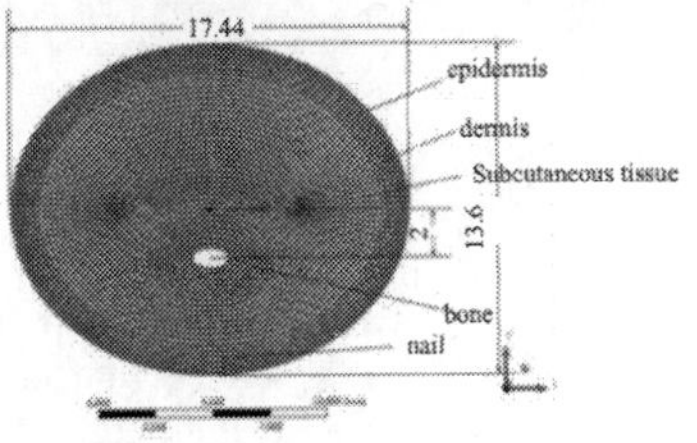

Figure 1. The proposed fingertip FE model.

A cross-section of two-dimensional FE model for the index fingertip is shown in Figure 1. The contour of a fingertip is measured using a reading microscope [8]. The geometry of the fingertip profile is reduced to be an ellipse with a long axis of 17.44mm and a short axis of 13.06mm. The hollow bone is simplified as an ellipse too. From the center of the hollow bone to the center of the outline, the distance is 2mm in the Y direction and 0mm in the X direction. Each layer has the axisymmetric tissue in the X direction. The thicknesses of the epidermis, dermis, and nail layer are assumed to be 0.2mm, 1.0mm and 1.2mm respectively.

This model is developed based on the universal software of ANSYS Workbench. The model is meshed using the hexahedron unit, with a size of 0.1mm on each side. Considering the fact that the discrimination threshold of the human finger is 2mm [17], the model has a high resolution. And we can infer that the depth of the model is only 0.1mm too. Hence it can be approximated as a 2D model. In the simulation, it is assumed that there is no relative sliding and separation between any two adjacent layers, and the degree of freedom (DOF), the motion law and the biomechanical responses of all points at the connection surface are consistent.

Since the model is symmetric along the X-axis, we only choose the right half as the analysis object. During the practical operation, the contact displacement produced by the load on skin is always less than 2.0mm. Therefore epidermis can be considered as linearly elastic. Both dermis and subcutaneous tissues are hyper-elastic and linearly viscoelastic materials. The constitutive equations of those materials are referenced from [18]. The bone and the nail are considered as a linearly elastic material with a large Young's modulus. The parameters of epidermis, bone and nail are based on [7, 8].

3 The Fingertip Finite Element Model Simulation

In this section, we will simulate the tactile biomechanics of skin under distributed load and sharp wedge load stimulation. In biomechanics, the following relations are expected to be found: the relation of the external pressure with the skin displacement, the distribution of pressure in the fingertip under distributed load, and the displacement of the fingertip surface under sharp wedge load. In electrophysiology, we will examine the relation between the biomechanical responses of the subcutaneous tissue stimulated by a T-shaped tool and the discharged rate evoked by the slowly adapting type I sensors (SA I).

3.1 The distributed-load stimulation experiment

Assuming that the displacement δ of the distributed load (Figure 2) is 2.0mm, 2.34mm, 2.5mm and 2.8mm, which are consistent with [2,3,19], the contact force of fingertip under a distributed load at different distance is determined and compared with the experimental measurements of Wu and Jindrich as shown in Figure 3.

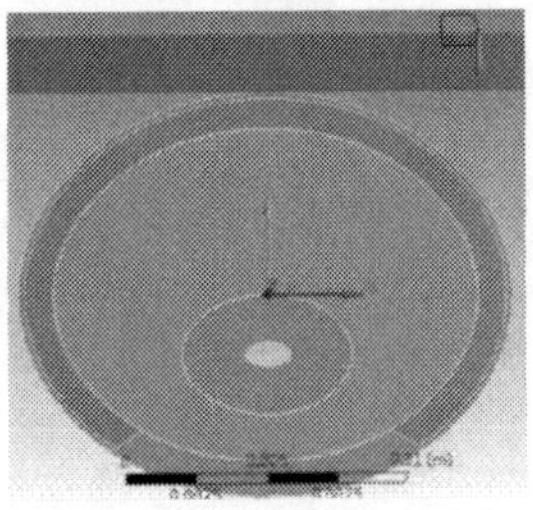

Figure 2. Distributed load applying to the skin surface.

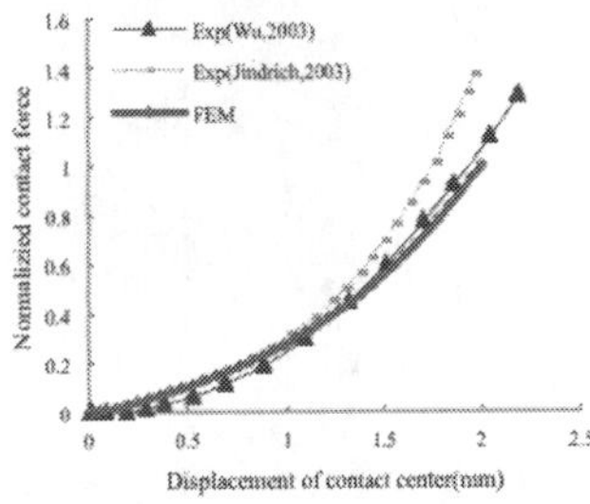

Figure 3. Comparison of predicted contact forces with [2,3].

From the figure 3, we can see that when the load is 40% of the maximum force, the size of deformation is up to 80% of the maximum value. With the increase of the pressure, the rate of change decreases gradually, which is identical to those of [2, 3].

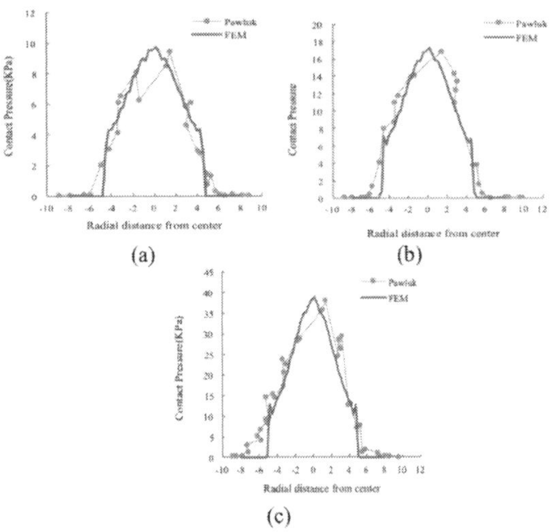

Figure 4. The predicted distributions of contact pressure under **a** δ=2.34mm, **b** δ= 2.5mm and **c** δ= 2.8mm.

The difference of distributed load may influence the distribution of contact pressure in the direct vicinity of the contact center as shown in Figure 4. The profile between the curves produced by FEM and reported by [19] are almost identical. We also can find that the curves of FEM are entirely symmetric while those of Pawluk and Howe's are right-oblique, i.e. the peak value of contact pressure of the experiments occurs at the point of X=2mm, not the point of X=0mm. This may be due to the fact that in the FEM, the index finger is assumed to be idealized symmetric.

3.2 The sharp wedge stimulation experiment

The sharp wedge is considered to be rigid, with the width of 0.05mm and 1.0mm on the tip and base respectively, and a height of 2.0mm. Applying the tip end on fingertip, when the displacement constraint is kept to 1mm, the contact model of finger is developed as Figure 5. After simulation with this model, the skin deformation is plotted in Figure 6.

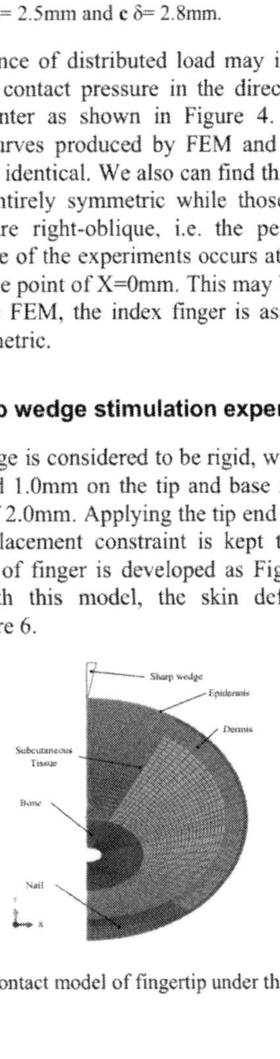

Figure 5. The contact model of fingertip under the sharp wedge load.

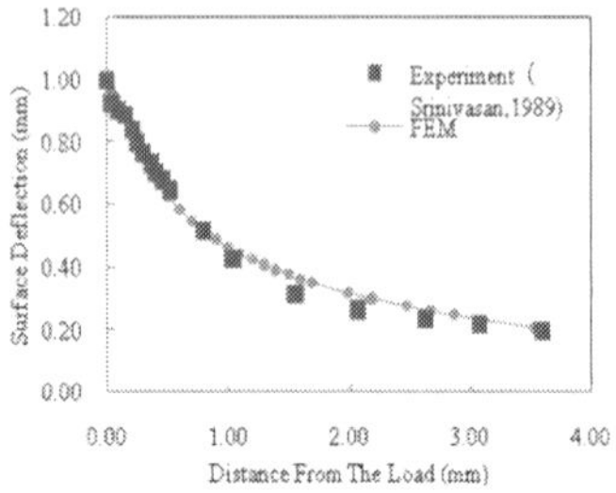

Figure 6. Comparison of predicted deformation with [20].

As Figure 6 shows the two curves match well in the all profiles. The farther the distance of the skin from the stimulated center, the less the deformation, and the rate of reduction decreases with the increase of the distance. When the distance is over 3mm, the decrement becomes almost negligible.

Literatures[12,21,22,23] suggested that the evoked discharge rate could be explained by SED. Figure 7 depicts the predicted distribution of the SED within the tissues.

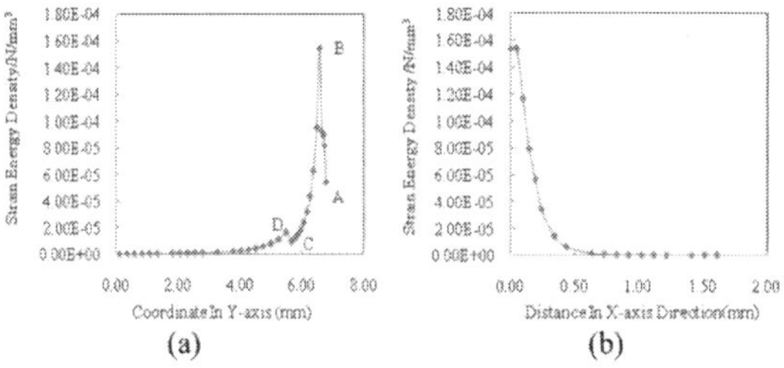

Figure 7. The predicted distribution of the SED within the tissues. **a** In the Y direction, **b** in the X direction.

In Figure 7(a), the letters A and B represent two different points located in the connection area between the epidemics and the dermis, at the depth of 6.7782mm and 6.57487mm from the un-deformed skin surface respectively. The letters C and D are two different points, both at the depth of about 5.5mm which are the connection area between the dermis and the subcutaneous tissue. As shown in the curve, B and D are two positive peak values of the SED, which just belong to different skin layers. In particular, B is the maximum value point of the curve. It is interesting to note that the SED of tissue presents a monotonous change in the magnitude within the same layer, while there is a turning point B in the magnitude of SED from one layer to the next one. The simulation results also indicate that the strain energy mainly accumulates in the range between A and D, where most of the mechanoreceptors are located. The predicted horizontal distribution of the SED of the tissue between the dermis and the subcutaneous is illustrated in Figure 7(b). It shows that at a given depth the tissues of different distance from the contact center present different horizontal SED. It is well known that different shape of the load will produce different strength sensed by the activated receptors and may produce different horizontal SED within tissues. In most cases, the profile of the horizontal SED can map the shape of the stimulator and may help us to obtain the outline of the stimulator.

When the displacement of the sharp wedge load is 1mm, the stress distribution within the skin tissue is also simulated as illustrated in figure 8. The predicted vertical stress and the horizontal stress are presented in figure 8(a-b) respectively. A comparison of figure 8(a) with 7(a) suggests that the predicted distribution of vertical stress agrees well with that of vertical SED. So does a comparison of figure 8(b) with 7(b). Figure 8(c) shows the predicted distribution of the Equivalent Von Mises stress, and we can see the deformation is not continuous.

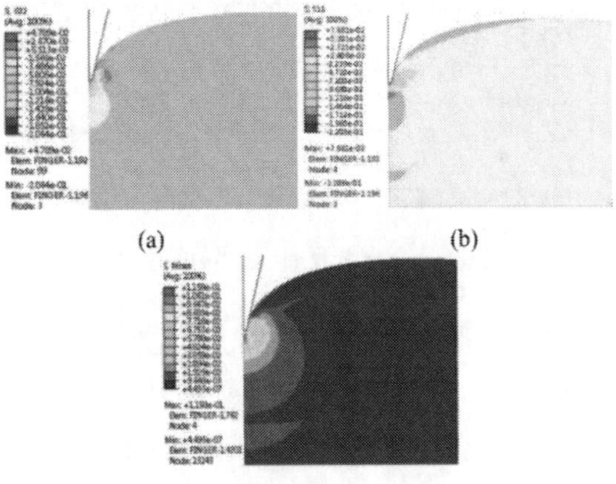

Figure 8. The predicted stress distributions. **a** the vertical stress, **b** the horizontal stress, **c** the equivalent von Mises stress.

The distributions of the maximum and minimum normal strains, shear strain, and strain energy density at depths of 0.20, 0.70, and 1.20 mm from the undeformed skin surface are derived over the width of the fingertip, and are shown in Figure 9(a-d), respectively, where the values are all represented by their logarithms.

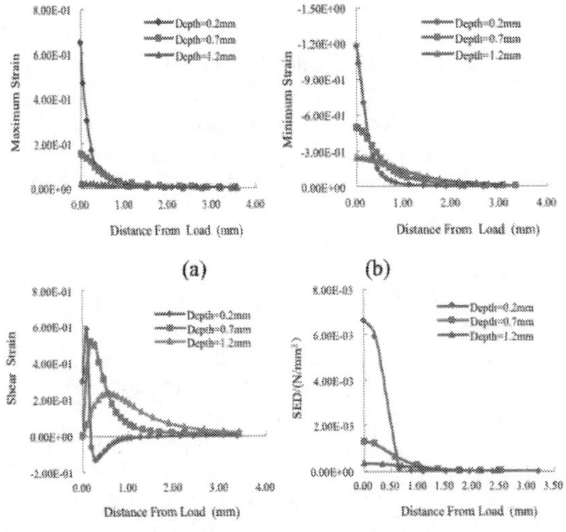

Figure 9. Predicted distributions of the fingertip indented using a sharp wedge (at depths of 0.20, 0.70, 1.20 mm). **a** The tensile strain, **b** the compressive strain, **c** the Shear strain, **d** the SED.

The maximum (tensile) strain of the tissue predominantly occurs in a direction parallel to the skin surface, while the minimum (compressive) strain mostly occurs in a direction normal to the skin surface. The maximum and minimum strain and the SED exhibit their maximal magnitudes in the immediate vicinity of the contact point, while the shear strain reverses its direction around the contact point.

From Figures 7-9, we may find that when skin is stimulated by a sharp wedge, the stress/strain of skin show a high gradient at the junctional plane between two layers. As we know, the mechanoreceptors are mainly located in the dermis layer, and very sensitive to the internal strain of the skin tissues. Therefore it makes sense to research the tactile sensation of the fingertip through investigating the strain distribution.

3.3 The electrophysiological experiment

During the simulation, it is assumed that: 1) The most sensitive point (MSP) of contact section is the center of the skin surface. 2) The deformation of soft tissue at a given depth represents the sensing intensity.

Based on the analysis above, it is known that these two assumptions are rational. In this experiment, a T-shaped hard metal tool with the convex width of 3.0mm or1.0mm and the edge chamfer radius of 0.1mm is served as the stimulator. The friction coefficient of T-shaped tool and skin surface is set as 0.3 according to [18].

The procedures of the experiment are as follows: 1) From the start point of the left edge through the MSP to the end point of the right edge, an action route parallel to the X-axis is determined. 2) At any action point, within 0.005s, the T-shaped tool is pressed vertically to 1.0mm deep into skin at a constant speed, and then it is withdrawn to the starting point at the same speed. 3) After that, it is moved to the right at a step of 0.2mm. And then, it will repeat the operation of 'press-withdraw-move right-press'. This is repeated until it reaches the end point on the right edge.

Figure 10(a) provides a comparison of the predicted SED of soft tissue at some given depths with the discharge rate of SA I , where the dotted line represents the discharge rate reported by [4], and the diamond line and triangle line illustrate the SED of soft tissue at 1.0mm and 0.75mm respectively. There are some deviations between the discharge rate and the two SEDs. In comparison, the one at 0.75mm matches better with the discharge rate than the one at 1.0mm. It indicates that the matching rate is in terms of the depth where the receptor locates. However, the existing deviation between them may result from the differences of the mechanical and geometric properties for the involved biological soft tissue. A comparison of the predicted maximum tensile strain (Ten), minimum compressive strain (Com), SED of soft tissue and the mean discharge rate at the depth of 0.75mm is drawn in Figure 10(b). The SED presents the high conformity with the discharge rate. The correlation coefficient between them is 0.94, which is more than that of other two mechanical responses with the discharge rate.

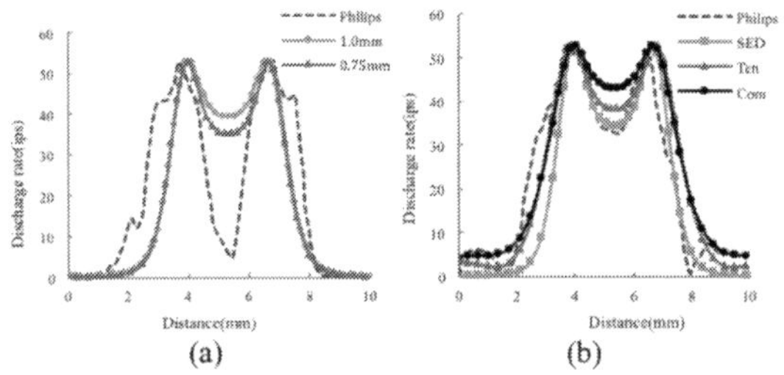

Figure 10. A comparison between the simulated mechanical responses and the discharge rate reported by [4] when the bar width of the T-shaped block is 3.0mm. **a** The predicted SED at the depth of 1.0mm and 0.75mm, **b** The predicted SED, the Ten, the Com of tissues at the depth of 0.75mm.

To observe the effect of bar width onthe match between the mechanical responses and the discharge rate, we also conduct the same simulation when the bar width of the T-shaped tool is 1.0mm, as shown in Figure 11. Compared with the Figure 10, it presents a better consistency between the simulation results and the data in literatures. Obviously, their deviations decrease with the reduction of the bar width. In addition, the predicted SED at the depth of 0.75mm still fits better with the public data than that at the depth of 1.0mm, which is identical to the condition of 3.0mm bar width.

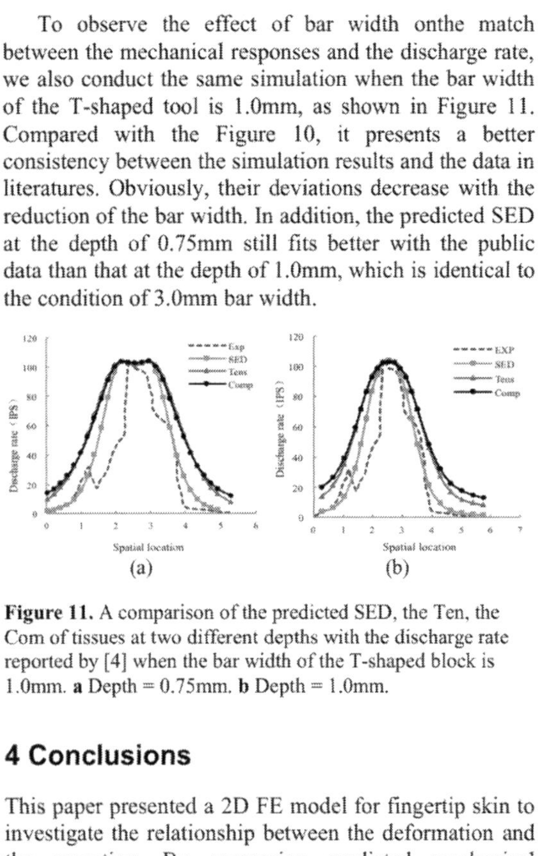

Figure 11. A comparison of the predicted SED, the Ten, the Com of tissues at two different depths with the discharge rate reported by [4] when the bar width of the T-shaped block is 1.0mm. **a** Depth = 0.75mm. **b** Depth = 1.0mm.

4 Conclusions

This paper presented a 2D FE model for fingertip skin to investigate the relationship between the deformation and the sensation. By comparing predicted mechanical responses using the proposed model with the results reported by literatures, they match well in the overall profile, the value of the key points, the changing tendency, and so on. All of the above can illustrate the validity of the model. In addition, the simulation results show that the soft tissues of the fingertips are very sensitive to the external stimulus, and the predicted distribution characteristics of the SED within soft tissues can map the evoked charging rate of mechanoreceptors effectively. Therefore, the model is examined to be valid and can be applied in the tactile research.

Acknowledgements

This work was supported by the National Natural Science Foundation of China (61175069), the Natural Science Foundation of Jiangsu Province of China (BK20131205) and 2012 Qing Lan Project of Colleges and Universities of Jiangsu Province.

References

1. M.A. Srinivasan, R.H. LaMotte, J. Neurosci. 7, 6: 1682-1697 (1987)
2. D.L. Jindrich, Y.T. Zhou, J.T. Dennerlein, J Biomech. 36, 4: 497-503 (2003)
3. J.Z. Wu, R.G. Dong, W.P. Smutz et al, Med. Eng. Phys. 25, 3: 397-406 (2003)
4. J.R. Phillips, K.O. Johnson, J. Neurophysiol. 46, 6: 1192-1203 (1981)
5. M.A. Srinivasan, K. Dandekar, J. Biomech. Eng. 118, 1: 48-55 (1996)
6. E.R. Serina, E. Mockensturm, C.J. Mote, D. Repmel, Journal of biomechanics, 31, 7: 639-646(1998)
7. T. Maeno, K. Kobayashi, N. Yamazaki. JSME 41, 1: 94-100 (1998)
8. J.Z. Wu, R.G. Dong, Proc. IMechE. 219, 9: 89-103 (2005)
9. Y.C. Fung, S.C. Cowin, Springer-Verlag, 2nd ed. New York (1994)
10. M.B. Rubin, S.R. Bodner, N.S. Binur, ASME J Biomech Eng, 120, 5: 686-9 (1998)
11. Y. Zheng, A.F. Mak, Biomedical Engineering IEEE Transactions on, 43, 9: 912-918 (1996)
12. A.B. Vallbo, R.S. Johansson, Hum. Neurobio. 3, 1: 3-14 (1984)
13. V.C. Mow, S.C. Kuei, W.M. Lai et al, ASME J. Biomech. Eng. 102, 1:73-84 (1980)
14. J.Z. Wu, R.G. Dong, S. Rakheja et al, Med. Eng. Phys. 26, 2: 165-175 (2004)
15. H. Yamada, F.G. Evans (1971). The Williams and Wilkens Company, Baltimore, 4(2): 297+x
16. L.H. Pan, L. Zan, F. Foster. Ustrasound Med. Biol. 24, 7: 995-1007 (1998)
17. K.B. Shimoga, IEEE Virtual Reality Annual International Symposium : 271-279 (1993)
18. J.Z. Wu, R.G. Dong, D.E. Welcome, Med. Eng. Phys. 28, (8): 816-826 (2006)
19. D.T. Pawluk, R.D. Howe, Biomech. Eng. 121, 2: 178-183 (1999)
20. M.A. Srinivasan, J. Biomech. 22, 4: 343–349 (1989)
21. K. Dandekar, B.I. Raju, M.A. Srinivasan, J Biomech Eng. 125, 5: 682-691 (2003)
22. K.O. Johnson, Curr. Opin. Neurobiol. 11, 4: 455-461 (2001)
23. K.O. Johnson, T. Yoshioka, F. Vega-Bermudez, J. Clin. Neurophysiol. 17, 6: 539-558 (2000)

MATEC Web of Conferences **45**, 02004 (2016)
DOI: 10.1051/matecconf/20164502004

Modeling and analysis of grinding force in ultrasonic honing considering the scale effect

Ye Linzheng , Zhu Xijing

Key Laboratory for AMT of Shanxi, North University of China, Taiyuan 030051, China

Abstract: To research the power ultrasonic honing mechanism at the micro scale, the scale effect is considered and the strain gradient plasticity theory based on the dislocation mechanism (MSG) is applied to establish the model of a whetstone grinding force, and the simulation analysis is conducted. Results show: the grinding force will increase when the scale effect is considered; the main influence parameter is honing depth on the grinding force; the grinding force increases nonlinearly with the continued reduce of honing depth after which decreases to 1.4 μm, which decreases slightly with the increase of the relative velocity of a whetstone. So the material becomes difficult to remove at the micro scale.

1 Introduction

Power ultrasonic honing is a new type of precision machining method, with some advantages such as small grinding force, low honing temperature and high processing efficiency, which is widely used in cylinder block and cylinder liner processing of cars and tanks, and has a broad application prospect.[1] In recent years, many scientists have observed the scale effect of materials at the micro scale extensively from experiments, that is, when the characteristic length of the non-uniform plastic deformation of the material is at the micron level, the material behavior is significantly different from that at the macro scale. In 1994, Fleck et al. observed the dimensionless torsional stiffness of copper wire increased almost two times when its diameter reduced from 170 μm to 12 μm in thin copper wire torsion test.[2] In1998, A.G.Evans and Stolken observed that the bending strength of the beam increased significantly with the decrease of the thickness of the thin beam in thin beam bending experiment.[3] A series of micro indentation experiments also showed that the hardness of material increased with the indentation decreasing to the micron level.[4,5]

There are having been many discussions about the grinding force model at home and abroad, but not much introduction about the grinding force model of ultrasonic honing. Li lijun divided the grinding force into the cutting force and the friction force, and established the mathematical model of grinding force.[6] Xie guizhi et al. studied the critical conditions for the removal of plastic deformation and the removal of brittle fracture in the process of engineering ceramic grinding, and set up corresponding grinding force model respectively.[7] Liu zhen et al. established the grinding force model of ultrasonic honing considering the strain rate of the material.[8]

In summary, materials processed will inevitably show scale effect when the honing depth of ultrasonic honing reduces to micron level, but some related researches have not yet been seen. Based on the ultrasonic honing processing mechanism, applying the strain gradient plasticity theory based on the dislocation mechanism(MSG theory) and the scale effect of material is considered, establishing the grinding force model and conducting its numerical simulation, and analyzing the scale effect' influence on the grinding force in this paper.

2 The mechanism of ultrasonic honing processing

There are three forms of vibration applied in ultrasonic honing: axial vibration, radial vibration and torsional vibration. Axial ultrasonic vibration is chosen in this paper. At the same time of the whetstone doing reciprocating and axial rotation motions, the axial ultrasonic vibration applied makes sinusoidal speed attached to the whetstone synthesis speed. Assuming the applied form of vibration is $A\sin 2\pi ft$, A and f represent the amplitude of axial vibration applied and the frequency of vibration respectively, the synthesis speed of the whetstone is:

$$v_{ea} = \sqrt{v^2 + (v_a + v_f)^2} \quad (1)$$

Where, v_{ea}, v, v_a and v_f are the synthesis speed, reciprocating speed, rotation speed and ultrasonic vibration speed of the whetstone respectively.

3 Scale effect of the power ultrasonic honing

Materials' scale effect should be considered when the honing depth of ultrasonic honing is a few microns. With the honing depth decreases, the hardness and the specific grinding energy of the material processed will increase nonlinearly, which is the so-called scale effect. Applying the strain gradient plasticity theory based on the dislocation mechanism, the shear strength of the material can be expressed by the dislocation density.

$$\tau = \alpha\mu b\sqrt{\rho} = \alpha\mu b\sqrt{\rho_s + \rho_G} \quad (2)$$

Where τ is the shear strength; α is an empirical constant between 0.2 to 0.5, here is 0.5; μ represents the shear modulus; b indicates the Burgers vector length, which is 0.29664 nm for the 45 steel; ρ, ρ_s and ρ_G are the total dislocation density, the statistics stored dislocation density and the geometrically necessary dislocation density, respectively. ρ_G is related to the strain gradient at the microscopic scale.

$$\eta = \frac{1}{\bar{r}}\rho_G b \quad (3)$$

Where $\bar{r}$ is the Nye factor, which reflects the ratio of the average density of geometrically necessary dislocation and the geometrically necessary density in the most effective arrangement, is 2 for a face-centered cubic polycrystal and 3 for a face-centered cubic single crystal.

The tensile flow stress in the crystal is shown in the formula (4).

$$\sigma = \bar{m}\alpha\mu b\sqrt{\rho_s + \rho_G} = \sigma_{ref}\sqrt{f^2(\varepsilon) + l\eta} \quad (4)$$

Where $\bar{m}$ is the Taylor factor, which is $\sqrt{3}$ or 3 for isotropic materials or anisotropic materials, since the model of the grinding force is studied from the microscopic view in this paper, $\bar{m}$=3; σ_{ref} is the uniaxial tension reference stress; l is the characteristic length of materials proposed by Fleck.

$$l = \bar{r}\left(\frac{\bar{m}\alpha\mu}{\sigma_{ref}}\right)^2 b \quad (5)$$

Nix and Gao developed a kind of dislocation model to estimate the geometrically necessary dislocation density under a conical indenter, and combined (4) the relationship between indentation hardness and indentation depth can be obtained.

$$\frac{H}{H_0} = \sqrt{1 + \frac{h^*}{h}} \quad (6)$$

Where H_0 is the indentation hardness without scale effect, which is 1.7052 Mp for 45 steel; H is the hardness when the indentation depth is h; h^* is a parameter of the length dimension, which can be expressed in the formula (7).

$$h^* = \frac{81}{2}b\alpha^2\tan^2\gamma\left(\frac{\mu}{H_0}\right)^2 \quad (7)$$

Where γ indicates the angle between the conical indenter surface and the surface of the material pressed. For standard Vickers indenter, the half angle of indenter tip is 68°, so γ is 22°.

4 Whetstone grinding force model

4.1 Single grain grinding force model

The grinding force consists of deformation force by cutting and friction, for a single grain:

$$\begin{cases} F_{\mathrm{gN}} = F_{\mathrm{gNc}} + F_{\mathrm{gNs}} \\ F_{\mathrm{gT}} = F_{\mathrm{gTc}} + F_{\mathrm{gTs}} \end{cases} \tag{8}$$

Where g represents a single grain, N and T indicate normal force and tangential force respectively, c and s indicate deformation force by cutting and friction respectively.

An abrasive on the whetstone is regarded as the Vickers indenter, as the following relationship:

$$P = \xi H R^2 \tag{9}$$

Where P represents the load imposed on a single abrasive, ξ is the indenter geometry factor, which is 2 for the Vickers indenter, H is the hardness of the material, R indicates the micro indentation size, and $R = a_p \tan\lambda$, a_p is the indentation depth, λ, the half angle of indenter tip, is 68°.

Abrasive may not always contact with the workpiece due to the fact that the whetstone grain is not same with the standard Vickers indenter completely and the empty cutting phenomenon in ultrasonic honing, so a dimensionless coefficient $\eta_0(0<\eta_0<1)$ is introduced, according to the specific processing environment and experimental experience, is taken 0.5, combined with (6) the normal cutting deformation force can be obtained.

$$F_{\mathrm{gNc}} = \eta_0 \xi H_0 a_{\mathrm{p}}^2 \tan^2\lambda \sqrt{1 + h^* / a_{\mathrm{p}}} \tag{10}$$

The tangential cutting force can be obtained according to the force analysis of a single grain and the geometrical conditions.

$$F_{\mathrm{gTc}} = \frac{\pi}{4} \eta_0 \xi H_0 a_{\mathrm{p}}^2 \tan\lambda \sqrt{1 + h^* / a_{\mathrm{p}}} \tag{11}$$

For a single grain, the grinding force owing to the friction is shown as follows:

$$\begin{cases} F_{\mathrm{gNs}} = S\overline{P} \\ F_{\mathrm{gTs}} = \overline{\mu} S \overline{P} \end{cases} \tag{12}$$

Where S represents the real contact area of a single grain and workpiece, due to the empty cutting phenomenon in ultrasonic honing, $S=\beta\pi(a_p\tan\theta+r_0)2+(1-\beta)\pi r_0^2$, β is the real contact area ratio of abrasive and workpiece, $0<\beta<0.5$ generally, which is taken 0.4 on the basis of past experience, θ is the half cone angle of abrasive and r_0 is grinding flat particle radius, $\overline{P}$ is the average pressure on a single grain, and $\overline{P} = \frac{P}{\pi R^2} = \frac{\eta_0 \xi}{\pi} H_0 \sqrt{1 + \frac{h^*}{a_{\mathrm{p}}}}$, $\overline{\mu}$ is related with the whetstone speed, and $\mu = A + Be^{-60Cv_{\mathrm{ea}}}$, A is the static friction coefficient 0.1 between grain and workpiece, B is sliding friction coefficient 0.08, C=0.014.

Grinding force model of a single grain can be obtained from the above as shown in formula (13).

$$\begin{cases} F_{\mathrm{gN}} = \eta_0 \xi H_0 \sqrt{1 + h^* / a_{\mathrm{p}}} [a_{\mathrm{p}}^2 \tan^2\lambda \\ \quad + \beta(a_{\mathrm{p}} \tan\theta + r_0)^2 + (1-\beta) r_0^2] \\ F_{\mathrm{gT}} = \eta_0 \xi H_0 \sqrt{1 + h^* / a_{\mathrm{p}}} \{ \frac{\pi}{4} a_{\mathrm{p}}^2 \tan\lambda + (A \\ \quad + Be^{-60Cv_{\mathrm{ea}}}) [\beta(a_{\mathrm{p}} \tan\theta + r_0)^2 + (1-\beta) r_0^2] \} \end{cases} \tag{13}$$

4.2 A whetstone grinding force model

The effective grain number on a whetstone can be estimated.

$$N_{\mathrm{d}} = \frac{4bl}{\pi d_{\mathrm{mean}}^2} \cdot p \tag{14}$$

Where b and l are the width and length of a whetstone respectively, d_{mean} is grain mean diameter, p indicates the percentage of abrasive scratching and cutting on a whetstone, some studies show that the grains numbers proportion which genuinely participating in scratching and cutting on the surface of the diamond whetstone were 3.84% and 0.52%.

A whetstone grinding force model can be obtained by making a single grain force and the effective grain number on a whetstone multiply, as shown in formula (15).

$$\begin{cases} F_{\mathrm{N}} = \eta_0 \xi H_0 \sqrt{1 + h^* / a_{\mathrm{p}}} \\ [a_{\mathrm{p}}{}^2 \tan^2 \lambda + \beta(a_{\mathrm{p}} \tan\theta + r_0)^2 \\ + (1-\beta) r_0{}^2] \cdot N_{\mathrm{d}} \\ F_{\mathrm{T}} = \eta_0 \xi H_0 \sqrt{1 + h^* / a_{\mathrm{p}}} \\ \{\frac{\pi}{4} a_{\mathrm{p}}{}^2 \tan\lambda + (A + Be^{-60Cv_{\mathrm{ea}}}) \\ [\beta(a_{\mathrm{p}} \tan\theta + r_0)^2 + (1-\beta) r_0{}^2]\} \cdot N_{\mathrm{d}} \end{cases} \tag{15}$$

5 Simulation analysis of grinding force model

In order to find out the influence of the scale effect on ultrasonic honing intuitively, and analyse the degree and trend of change on grinding force, applying Matlab to doing simulation analysis of the grinding force model established. The material of a workpiece is taken the 45 steel, for the diamonds grain, the grain size was 150# and the concentration is 100%, the half cone angle is π/4, the grinding flat particle radius is 0.015 mm, the structure size of a whetstone is 6×6×100 mm, the simulation parameters of power ultrasonic honing is shown in table 1.

Table 1. Parameters of power ultrasonic honing

rotation speed	80、125、200、315 r/min
reciprocating speed	50、100、150、200 mm/s
Honing depth	1~10 μm
Ultrasonic frequency	18.6 KHz
Ultrasonic amplitude	0.01 mm

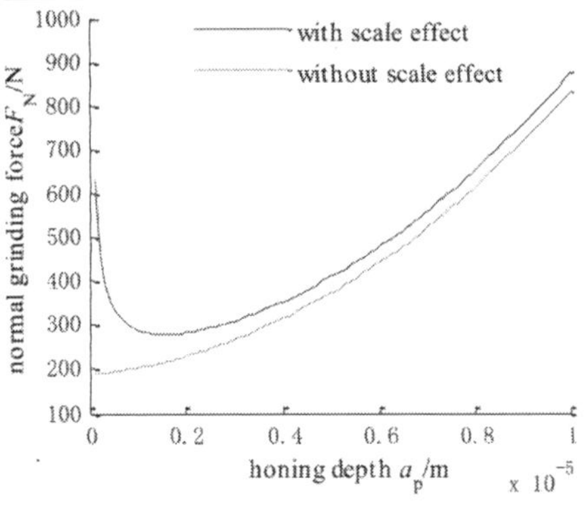

(a) n=200 r/min, v_a=100 mm/s

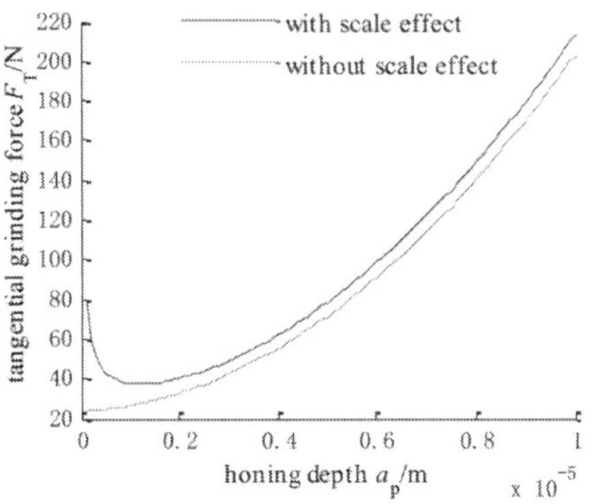

(b) n=200 r/min, v_a=100 mm/s, t=1/f

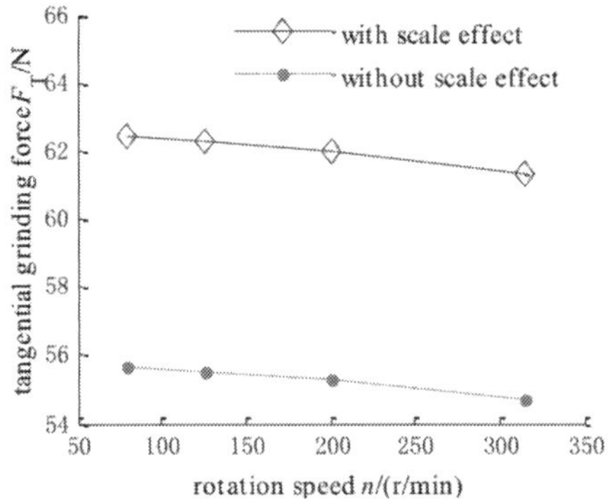

(c) a_p=4 μm, v_a=100 mm/s, t=1/f

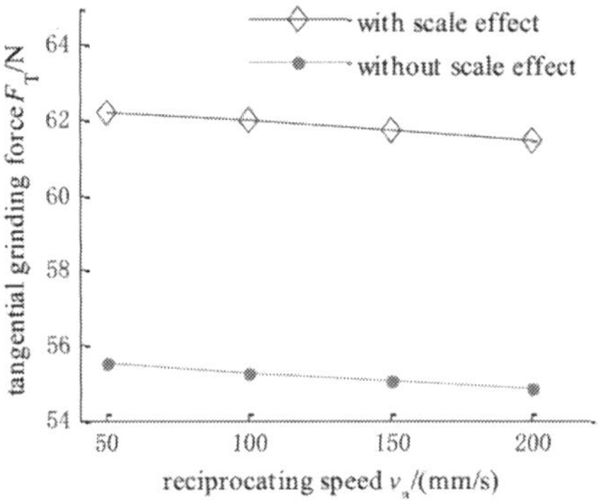

(d) a_p=4 μm, n=200 r/min, t=1/f

Figure 1. Curves of Grinding Force Change

Simulation results are shown in figure 1. As shown in (a) (b) (c) (d), honing depth is the main parameter influencing the grinding force, reciprocating and rotation speed is relatively small, compared with no considering the scale effect, the normal and tangential grinding force all increase under the influence of the scale effect. From (a) and (b) can see, the normal and tangential grinding force all change nonlinearly with the reduce of honing depth, which decrease with the decrease of honing depth, when the honing depth reduces to a certain value (about 1.4 μm), the grinding force reaches the minimum, and if honing depth continues to decrease, the grinding force will increase rapidly, so that material is difficult to remove. This is due to some changes of material characteristic at micro scale, the lattice dislocation needs to be fully considered, the hardness, shear strength and tensile flow stress of the material increase and the grinding force increases. When the honing depth is greater than the certain value, gradually tending to a macroscopic cutting, grinding force increases with the increase of honing depth. The grinding force reduces slightly with the increase of the reciprocating or rotation speed, because the increase of whetstone speed will reduce the dynamic friction coefficient so as to decrease the grinding force.

6 Conclusions

The scale effect should be considered when the honing depth of power ultrasonic depth is a few microns, which is essential for the microscopic mechanism of ultrasonic honing. The grinding force established is related with the microscopic parameters of the material processed, such as the Burgers vector length. When the scale effect is considered, the grinding force increased obviously, the normal and tangential grinding force will increase nonlinearly with the reduce of honing depth when the honing depth is less than 1.4 μm, and the grinding force decreases slightly with the increase of the relative velocity of a whetstone.

References

1. Zhu Xijing, Wang Jianqing, Cheng Quan, et al. Radial power ultrasonic vibration honing friction analysis[J]. Solid State Phenomena, 2011, 175: 183-186.
2. Fleck N A, Muller G M, Ashby M F, et al. Strain gradient plasticity: theory and experiment [J]. Acta Metallurgica Et Materialia, 1994, 42(2):475-487.
3. Stolken J S, Evans A G. A microbend test method for measuring the plasticity length scale [J]. Acta Materialia, 1998, 46(14):5109–5115.
4. Poole W. J., Ashby M. F., Fleck N. A. Micro-hardness of annealed and work-hardened copper polycrystals[J]. Scripta Materialia, 1996, 34(4):559–564.
5. Mcelhaney K. W., Vlassak J. J., Nix W. D. Determination of indenter tip geometry and indentation contact area for depth-sensing indentation experiments [J]. Journal of Materials Research, 1998, 13(05):1300-1306.
6. Li Lijun, Fu Jiecai. A Study of Grinding Force Mathematical Model [J]. Journal of Hunan University, 1979, 3:44-55.
7. Xie Guizhi, Shang Zhentao, Sheng Xiaomin, et al. Grinding Force Modeling for High-speed Deep Grinding of Engineering Ceramics [J]. Journal of mechanical engineering. 2011, 47(11): 167-176.
8. Liu Zhen, Zhu Xijing, Wang Jianqing, et al. Modeling and Simulation of Grinding Force on Axial Power Ultrasonic Honing [J]. Mechanical Science and Technology for Aerospace Engineering. 2014, 33(6):858-86

MATEC Web of Conferences **45**, 02005 (2016)
DOI: 10.1051/matecconf/20164502005

Stiffness Equivalent Finite Element Modelling of a Physical Assembly by Structural Optimization Method

Engin Metin Kaplan, Kamuran Kamil Yeşilkaya and Kemal Yaman

Defence Industries Research and Development Institute, Turkish Scientific and Technical Research Council, (TÜBİTAK-SAGE), 06261, Ankara, Turkey

Abstract. In this study, stiffness equivalent finite element modelling (EFEM) technique of a physical assembly which has several detailed electronic, mechanic and hydraulic parts is presented. Physical assembly can be evaluated as a subsystem. Outer geometry of the detailed subsystem is meshed by shell elements. Beam elements, of which cross sectional dimensions are found by structural optimization, are used at the inner section of the subsystem. To reach structural integrity, beam elements are connected to the outer shell with rigid elements. The deflection results of the detailed finite element model (DFEM) which is constructed by 3D solid elements are investigated under unit force laterally. The deflection results are defined as objective function in structural optimization of the EFEM. Design parameters are selected as rectangular cross section dimensions of the beam elements. The deflection results of the EFEM with optimized beam cross sections and DFEM indicate good agreement. Moreover EFEM is computationally efficient for the deflection analysis. It is shown that EFEM solution time is %1 of the DFEM.

1 Introduction

Aerospace and Defence industry is one of the most attractive market in the world [1]. Most of the worldwide and innovator companies do their designs according to their specialty, as coordinaly. Companies, which are responsible for subsytem design, do their desing about their specialty and deliver it to the system designer company. In delivery process, subsystem companies do not prefer to transfer their know-how to the system designer company. They prefer to deliver their subsytem in closed form and give information about input and output relationship of subsytem with the system. On the other hand, structural strength of the subsytem is important for the overall system strength. All detailed part of the subsytem affects the total system stiffness and must be defined as it is.

In this study a subsytem, of which outer geometry is known, and involves some detailed parts like electronic, mechanic and hydraulic are equivalently modelled with real detailed model concerning mechanical stiffness, by shell and beam elements. To reach the accurate results, firstly, real detailed 3D model of the subsytem is solved under unit forces. Deflection results of the track points are determined. These deflection result are seeked in the structural optimization process of the basic shell-beam equivalent model by sizing the beam cross section dimensions.

The solution time of EFEM and DFEM input files are also compared in this study.It is seen that EFEM is pretty short accoring to DFEM. Since two models are equal in stiffness, EFEM can be used for multiple parametric loading cases.

2 Literature Survey

Equivalent modelling by using optimization technique is useful and computationally efficient process. Various studies are available for the solution in literature.

Several researchers have worked on equivalent modelling technique. Yu Wang et all developed three-step optimization strategy. They use NASTRAN for their application. They performed their study on the wing-type structure made of composite material. Researchers interest not only static behaviour, but also dynamic, buckling and aeroelastic behaviour. Their static displacement results of the equavalent modelling with the detalied modeling are in a good agreement. In addition to that, their modal analyses results show that first ten elastic modes are equally modelled in %5 range with detailed finite element model by optimization technique. Moreover, first two buckling mode patterns are so similar. Lastly, flutter behaviour results of the both modelling are very close. The results show that the optimization technique for the composite-wing systems is adventageous, and the equivalent finite element model prepared with the rapid modeling is appropriate for the initially structural optimization with the static and dynamic constraints [2].

Jasson Gryzagoridis et all are studied on estimation flexural rigidy of the composite panels. They also performed experimental varification about their analytical

Article available at http://www.matec-conferences.org or http://dx.doi.org/10.1051/matecconf/20164502005

approach. Analytical approach is simply based on the euler-bernouli beam theory. They used cantilever beam in calculations. Their experimental results show that, analytical approach predicts the behaviour of the sandwich composite panel[3].

A. Moroncini et all built simple finite element model which is equivalent to the detailed model by using beam and shell elements only. They used NASTRAN both finite element analyses and structural optimization. Beam cross-section parameters are defined as design region in their analyses. The researchers emphasize that this technique is useful for initially design process, for predicting best alternative solution on vibration and acoustic behavior of the structure [4].

Fenghe Wu et all worked on optimization of the ram, which is like beam structure by using equivalent elastic modulus method. They used this method to super-heavy-duty CNC floor-type milling and boring machine. Instead of defining the complex constrains and contact parameters of the detailed finite element model, simple optimized model is used. In optimization technique, objective function is minimum volume under static loading. Furthermore, displacement and stress values are defined as constraints. Accoring to the results, basic method of the equivalent elastic modulus can effectively control the scale of optimization calculation and provide optimization going on without problem [5].

David J. Malcolm et all studied on blades modelled by equivalent beam elements for aeroelastic analysis. They use ANSYS for finite element modelling and MATLAB for equivalent model calculations. Since, their concern is dynamic equivalent modelling of the blades, their model must be also mass equivalent. Location of the elastic axes, orientation of principle axes and coupling between terms of the equivalent model should reflect 3D detailed blade parts. In conclusion, their equivalent beam modelling method demonstrates significant impression for easily involving all properties into beam elements for dynamic simulations [6].

T. R. Baumann studied on formulation of equivalent stiffness by experimental method. Researcher builds the stiffness matrices in form of two by two. He used beam elements in equivalent system and concentrated on shear and bending stiffness values. Author claims that, the equivalent model can be used in dynamic analysis. Researcher also points out that some singularity issues can occur in the model and it must be controlled. Results are acceptable for arbitrary length beams [7].

J. Sun et all worked on determination of elastic properties and lateral stiffness of multi-ribbed walls that is use in load-resistant elements in recent years. They used Mori-Tanaka method to derive analytical expression on equivalence elastic property determination. Furthermore, they verified their approach by both experimentally and numerically. They did static tests on six wall specimens, and performed finite element analysis of detailed model [8].

3 Theory

NASTRAN is used for the finite element analysis. The governing equation of the linear static finite element analysis is given below.

$$[K]\{u\} = \{P\} \qquad (1)$$

Here K represents system stiffness matrix. P is load vector and u is displacement of the nodes. Displacements are unknown in the Equation 1. Displacements is obtained by inverting the stiffness matrix and multiplying it by the force value [9] .

GENESIS is used for the structural optimization. Basically, numerical optimizations solve the problem: Seek the set of the variable OF_i, i=1, N included in the vector OF that minimizes F(OF).
Subjected to;

$$g_j(OF) \le 0 \quad j = 1, M \qquad (2)$$

$$h_r(OF) = 0 \quad r = 1, S \qquad (3)$$

$$OF_i^L \le OF_i \le OF_i^U \quad \text{i=1, N} \qquad (4)$$

Here OF is the objective function. It is calculated by design variables. g_j and h_r are inequality and equality constraints, respectively. OF_i^L and OF_i^U are the lower and upper region bounds of the design variable.

In sizing optimizations, element cross sectional parameters (height, width, etc.), section parameters (area, area moment of inertia, etc.), structural parameters (length, span, etc.), and responses (strains, stresses, eigenvalues) can be studied. The cross sectional parameters, section parameters, and node locations construct the optimization model. The designer interests design variables that, generally, have some sort of nonlinear relationship with the definition of the analysis model, and constraints on the produced by the analysis model. The sum of design variables, constraints, and the objective function construct the design model. The fundamental design capabilities are used to communicate the analysis model to a basic design model [10].

4 Calculations

Syhentesis steps of the methodology is given Figure 1 below.

STEP 1
Performing finite element analysis for detailed physical system by 3D solid element and obtain deflection results

↓

STEP 2
Mesh the outer geometry of the physical system by 2D shell element. Construct 1D elements on centerline and tie them with 2D shell elements by rigid beams.

↓

STEP 3
Start structural optimization by same B.C with step 1. Select beam cross section properties as design region. Define deflection results of Step 1 as objective function.

Figure 1. Syhentesis steps of the methodology

The physical model can be evaluated as fixed-free beam problem. The degree of freedom of the root of the kit is constraint in all directions. The unit force is applied to the kit at the free end in both y and z direction. The schematic view is given in Figure 2 below.

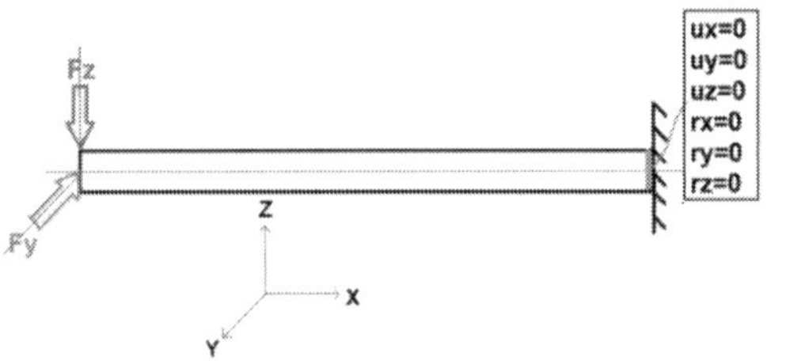

Figure 2. Boundary Conditions of the system

The number of entities of the detailed finite element analysis is given Table 1. The schematic view of the detailed finite element model is given in Figure 2.

Table 1. Number of entities of detailed finite element model

# of Nodes	# of 3D Elm's	# of MPC's
635789	365114	90

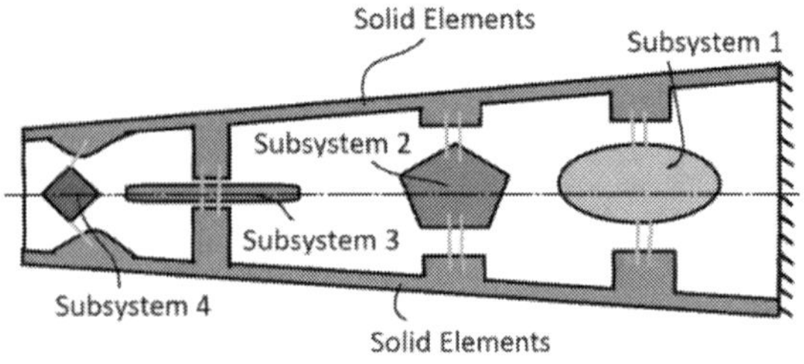

Figure 3. Schematic view of the detailed finite element model

Subsystems are modelled as dimensionless points elements which has same mass and inertial properties with real systems. The connection of the subsytem to the physical model is provided by rigid elements.

The number of entities of the stiffness equivalent finite element analysis is given Table 2. The schematic view of the detalied finite element model is given in Figure 4.

Table 2. Number of entities of stiffness equivalent finite element model

# of Nodes	# of Elm's		# of MPC's
	1D	2D	
23064	295	22860	7

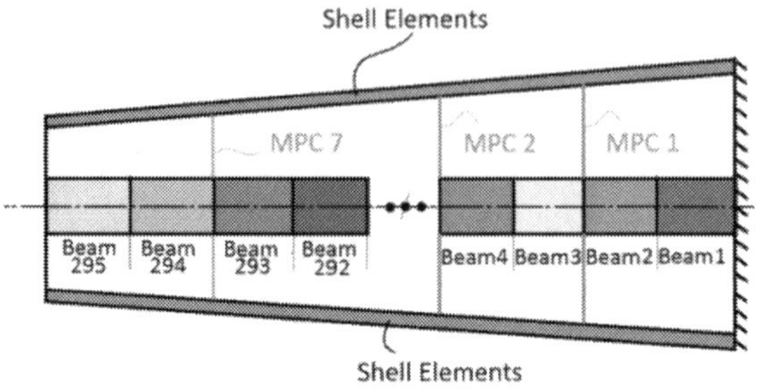

Figure 4. Schematic view of the stiffness equivalent finite element model

Material properties used in the analyses are defined in Table 3.

Table 3. Material properties of the structures used in the analyses

Material	Elastic Modulus (GPa)	Poisson Ratios	Density(kg/m^3)
Structural Steel	200	0.3	7800

In structural optimization, The design region is selected as closed rectangular cross section parameters of beam elements as shown in Figure 5.

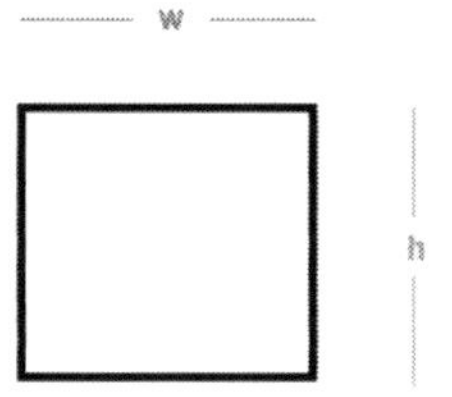

Figure 5. Design region of the beam used in structural optimization

The physical model is divided into 8 parts to catch the deflection pattern of the 3D detailed finite element analysis results. Here 9 different beam cross section parameters are optimized. Since two parameters as w and h are exist in a part, totally eigthteen design parameters are studied.

The design region bounds of the beam regions are given in Table 4.

Table 4. Optimization parameters of the beams

Parts	Initial Value	Lower Bound	Upper Bound
Beams (w,h)	(A_w), [A_h]	(%10A_w), [%10A_h]	(%1000A_w), [%1000A_h]

5 Results

There are four track points in the both 3d detailed finite element analysis and basic 1d-2d equivalent finite element analysis. These points are given in Figure 6.

According the results percentage errors of these points are given Table 5.

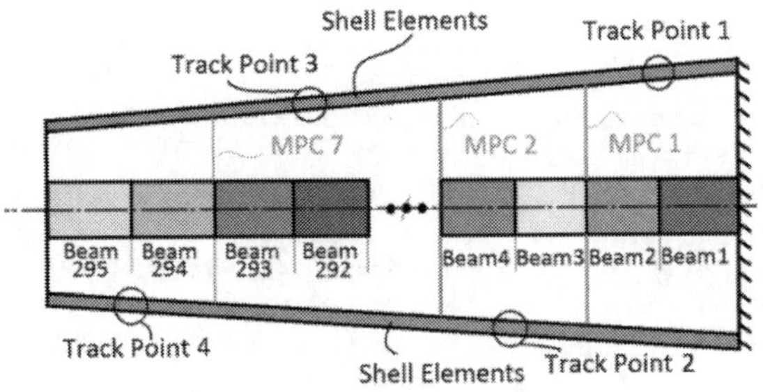

Figure 6. Schematic view of the stiffness equivalent finite element model

Table 5. Track point deflection errors of the stiffness equivalent model wrt detailed model

Track Points	Percentage Deviation of 1d-2d Equivalent Model Results wrt 3d Model	
	y axis	z axis
Point 1	%2.9	%0.12
Point 2	%1.01	%3.09
Point 3	%4.5	%3.2
Point 4	%0.9	%1.5

The cross section of the beams in y-z axes are shown in Figure 7 after structural optimization. It can be noted that, beam dimensions are changed differently. Figure 7 and Figure 8 present beam length results in y axis and z axis respectively. The results are different due to having dissimilar stiffness characteristics in each axis.

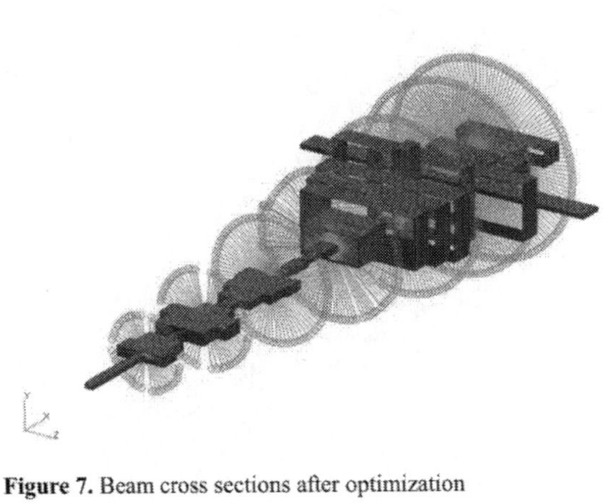

Figure 7. Beam cross sections after optimization

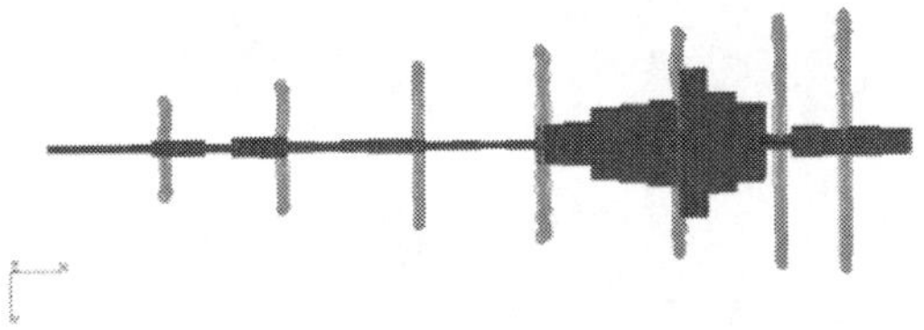

Figure 8. Beam dimensions in "y axis" after optimization

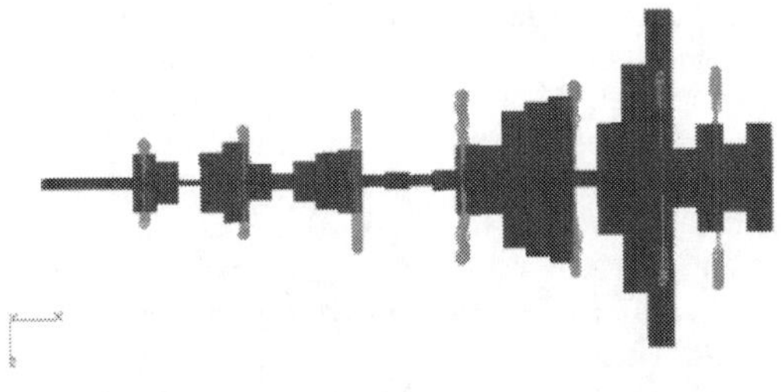

Figure 9. Beam dimensions in "z axis" after optimization

EFEM is also computaionally cheaper than DFEM. Total analysis times are compared and it is found that solution of EFEM takes 1 minutes. Moreover, solution of the DFEM takes about 98 minutes.

6 Conclusion

Equivalent stiffness finite element model of the subsystem which is constructed by 1d beam elements and 2d shell elements is investigated in this study. Firstly, linear static finite element analysis of the subsytem is performed with all detail parts. Deflection results of the several points in both y and z directions are noted. Then, equivalent finite element model is prepared by beam and shell elements. Beam elements are placed in the center of the sections along the part. Beam elements are connected in several regions to the shell elements by rigid beams.

Structural optimization is used in the second stage of the technique. Deflection results of the DFEM is defined as objective function both on y and z directions. Cross section dimensions of the beam is selected as design region. Lastly optimization algorithm determines the beam cross section dimensions.

As it can be seen from the Table 5, percentage error of the EFEM technique is so small with respect to DFEM. It can be said that EFEM of the subsytem can be used in the linear static analysis of the system instead of DFEM Thus, no detailed information about electronics, mechanics, hydraulics etc. of subsytem is delivered to the system design authority.

EFEM is also computaionally efficient. As it can be seen from Table 1 and Table 2, finite element quantity of the EFEM is lower than DFEM. Thus, computationally time solution of EFEM input file is approximately %1 of the DFEM. EFEM can be used for deflection analysis for several loading cases efficiently.

7 Future Work

Numerical studies mentioned above will be checked with the experimental setups using load cells, strain gages and laser displacement sensors to measure deflection values.

References

1. Vision 2023 Project, TUBITAK Conference Report, Defence and Aerospace Industry Report-Additional Part 3, (2003)

2. Y. Wang, X. Ouyang, H. Yin, X. Yu, Structural-Optimization Strategy for Composite Wing Based on Equivalent Finite Element Model, Nanjing University of Aeronautics and Astronautics, 210016 Nanjing, People's Republic of China, (2014)
3. J. Gryzagoridis, G. Oliver, D. Findeis, On the Equivalent Flexural Rigidity of Sandwich Composite Panels Cape Peninsula University of Technology, South Africa
4. A. Moroncini, Dr. L. Cremers, Dr. M. Kroiss, NVH Structural Optimization using Beams and Shells FE Concepts Models in the early Car Development Phase at BMW, Isma (2010)
5. F. Wu, X. Xu2, W. Jun, J. Fan, Static Stiffness Optimization of Large Ram Based on Equivalent Elastic Modulus, Key Engineering Materials Vol. **621** (2014)
6. D. J. Malcolm, D. L. Laird, Modelling of Blades as Equivalent Beams for Aeroelastic Analysis, AIAA Reno Jan, **#0870**, (2003)
7. T. R. Baumann, Formulation of Equivalent Uniform Beam Elements from Structual Stiffness Matrices, AIAA-92-**#2418**-CP(1992)
8. J. Sun, Y. Jia Y. L.Mo, Evaluation of Elastic Properties and Lateral Stiffness of Multi-Ribbed Walls Based on Equivalent Elastic Model., Elsevier Engineering Structures, **#72** 92-101, (2014)
9. MSC Software, Linear Static Analysis User's Guide, MD/MSC Nastran (2010)
10. GENESIS Design Manuel, Volume 2, Vanderplaats Research & Development, Inc., Colarado-USA(2012)

MATEC Web of Conferences **45**, 03001 (2016)
DOI: 10.1051/matecconf/20164503001

A capacitor-clamped inverter based torsional oscillation damping method for electromechanical drivetrains

Viknash Shagar, Doupadi Bandara, Shantha Jayasinghe, and Hossein Enshaei

Australian Maritime College, University of Tasmania, Australia

Abstract. A typical electromechanical drivetrain consists of an electric motor, connecting shafts and gears. Premature failures of these shafts and gears have been reported which are mainly due to fatigue caused by extreme loads and torsional oscillations. Overdesign and passive damping are the common approaches taken to increase the fatigue life. Nevertheless, they increase the system cost, weight and volume. Alternatively, active damping through advanced inverter control of the motor drive has been identified as a promising solution that does not require overdesign or alterations to the existing system. Even with the active damping control, oscillations propagate into the dc side of the power converter and subsequently to the upstream power bus. Generally, a large capacitor or an additional energy storage system is placed to suppress these oscillations. This paper proposes to use the clamping capacitors of the capacitor-clamped inverter as energy storage elements and thereby eliminate the need for a large dc side capacitor or an additional energy storage system. The efficacy of the proposed method has been verified with computer simulations. Simulation results show that the clamping capacitors are capable of containing torsional oscillations within the inverter without passing them to the upstream power bus.

1 Introduction

More electric technologies (METs) play an important role in meeting ever growing demands for energy efficiency and emission reduction in the transportation sector [1]. Improved performance and control flexibility are the other advantages that make METs stand out amongst predominantly mechanical technologies. The automobile industry is the first to incorporate METs into vehicles whereas the maritime and aviation industries are catching up fast. As a result, conventional mechanical transmission based propulsion and drive systems in ships and aircrafts are gradually being replaced with electromechanical drive systems. The most prominent MET in ships is the electric propulsion azimuth thrusters or pod propulsion systems [2].

In a typical electromechanical drive system, there is a power converter to convert the fixed voltage-fixed frequency electrical supply into a variable voltage-variable frequency output to feed the motor. The load is connected to the motor through shafts and gearboxes. The L-drive type azimuth thruster is a good example for this configuration where the motor is connected to the propeller through a vertical shaft, a horizontal shaft and a bevel gear. Depending on the application, the shafts may require being long and thin which results in a certain degree of elasticity. The elasticity leads to torsional oscillations in transient conditions. Unless appropriate control measures are taken, repetition of such oscillations over a period of time leads to fatigue and result in driveline failure [3]. The easiest way to avoid such premature failures is the overdesign with bigger and stiffer shafts and gears. But it increases the weight, volume and cost of the system. Passive damping is another solution that uses additional hardware to suppress oscillation. This also increases the weight, volume and cost of the system.

Alternatively, active damping through advanced power converter control has been identified as a promising solution that can suppress torsional oscillations and thereby increase the fatigue life without hardware modifications [4]. Therefore, the size and weight of the drivetrain can be kept at a reasonable level eliminating the need for overdesign.

Even though active damping reduces torsional oscillations, the remaining oscillations in the shaft propagate into the dc-side of the motor drive. These oscillations can propagate further into the main power system of the ship as well causing damages to the attached equipment. Increasing the dc-link capacitance or adding an energy storage is the common solution to prevent the propagation of oscillations into the power system. However, they require large capacitors or additional hardware. Therefore, as an alternative, this paper proposes to use clamping capacitors of the capacitor-clamped inverter to absorb oscillations and thereby eliminate the need for large dc-link capacitor or an additional energy storage system.

The proposed concept has been verified with computer simulations considering an L-drive azimuth thruster system as an application. The simulations were carried out for three different scenarios. Firstly, the system has been simulated for the normal operation without active damping or clamping-capacitor support to contain oscillations. The results show significant oscillations in the shaft, dc voltage and input current. Secondly, the active damping was added to the control which reduced the oscillation in the shaft. However, the remaining oscillations get propagated into the inverter and appeared in the dc voltage and the input current. Thirdly, both active damping and clamping capacitor support were added to the controller which in turn reduced oscillations in the shaft and smoothened dc voltage and input current. This concept is thus proven to be feasible as the torsional oscillations can be contained within the converter without passing them on to the shipboard power system.

2 Proposed system

The proposed system is shown in Figure 1 where a capacitor-clamped inverter is used to drive the propulsion motor of an L-drive type azimuth thruster. In this study, a permanent magnet synchronous motor (PMSM) is used as the propulsion motor. The motor is connected to the propeller through a vertical shaft, horizontal shat and a bevel gear. For simplicity, in the simulation model, an ideal gear was assumed and the two shafts were represented by a single shaft. Moreover, low values were assigned for the shaft stiffness, K_s, and the damping coefficient, B in order to have a more flexible shaft and hence getting a better illustration of the torsional oscillations. The block diagram of the drivetrain model is shown in Figure 2 where θ_m is the angle at the drive end, θ_p is the angle at the load end and J_P is the inertia of the load. The torque transmitted through the shaft, T, varies with the difference in the twisting angle of the shaft at the drive end and the load end. The corresponding mathematical expression for T can be expressed as in (1).

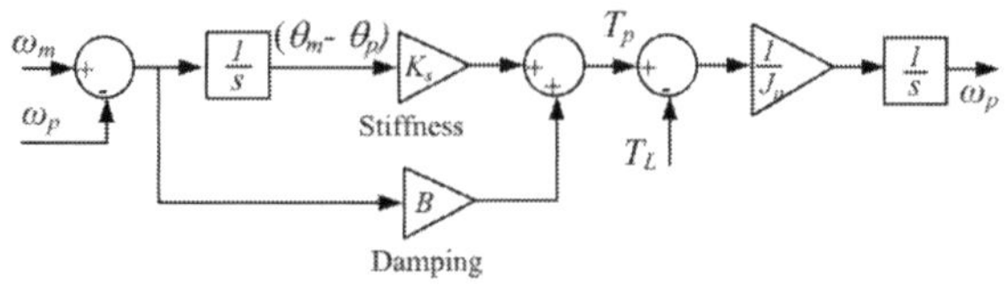

Figure 2. Drivetrain model

$$T_p = K_s \int (\omega_m - \omega_p) dt + B(\omega_m - \omega_p) \quad (1)$$

3. Control strategy

3.1 Motor controller

A block diagram of the motor controller is shown in Figure 3. In the steady state operation, the controller compares the actual motor speed, ω_m, with a reference speed, ω_m^*, and the error is then passed through PI controller. The output of the PI controller is the reference torque from which the q axis current reference, i_q^*, can be derived. The d axis current reference, i_d^*, is set to zero. The actual dq axes currents are compared with the references and the errors are passed through PI controllers which in turn produce references for dq axis voltages as shown in Figure 3. Finally, using the voltage values of the dq axes, the amplitude and angle of the inverter output voltage are generated in the abc frame. The PI controller at the bottom left of Figure 3 also compares the speed difference between the drive end and the load end of the drivetrain in order for the motor to vary its torque so as to reduce the amount of twisting in the shaft that can occur during wildly fluctuating load conditions. This is the control action that implements active damping. Therefore, a more flexible shaft can have its mechanical stress converted to the electrical equivalent by increasing the gain of the corresponding PI controller. The clamping capacitor charge/discharge controller and modulation technique described below can then help to absorb and hence damp down the voltage and current oscillations in the inverter.

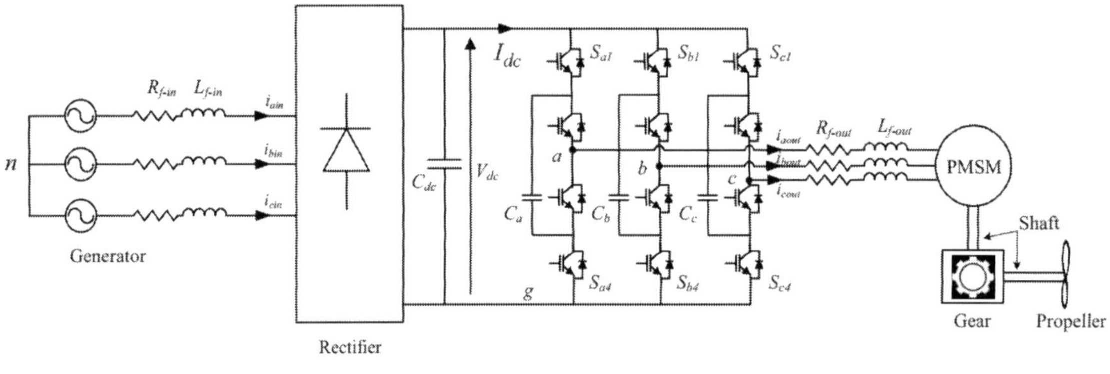

Figure 1. Capacitor clamped inverter based azimuth thruster drive system

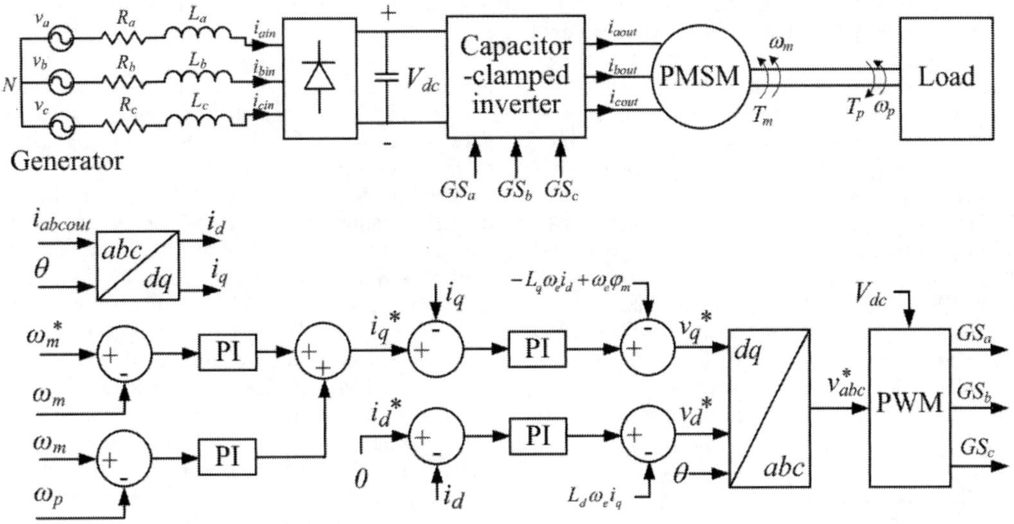

Figure 3. Motor speed controller

3.2 Inverter modulation

Modulation techniques for the capacitor-clamped multilevel inverter have been extensively covered in literature [5-9]. However, most of these techniques require balanced capacitor voltages for the operation. In the approach taken in this paper, the goal is for the clamping capacitors to absorb any oscillation in the dc-link voltage caused by loading changes. Therefore, the capacitor voltages cannot be assumed to be balanced in such operating conditions. This means that the conventional modulation (PWM) techniques such as carrier based PWM and space vector PWM methods would not be suitable in these capacitor clamped inverters. Hence, the modified PWM method presented in [10] has been adopted as the modulation method for this study.

3.3 Clamping capacitor charge/discharge controller

The charging and discharging of the clamping capacitors serve to absorb the oscillations in the dc-link voltage of the motor drive and thus prevent them from being transferred to the upstream power bus. The clamping capacitor charge/discharge control strategy presented in [10] has been adopted in this study. The corresponding controller block diagram for the capacitor, C_a, is shown in Figure 4 where the PI controller at the top of the diagram helps bring the capacitor voltage to half the dc-link voltage value. The second PI controller from the top left corner of Figure 4 compares the square of the actual dc-link voltage against a reference generated by passing the same signal through a low pass filter. This reference represents a smooth change in the dc-link voltage and thereby removes the oscillation that occurs during transients. The two PI controllers are designed in a way that the First PI controller is dominant during steady state

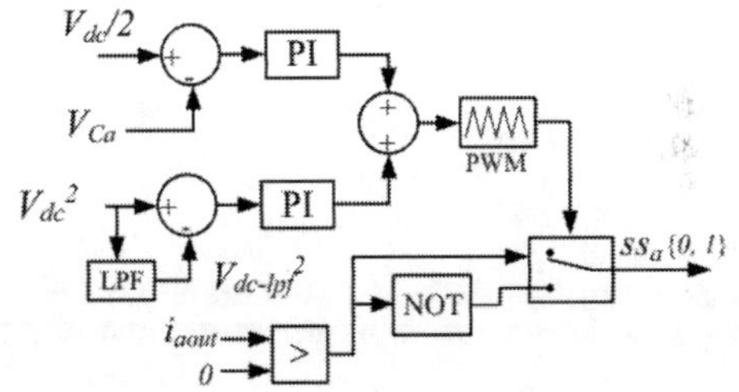

Figure 4. Charge/discharge controller for the capacitor C_a attached to the leg 'a' of the inverter

conditions while the second PI controller is dominant during transient conditions. For instance, if the electrical system experiences more transients reflected as oscillations in the dc-link voltage one way to make it smooth is to make the second PI controller to act faster. However, as the effect of the second PI controller is present only during transients it does not bring the capacitor voltage to the balanced condition of half the dc-link voltage after the transient. The first PI controller becomes prominent after the transient and thus it brings clamping-capacitor voltages to the balanced condition. The output function SS_a of the charge/discharge controller shown in Figure 4 selects one of the two redundant switching states depending on the polarity of the phase current at the output [10].

4 Simulation results

Three sets of simulations were carried out on the MATLAB/Simulink platform to verify the efficacy of the proposed torsional oscillation suppression method. Both active damping and clamping capacitor support were not included in the first simulation. Therefore, the results shown in Figure 5 correspond to the transient response of the normal system. The load changes introduced to create transients are shown in Figure 5(a). These transients excite torsional oscillations in the shaft as shown in

Figure 5(b). The shaft has a certain damping coefficient and thus these oscillations get damped slowly. Amidst this natural damping, oscillations continue for a significant period of time during which the shaft experiences stresses and resultant fatigue. The output current of the inverter is shown in Figure 5(c) where the envelop shows similar oscillations. This is an indication that the oscillations get propagated into the electrical side. The dc side voltage and input current waveforms shown in Figures 5(d) and 5(e) confirms the propagation of oscillations into the dc-side and upstream power bus respectively. The clamping capacitor voltages are kept at balanced conditions as shown in Figure 5(e) and thus they do not absorb oscillations.

In the second simulation, the active damping was enabled and thus the oscillation in the shaft got damped quickly as shown in Figure 6(a). As shown in Figures 6(b), 6(c) and 6(d) the output current, dc-link voltage and the input current show the same oscillation, but for a relatively short period of time. As evident from these results active damping is capable of reducing torsional oscillations. Nevertheless, the remaining oscillations still get propagated into the dc side of the inverter and then to the power system.

The proposed method absorb these oscillations through clamping capacitors of the inverter. The third simulation was carried out to test the performance of this method and thus both active damping and clamping-capacitor support were enabled in the controller. Due to the presence of the active damping controller the oscillations in the shaft got damped quickly as shown in Figure 7(a). This graph and the output current waveforms shown in Figure 7(b) are very much similar to the corresponding graphs in Figure 6(a) and 6(b) respectively. Therefore, torsional oscillation damping on the shaft depends entirely on the active damping controller. In other words, the clamping-capacitor support does not have a significant effect on shaft oscillations.

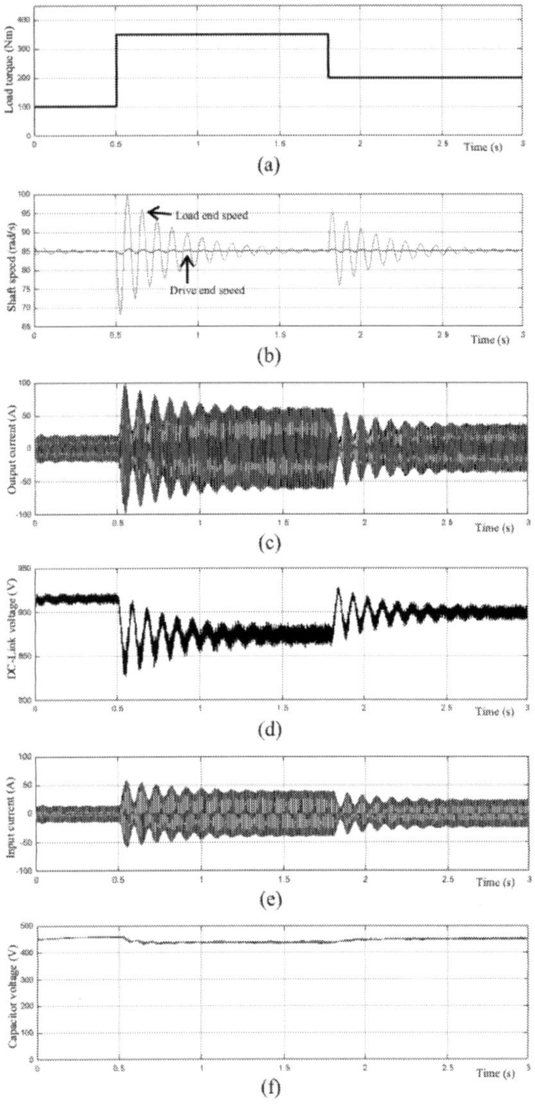

Figure 5. Normal operation (a) load torque, (b) shaft speed at the drive end and load end, (c) output current, (d) dc-link voltage, (e) rectifier input current, (f) clamping capacitor voltage

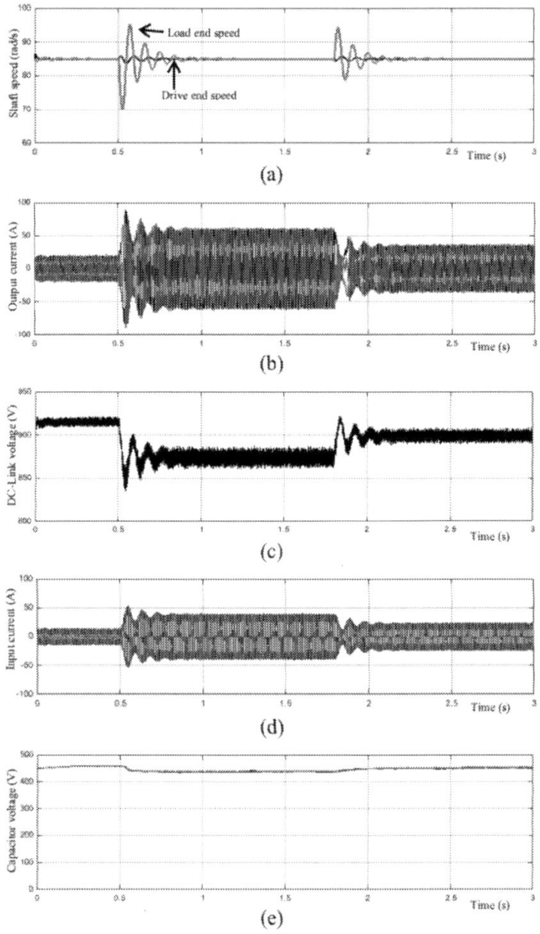

Figure 6. Proposed operation with active damping (a) shaft speed at the drive end and load end, (b) output current, (c) dc-link voltage, (d) rectifier input current, (e) clamping capacitor voltage

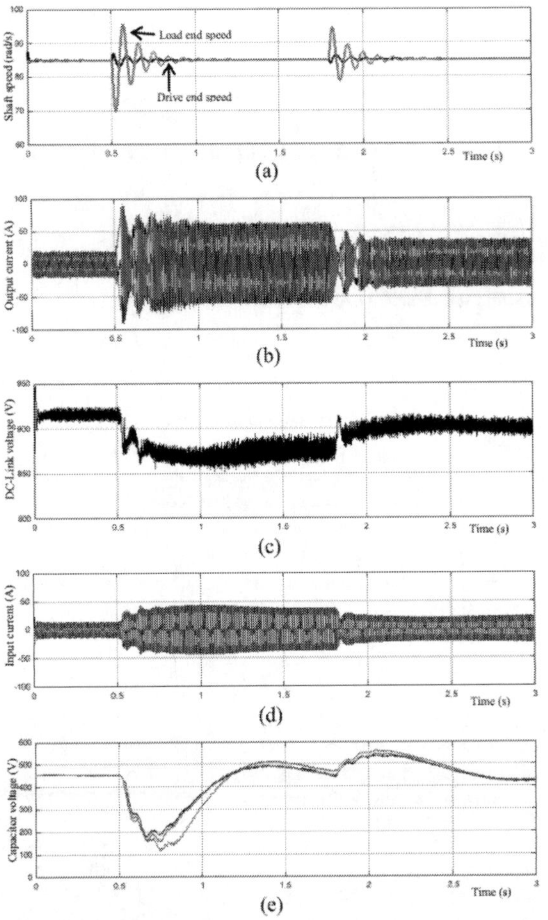

Figure 7. Proposed operation with active damping and clamping capacitor support (a) shaft speed at the drive end and load end, (b) output current, (c) dc-link voltage, (d) rectifier input current, (e) clamping capacitor voltage

As evident from the smooth transients in the dc-link voltage and input current waveforms shown in Figure 7(c) and 7(d) respectively, the clamping-capacitor support makes a significant contribution to contain the oscillations within the inverter without passing them to the upstream power bus. The variations in the clamping-capacitor voltage, as shown in Figure 7(e), verifies this claim. When the load gain occurs at 0.5s, the clamping capacitors get discharged fast to supply the sudden increase of the load power. Moreover, they absorb the oscillations in the dc-link voltage as well. Therefore, the dc-link voltage changes smoothly. When the steady state occurs, capacitor voltage rises slowly to the balanced condition. The opposite happens at the load drop.

5 Conclusions

This paper demonstrates the effective use of short-term energy storage capability of the capacitor-clamped inverter to contain oscillations within the inverter. Moreover, an active damping controller is used to suppress the torsional oscillations present in electromechanical drivelines with long and flexible shafts. An azimuth thruster with long flexible shafts is used as the example application to assess the efficacy of the proposed method. The results show that the proposed method is feasible and the shaft oscillations and vibrations can be damped quickly with the active damping controller. Moreover, the results show that the clamping capacitors are capable of absorbing the resulting transients arising from the shaft oscillations, without passing them on to the upstream power bus.

References

[1] S.D.G. Jayasinghe, G. Lokuketagoda, V. Shagar, D. Ranmuthugala, and H. Enshaei, "Electro-technologies for Energy Efficiency Improvement and Low Carbon Emission in Maritime Transport," in *Proc.* IAMU AGA, pp.119-123, (2015)

[2] C.A. Reusser, and H. Young, 'Full electric ship propulsion based on a flying capacitor converter and an induction motor drive,' in *Proc. International Conference on Electrical Systems for Aircraft, Railway, Ship Propulsion and Road Vehicles* (ESARS 2015), pp.1-6, (2015)

[3] M. Fonte, L. Reis, and M. Freitas, "Failure analysis of a gear wheel of a marine azimuth thruster," *Engineering Failure Analysis*, (**18**), pp.1884-1888, (2011)

[4] H. Geng, D. Xu, Bin Wu, and Geng Yang, "Active Damping for PMSG-Based WECS With DC-Link Current Estimation," *IEEE Trans. Industrial Electronics*, (**58**), pp.1110-1119, (2011)

[5] L. Ziyou, A. I. Maswood, and G. H. P. Ooi, "Modular-Cell Inverter Employing Reduced Flying Capacitors With Hybrid Phase-Shifted Carrier Phase-Disposition PWM," *IEEE Trans. Ind. Electron.*, (**62**), pp.4086,4095, (2015)

[6] V. Dargahi, A. K. Sadigh, M. Abarzadeh, S. Eskandari, and K. A. Corzine, "A New Family of Modular Multilevel Converter Based on Modified Flying-Capacitor Multicell Converters," *IEEE Transactions on Power Electronics*, (**30**), pp.138,147, (2015)

[7] W. Li, Q. Luo, Y. Mei, S. Zong, X. He, and C. Xia, "Flying-Capacitor Based Hybrid LLC Converters with Input Voltage Auto-Balance Ability for High Voltage Applications," *IEEE Trans. Power Electron.*, (31), pp.1908-1920, (2015)

[8] A. Ashraf Gandomi, K. Varesi, and S. H. Hosseini, "Control strategy applied on double flying capacitor multi-cell inverter for increasing number of generated voltage levels," *IET Power Electronics*, (**8**), pp.887,897, (2015)

[9] P. R. Kumar, R. S. Kaarthik, K. Gopakumar, J. I. Leon, and L. G. Franquelo, "Seventeen-Level Inverter Formed by Cascading Flying Capacitor and Floating Capacitor H-Bridges," *IEEE Trans. Power Electron.*, (**30**), no.7, pp.3471-3478, (2015)

[10] D.M. Vilathgamuwa, S.D.G. Jayasinghe, and U.K. Madawala, "Battery clamped three-level inverter for renewable energy systems," in *Proc. IEEE Ind. Electron. Society Conf.*, pp.3105-3110, (2011)

MATEC Web of Conferences **45**, 03002 (2016)
DOI: 10.1051/matecconf/20164503002

Performance and Emission Studies of a SI Engine using Distilled Plastic Pyrolysis Oil-Petrol Blends

Kareddula Vijaya Kumar[1], Ravi Kumar Puli[1], A. Swarna Kumari[2] and P. Shailesh[3]

[1] Dept. of Mechanical Engineering, National Institute of Technology, Warangal, TS-506004, INDIA
[2] JNTUK/MED, Kakinada, INDIA
[3] MCET, OU/MED, Hyderabad, TS-500001, INDIA

Abstract. In the present work, an experimental investigation is carried out to evaluate the use of plastic oil derived from waste plastic which used in a Spark Ignition engine. Experiments are conducted, the measured performance and emissions of plastic oil blends at different proportions are compared with the baseline operation of the SI engine running with gasoline fuel. Engine performance and exhaust gas emissions such as carbon monoxide, total unburned hydrocarbons, carbon dioxide and oxides of nitrogen are measured. From the experiments it is observed that 50% Distilled Plastic Pyrolysis Oil (50%DPPO) exhibits the substantial enhancement in brake power, brake thermal efficiency and reduction in brake specific fuel consumption running at full load conditions among different blends and pure petrol. There is also noticed decrement of carbon dioxide and unburned hydrocarbons emissions at the same blend. The experimental result shows that plastic oil shall conveniently be used as a substitute to gasoline in the existing SI engines without any modifications.

1 Introduction

Increasing industrialization, urbanization and changes in the pattern of human life, will accompany the process of economic growth but at parallel causes for increasing quantities of environmental wastes leading to increased threats to it. Developing countries like India depends heavily on oil import. From the last two decades research around the world is concentrating on alternative energy generation techniques, it is the high time to generate the energy by alternative sources. Alternative fuels for both spark ignition and compression ignition engines have become very imperative owing to increased environmental protection concern, need to reduce dependency on petroleum products and even socioeconomic aspects. The past three and half decades are witnessed for an explosive growth of the plastic utilization across world. The production of synthetic polymers represented by polyethylene (PE), polypropylene (PP), polystyrene (PS), and polyvinyl chloride (PVC). These plastics are widely used in many important day to day applications such as clothing, household appliances and in automotive products and aerospace. While we enjoying the conveniences that plastics can provide, the treatment of waste plastic becomes an unavoidable and imminent issue.

In this context, it is urgent need to search a safety and effective ways to recycle waste plastics. Recently new ways of environmental friendly waste plastic recycling have been of interest, and among them, the use of waste plastics as a supplementary fuel has attracted interest of many researchers. Mehdi et al. [1] reviewed the PVC wastes and suggested chemical recycling is the best process for the environmental pollution problems and addressed several recent separation and recycling problems. Mitsuhara et al. [2] attention is also focused on using oil derived from waste plastics in diesel engines. Williams et al. [3] investigated the diesel engines are the most preferred power plants due to their excellent driveability and higher thermal efficiency. Naima et al. [4] had investigated alternative fuels for internal combustion engines by using waste plastics, waste engine oil and waste cooking oil. The properties of the oil derived from these three types of oils were analysed and compared with the diesel fuel. Moinuddin Sarker, et al. [5] used stainless steel reactor to produced hydrocarbon vapors from waste thermoplastics. Plastics are non-biodegradable polymers mostly containing carbon, hydrogen and few other elements which were the root cause for the plastic waste management and environmental issues [6-8]. Many researchers are investigated the usage of waste plastic oil and waste tyre oil as an alternative fuel for diesel engine. Shanmuga Vadivel et al. [9] are investigated the performance and emission characteristics of Nano engine running on plastic oil with gasoline blends. Their results shows an increase in brake thermal efficiency for the blends when compared to that of gasoline fuel. [10-12] were run the diesel engine blended with plastic oil to evaluate the

Article available at http://www.matec-conferences.org or http://dx.doi.org/10.1051/matecconf/20164503002

performance and emission characteristics without any engine modifications, their investigation clears that the plastic oil is alternate fuel for diesel engine. [13-17] are experimentally investigated on a diesel engine using waste plastic oil and tyre oil blends with exhaust gas recirculation. Their results shows that plastic and tyre oil blended engines suffers with high NO_x emission, which can deduced by using exhaust gas recirculation. Vijaya Kumar et al. [18] had investigated the performance and emission characteristics of diesel engine with three hole injector by using tyre oil blends and their results shows that 20% tyre oil blend was given better performance compared with diesel and other blends at all injections pressures. [19-22] are conducted experiments on diesel engine fuelled with tyre pyrolysis oil blended at different proportions to evaluate the performance and emissions with and without ethanol. [23, 24] are investigated the performance, emission and combustion studies using distilled tyre pyrolysis oil blends on diesel engine without any engine modification. However, very little work has been done to test their use in high-speed diesel engines. The main objective of this research is derived from the literature review, analyze the engine performance, combustion and emission characteristics of petrol engine fuelled with distilled waste plastic oil blends with petrol fuel. The crud waste plastic oil prepared from the pyrolysis process was carried out the batch distillation apparatus. This distilled plastic oil blended with gasoline at different volume fractions and the performance and emission characteristics was carried out on 4stroke single cylinder kerlosker petrol engine without any engine modifications.

1.1 Preparation of Plastic Pyrolysis Fuel

The production method for the conversion of plastics to liquid fuel is based on the pyrolysis of the plastics and the condensation of the resulting hydrocarbons are shown in Fig. 1. Pyrolysis refers to the thermos-chemical process for conversion of waste plastic by heating the feed stock at high temperature in absence of air which produces gaseous products which is them condensed to give liquid fuels consisting of pyrolytic oil or liquid oil.

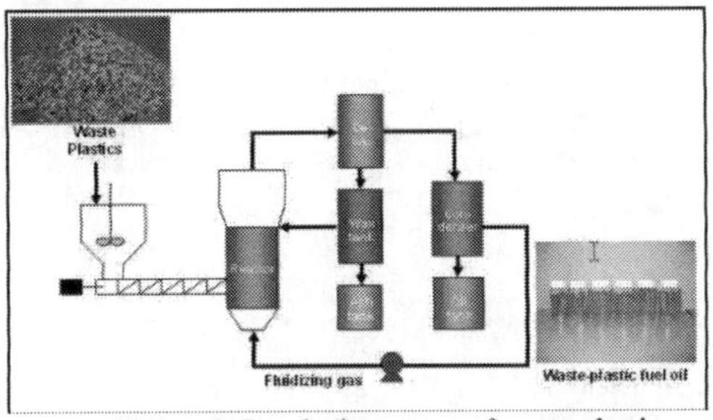

Figure 1. Pyrolysis process of waste plastics

Waste plastic is treated before feed in to the reactor. The feed stock passed to hopper through mills and send in to reactor by the hot screw conveyor. In reactor waste plastic is subjected to high temperature and low pressure by the hot gasses coming from combustion chamber where they will decompose at 450°C to 550°C. Vapors, gasses and char products are generated and they passed through cyclone filter. The char is collected at the ash tank and gasses are send to condenser. In the condenser heat from this gasses is utilized to generate steam. The steam can used to generate the power as by product. So, that the plastic vapors are gets condensed in the condenser. The condensate oil is separated and used to further transformation, the non-condensable gasses are feed back to the reactor as fluidized gasses. Almost 70-80% of gaseous products are converted in to liquid fuels providing a good quantity of liquid fuel which then can be enhanced.

1.2 Reducing Viscosity of Plastic Pyrolysis Oil

The plastic oil which obtained by the pyrolysis method has some disadvantages such as high viscosity, low volatility, gumming effect and carbon residue. So, that the demerits impose a negative impact on the engine performance, emissions durability. Therefore, before using they as fuels the viscosity of the plastic oil have to be reduced. One interesting method to reduce the viscosity of the plastic oil and also to improve the spray formation so as to use them in the spark ignition engines are distillation. This process applied for petrol and diesel grade fuel production process. Fig. 2, distillation apparatus is used for to remove the impurities from basic fuel and increase the volatility of fuel. After distillation volatility of fuel is increase, it helps to homogenization of air fuel mixture in case of SI engine carburetor itself and also avoid logging of fuel impurities in filter or in nozzle holes.

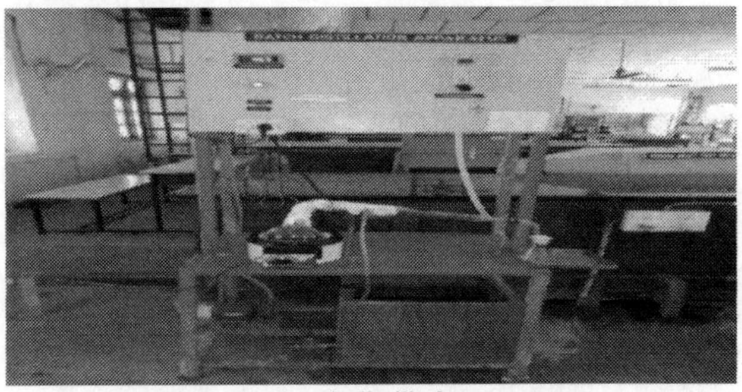

Figure 2. Batch distillation apparatus

Waste plastic to fuel was use for further distillation process and making petrol grade fuel. Distillation column was use for distillation process. Distillation process set up different columns with different temperature profile like low boiling point fuel to high boiling point fuel. Petrol grade fuel collected from fractional column and temperature range was 90°C to 160°C. This fuel hydrocarbon compound also heavier and this fuel are not igniting. Collected petrol grade fuel percentage was about 42% and rest of all other fractional fuel percentage was 58% including diesel grade fuel and light gas also. Fractional distillation process was also generating some light gases. Plastic pyrolysis fuel to different fuel by using fractional distillation column used for heat applied with different column temperature wise and fuel break down into shorter into longer chain wise and come out into different fraction column then collected into separate container.

In the present study distillate of the plastic oil is obtained in the same boiling range of petrol (90-160°c). The oil is obtained by pyrolysis come distillation is shown in the below Fig. 3.

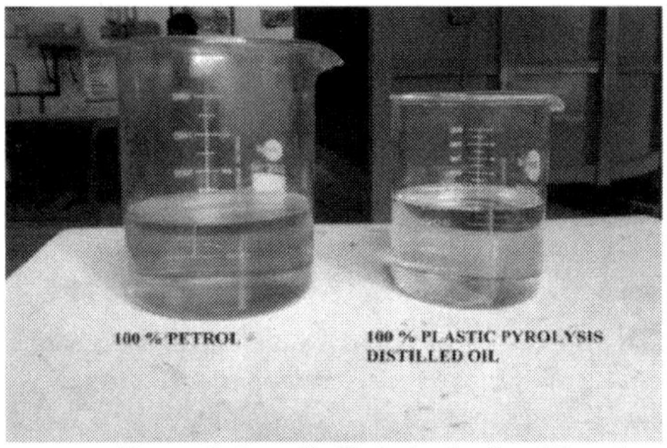

Figure 3. Pure distilled plastic oil and Petrol

Table 1: Basic properties of fuels

S. No.	Characteristics	Petrol	PPO	DPPO
1	Specific Gravity @ 40°C	0.720	0.820	0.780
2	Kinematic Viscosity m^2/sec @ 40°C	0.5 x 10^{-6}	2.52 x 10^{-6}	0.7 x 10^{-6}
3	Calorific Value (kJ/kg)	43953	42808	46817
4	Density @ 40°C (kg/m^3)	720	820	780

As the derived plastic fuel from the waste plastic after pyrolysis and distillation catches fire in room temperature itself, so flash and fire point is not valid for the arrived fuel as same like as petrol. And the extraction from the distillation i.e. plastic fuel slightly having more calorific value compare to that of gasoline, and also having density more than that of the gasoline. Some of the properties of the petrol, plastic pyrolysis oil (PPO) and distilled plastic pyrolysis oil (DPPO) are compared in table 1.

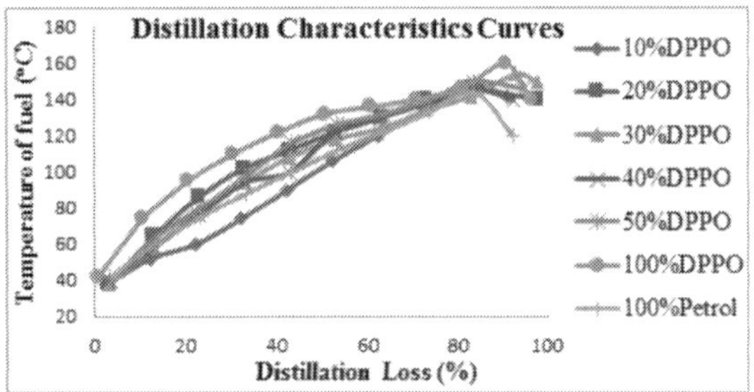

Figure 4. Volatility variation in petrol and DPPO blends

The tendency of a fuel to vaporize is also characterized by determining a series of temperatures at which various percentages of the fuel have evaporated (boiling temperatures), as described in ASTM D86, Test Method for Distillation of Petroleum Products. The temperatures at which 10 percent, 50 percent, and 90 percent evaporation occurs are often used to characterize the volatility of gasoline.The distillation characteristics curves of the petrol and plastic fuel as well as the different blends are shown in Fig. 4. From Fig. 4, it can be observed that the volatility property is more in the plastic pyrolysis distilled oil blends compared to the 100% petrol.

1.3 Preparation of Plastic Oil Blend

The basic properties of DPPO-PF were measured and compared with conventional petroleum fuels. The Carbon content are higher for DPPO than petrol fuel. In the present work 10%DPPO, 20%DPPO, 30%DPPO, 40%DPPO and 50%DPPO of DPPO was blended with petrol fuel on volume basis and observed for 15 days to check for any separation. No such separation was noticed. DPPO blended with PF is indicated as 10%DPPO. For example, 10%DPPO blended with 90% PF is denoted as 10%DPPO.

1.4 Experimental Setup

The Test rig comprises of the following apparatus and the layout is shown in Fig. 5,

- Petrol engine
- Electrical Dynamometer
- Burette setting with fuel consumption pipe and fuel tank
- Air and fuel supplied to engine
- Gas analyzer and Display

The test rig consists of single cylinder, 4-stroke, air-cooled petrol engine coupled to a electrical dynamometer, which acts as a loading device. A six channel temperature indicator provided to measure the temperature at various points of the test rig.

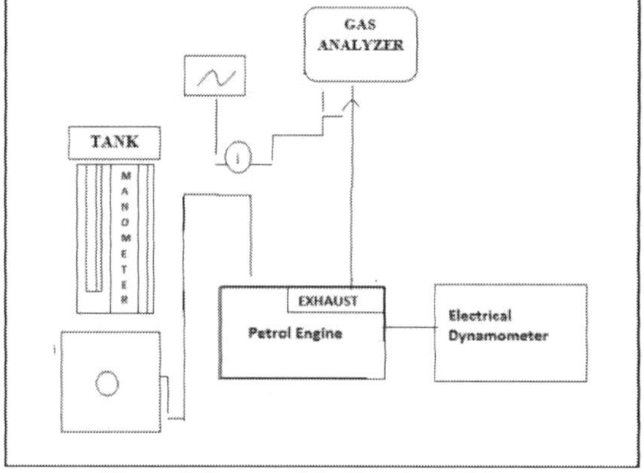

Figure 5. Layout of the Experimental Set up

The engine is complete with carburetor, governor, exhaust silencer, etc... a 2.2 KVA capacity, single phase, DC Generator is coupled on a strong iron base plate with bank (fin type resistor) arrangement a panel board consisting of main switch, voltmeter, ammeter to measure the power output. A fuel tank of about 5 liters capacity, a 100cc x 0.1cc burette and a three way cock arrangement. A silencer is fitted to the exhaust pipe to reduce the noise level to a minimum. A single phase panel board arrangement consisting of a digital ammeter and voltmeter and single phase main switch. All this above instruments with panel board are mounted on the M.S. Fabricated stand. The speed of the engine is measured by the use of non-contact type digital tachometer.

1.5 AVL Gas Analyzer

The purpose of the AVL gas 444 is to measure the relative volumes of certain gases constituents in the exhaust gas of motor vehicles has shown in Fig. 6. These gases are carbon monoxide, carbon dioxide, hydrocarbons, and oxygen nitric oxide. The air fuel ratio is calculated from the CO, CO_2, HC and O_2, constituents and displayed. The Liquid crystal display shows the constituents of the gas once the gas is allowed to flow in. It is a standard output device for the AVL gas 444. The functional keys serves for various purposes such as to fine filter helps in removal or separation of water which enters into meter with tested gas. This prevents the damage of the machine. Frequently the filter has to be replaced depends on the manufacturer catalogue. The water inlet allows the gas from the engine exhaust into the machine. The exhaust gas probe takes a sample of gas from the exhaust tail pipe of the vehicle and carries it to the gas analyzer via exhaust gas hose. The important elements that are present on the back side are condensate outlet, ventilator and fan, sensors rpm and oil temperature RS 232 connection and power supply.

Figure 6. AVL Gas Analyzer

2 Results and Discussion

2.1 Engine Performance Characteristics

2.1.1 Brake Specific Fuel Consumption

Brake specific fuel consumption of an engine is defined as the rate of fuel consumption per unit power produces. This is an important performance parameter as it determines the mileage of the vehicle. It can be seen from the Fig. 7, that the BSFC of the engine decreased with increase in load and also BSFC decreased with increase in the proportion of plastic oil in the blend. Density and calorific values of the blend increases with increases in the plastic proportion as the plastic oil is denser than petrol and also have higher calorific value.

2.1.2 Brake Thermal Efficiency

Fig. 8, shows the variation of brake thermal efficiency with load for various blends. Brake thermal efficiency increased with increase in load with the rate of variation decreasing with increase in load acting. For the 1/6th, 1/3rd and 1/2 load conditions the brake thermal efficiency of 50%DPPO is more than the pure Petrol, after half load condition the brake thermal efficiency is fall and again raised at the full load conditions.

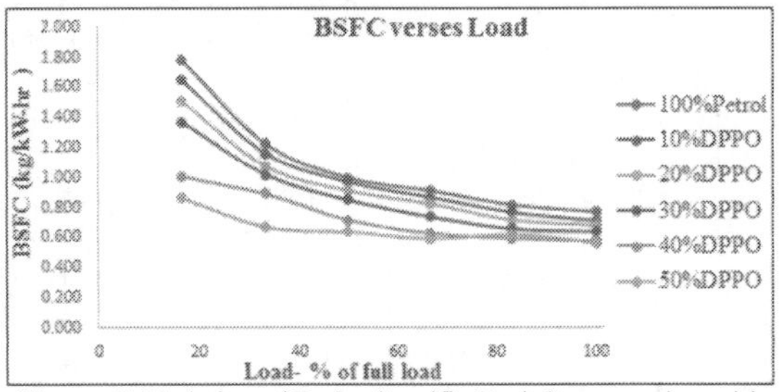

Figure 7. Variation of Brake Specific Fuel Consumption with Load

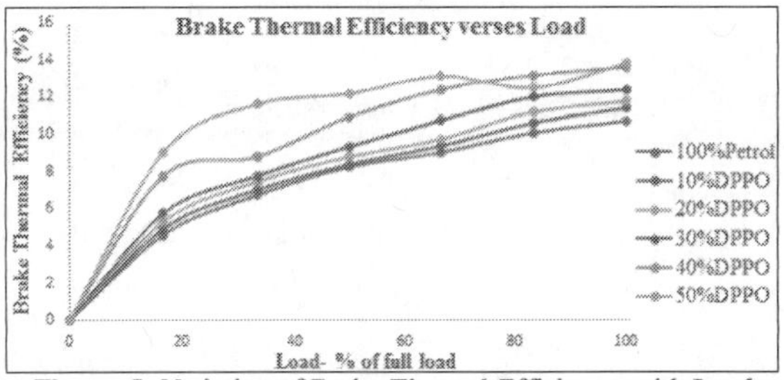

Figure 8. Variation of Brake Thermal Efficiency with Load

2.2 Engine Emission Characteristics

2.2.1 Carbon Monoxide

From Fig. 9, it can be observed that, CO emissions increased with increase in load reached the maximum at 4th or 5th load and there after slightly decreased. CO emissions are due to incomplete complete combustion of fuel either due to inadequate oxygen or flame quenching. In the present study as the engine is equipped with a carburetor to supply fuel to the engine cylinder, fuel air mixture might not be homogenous throughout the cylinder. CO emissions increased marginally with increase in the proportion of the plastic oil in bend.

2.2.2 Oxides of Nitrogen

NO_x emissions increases with increase in the peak cylinder temperature. Peak cylinder temperature increases with increase in load. NO_x emissions also varies with air fuel ratio, but in petrol engine air fuel ratio does not vary much with load. From Fig. 10, it can be observed that

NO_x emissions increased with increase in load as predicted.

2.2.3 Unburnt Hydro Carbons

Fig. 11, shows the variation of unburnt hydro carbon emissions with load for all blends. Unburnt hydro carbon emissions decreased with increase in the proportion of plastic oil in bend. Whereas CO emissions increased with increase in the plastic oil content in blend. In-cylinder temperature is a key factor in determining unburned HC and CO emissions characteristics. Lower temperatures generally cause higher emissions of unburned HCs and CO, however there are certain temperature ranges where HC emissions increase while CO decreases. An important consideration for the HC and CO emissions characteristics is the hot ignition and CO to CO_2 Oxidation threshold temperatures. Prior research has shown that the rapid breakdown of H_2O_2 leading to hot ignition for HC fuels begins at roughly 1000 K. CO-to-CO_2 oxidation requires temperatures of at least 1500 K.

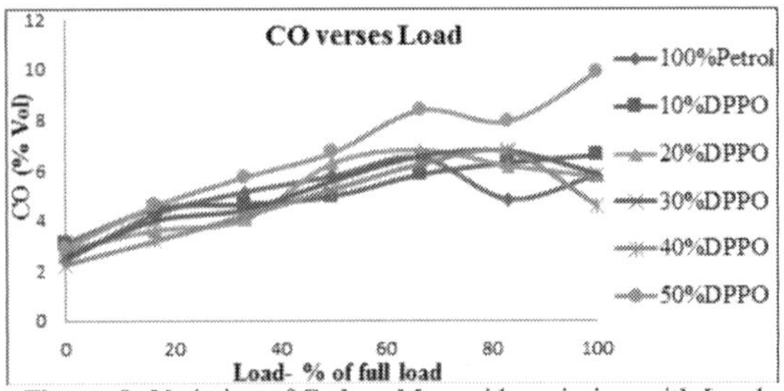

Figure 9. Variation of Carbon Monoxide emission with Load

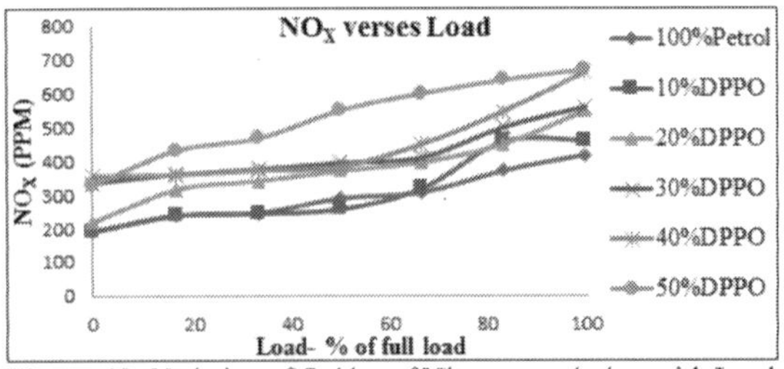

Figure 10. Variation of Oxides of Nitrogen emission with Load

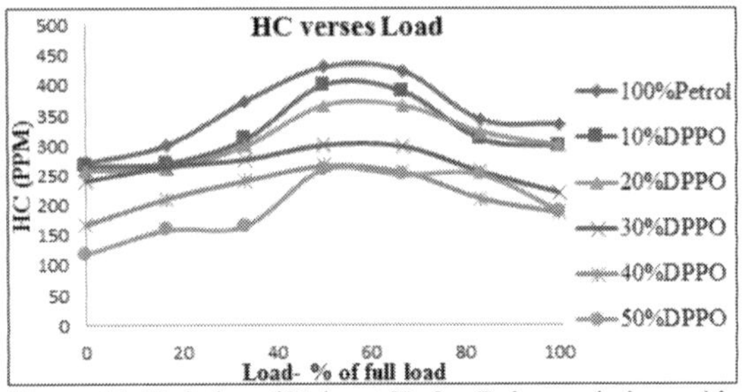

Figure 11. Variation of Unburnt Hydro Carbon emissions with Load

In the present study as the plastic oil has more calorific value than petrol, engine temperatures might be higher when the engine is operated with plastic oil. The HC and CO emissions exhibited an opposite trend with increase in plastic oil content in blend. This might be due to in-cylinder temperatures being near or below the H_2O_2 breakdown temperature, where radicals released from rapid H_2O_2 breakdown cause HC molecules to break down to form CO . When the temperatures are near or below 1000 K, partial oxidation of HCs into CO does not occur thereby causing higher emissions of HCs and lower emissions of CO.

3 Conclusions

Plastic pyrolysis distilled fuel is tested on the SI engine and the performance and emissions characteristics are compared with the different blends and the following observations are made.

- Plastic oil obtained from the pyrolysis of a plastic waste can be used as a fuel in petrol engine by obtaining the distillate in the temperature range of petrol.
- Petrol and plastic pyrolysis distilled fuel can be readily blended without any precautions or the methods as both their viscosity and density are of in the same range.
- It is observed that the distillation characteristics of both petrol and plastic pyrolysis distilled fuel are more or less similar.
- The basic specific fuel consumption of the engine fuelled with 50%DPPO is 26.84 % lesser than that of the engine fuelled with sole petrol at full load.
- The brake thermal efficiency of the engine with 50%DPPO is 19% higher than that of the sole petrol operation.
- The CO emissions in 50%DPPO are 1.7 times more than that of the engine fuelled with the 100% Petrol.
- The unburnt hydrocarbons are 43.58% lesser in the 50%DPPO engine operation compared to the sole petrol operation.
- The NO_x emissions are 1.6 times more in the 50%DPPO compared to that of the engine with 100% petrol as fuel.

In the view of performance, experiments can run beyond 50% distilled plastic pyrolysis oil without any engine modification, but while considering in NO_x emissions, they are proportionately increasing with the percentage of distilled plastic pyrolysis oil compared to 100% petrol. Which should be consider as per EPA regulation.

References

1. Mehdi Sadat-Shojai, Gholam Reza Bakhshandeh, Recycling of PVC wastes, Elsevier, Polymer Degradation and Stability, **96**, 404-415, (2011).
2. Mitsuhara Y, Soloiu VA, Nakaishi Y, Application of new fuel produced from waste plastics and heavy oil to diesel engine, Transaction of the Japan Society of Mechanical Engineers, **67**, 2618-2624, (2001).

3. Williams PT, Williams EA, Interaction of plastics in mixed plastics pyrolysis, J. Energy Fuels, **13**(1), pp 188-196, (1999).
4. Naima K and Liazid A, Waste oils as alternative fuel for diesel engine: A review, Journal of Petroleum technology and Alternative fuels, Vol. **4**(3), pp. 30-43, (2013).
5. Moinuddin Sarker, Mohammad Mamunor Rashid, Mohammad Molla and Ashiquz Zaman, Conversion of Municipal Waste Plastic into Liquid Hydrocarbon Fuel Using a Stainless Steel Reactor, JESE A 1, 721-726, ISSN: 1934-8932, (2012).
6. Mani M, Nagarajan G, Influence of injection timing on performance, emission and combustion characteristics of a DI diesel engine running on waste plastic oil. Energy, **34**, 1617-1623, (2009).
7. Soloiu A, Yoshihara Y, Hiraoka M, Nishiwaki K, Mitsuhara Y, Nakanishi Y, The investigation of a new diesel produced from waste plastics, Eco-Technology Research Center, ISME, (2000).
8. Hai VP, Nishida O, Fujita H, Harano W, Toyoshima N, Iteya M, Reduction of NOx and PM from diesel engines by WPD emulsified fuel, SAE 2001-01-0152, ISSN: 01487191, (2001).
9. T. Shanmuga Vadivel, P. Balashanmugam, T Sivakumar, Performance and emission characteristics of Nano engine running on Plastic oil extract, International Journal of Scientific Research (IJSR), Vol. **3**, Issue 7, ISSN: 2277-8179, (2014).
10. Jane Pratoomyod, Krongkaew Laohalidanond, Performance and Emission Evaluation of Blends of Diesel fuel with Waste Plastic Oil in a Diesel Engine, International Journal of Engineering Science and Innovative Technology (IJESIT), Volume **2**, Issue 2, ISSN: 2319-5967, (2013).
11. M. Mani, C. Subash, G. Nagarajan, Emission and combustion characteristics of a DI diesel engine using waste plastic oil. Applied Thermal Engineering, **29**, 2738-2744, (2009).
12. Senthilkumar Tamilkolundu, Chandrasekar Murugesan, The Evaluation of blend of Waste Plastic Oil- Diesel fuel for use as alternate fuel for transportation. 2nd International Conference on Chemical, Ecology and Environmental Sciences (ICCEES'2012), April 28-29, Singapore (2012).
13. Mani M, Nagarajan G, Sampath S, An experimental investigation on a DI diesel engine using waste plastic oil with exhaust gas recirculation. Fuel **89**, 1826-1832 (2010).
14. M Paul Daniel, Kareddula Vijaya Kumar, B Durga Prasad, Ravi Kumar Puli, Performance and Emission Characteristics of Diesel Engine Operated on Plastic Pyrolysis Oil with exhaust gas recirculation, IJAE, DOI: 10.1080/01430750.2015.1086677.
15. M Paul Daniel, Kareddula Vijaya Kumar, B Durga Prasad, Performance and Emission Characteristics of Diesel Engine Operated on Tyre Pyrolysis Oil with exhaust gas recirculation, IJAE, May 2015, DOI: 10.1080/01430750.2015.1023837.
16. Kareddula Vijaya Kumar, Ravi Kumar Puli, Monika Dixit, D Ravi Chandra, Performance and Emission Evaluation of Tyre Oil blended with Diesel Fuel in Compression Ignition Engine, ICEE-2014, December 2014, JNTUH Kukatpally, Hyderabad, India.
17. Kareddula Vijaya Kumar, Ravi Kumar Puli, A Veeresh Babu, J.A. Ranga Babu, Performance and Emission Evaluation of Plastic Oil blended with Diesel Fuel in Compression Ignition Engine, ICEE-2014, December 2014, JNTUH Kukatpally, Hyderabad, India.
18. K. Vijaya Kumar, Dr. Ravi Kumar Puli, D. Ravichandra, K. Abhishek, Investigation on a CI Engine Fuelled with blends of Waste Tyre oil, Third International Conference on Advances in Mechanical, Aeronautical and Production Techniques (MAPT), April 2015, Malysia.
19. Mr. Hirenkumar M Patel, Prof. Tushar M Patel, performance analysis of single cylinder diesel engine fuelled with pyrolysis oil- diesel and its blend with Ethanol, International Journal of Engineering Research & Technology (IJERT), Vol. **1**, Issue 4, ISSN: 2278-0181, (2012).
20. Jinang M.Patel, Krunal J. Patel,Vatsal V. Patel, Kalpesh V. Vaghela, Investigates that Performance studies of Tire Pyrolysis Oil blends with Diesel Fuel, International Journal of Engineering and Advanced Technology (IJEAT), Volume-**3**, Issue-2, ISSN: 2249–8958, (2013).
21. Murugan. S, M.C. Ramaswamy, G. Nagarajan, Assessment of pyrolysis oil as an energy source for diesel engines, Fuel Processing Technology, **90**, 67-74, (2009).
22. Merchant, A.A., and Petrich, M.A., Pyrolysis of Scrap Tires and Conversion of Chars to Activated Carbon, American Institute Chemical Engineering Journal, Vol. **39**, 1370-1376, (1993).
23. Murugan. S, M.C. Ramaswamy, G. Nagarajan, Performance, emission and combustion studies of a DI diesel engine using distilled tyre pyrolysis oil-diesel blends, Fuel Processing Technology, Elsevier, **89**, 152-159, (2008).
24. Murugan. S, M.C. Ramaswamy, G. Nagarajan, A comparative study on the performance, emission and combustion studies of a DI diesel engine using distilled tyre pyrolysis oil–diesel blends, Fuel **87**, 2111-2121, (2008).

MATEC Web of Conferences **45**, 03003 (2016)
DOI: 10.1051/matecconf/20164503003

Performance Evaluation of an Ultra-Lean Combustion Studies in IC Engines

Ravi Kumar Puli[1], P Gandhi[2], K Vijaya Kumar[1], Damera Babitha[3] and Manoranjan murthy Kumar Puli[4]

[1]*Dept. of Mechanical Engineering, National Institute of Technology, Warangal, TS-506004, India*
[2]*JNTU/ MED, Anantapur, A.P-515002, India*
[3]*JNTU/ MED, Hyderabad, TS-500085 India*
[4]*OU/ MED, Hyderabad, TS-500007 India*

Abstract. An investigation was made to determine the effects of ethanol at ultra-lean effective systems utilizing an experimental study. The present study is planned to develop and evaluate the performance, emission and combustion characteristics of the ultra-lean burn Internal Combustion engine was modified to operate on ethanol blends fuels. The study looked at part throttle, constant speed operation of 0%, 10%, and 15% ethanol blends with gasoline mixtures operating in ultra-lean operating systems. In experiment practice the use of homogeneous lean mixtures in engine has been handicapped by several difficulties. The most serious one is that the flame propagation through mixtures becomes gradually slower as the mixture becomes leaner. The mixture distribution in a multi-cylinder engine is a problem because even small variation in mixture ratio on the linear side will strongly effect power output. Enhancement of lean combustion of homogeneous mixtures can be achieved by (i) using ethanol blends with gasoline (ii) using high-ignition energy (iii) providing high compression ratios (iv) creating high swirl in the combustion chamber.

1 Introduction

Clean, high efficiency combustion technology is one of the most realistic and effective remedies for the current global environmental and energy problems since fossil fuels account for a large proportion of the world's total primary energy resources. Combustion procedure are typically "fast" compared to other types of flow difficulties, requiring only milliseconds for ignition of stationary reactants [1] once adequate energy is provided to allow chemical shifts. In a combustion procedure with a homogeneously mixed reacting flow, the limiting factor is the reaction rate of the fuel and oxidizer. Reaction rates are a function of specific to the chemical composition and are temperature of the fuel and oxidizer. In particular, lean premixed combustion has attracted much attention for its low emissions of NOx and high energy efficiency [2]. A model study [3], for a packed bed porous burner operates on the principles of a one-dimensional model, including gas-phase transport, solid conduction, interphase heat exchange, and radiation. It is very important to note, the model produced assumes not an isothermal solid, but rather thermal equilibrium between the gas and solid phases. They also assume the solid is only used as a heat recirculation tool and does not influence the chemical reactivity. Based on their research sought to explicitly inspect the validity of modeling ultra-lean conditions. In an attempt to reduce fuel consumption and engine-out emissions from small air cooled engines, lean burn technology had already existed investigated by small motorcycle engine manufacturers [4-8]. Some of these expertise include use of oxygenated or emulsified fuels, air-assisted fuel injection for better atomization of the fuel droplets, and use of catalytic material in the combustion chamber to accelerate combustion. While some technologies may appear overpoweringly expensive for less cost utility engines, they prepare the feasibility of lean operation of small engines. The application of leaner AFRs and a suitable catalytic converter have already been shown to decrease the CO, (HC + NO_x) emissions significantly [9]. Emissions from engine-out are a strong function of fuel supply control, and the best way to achieve accurate fuel quantity metering is by injection fuel. Motorcycle engines have already applied injection fuel over carburetion and now this route is slowly gaining approval with small useful engines.

Because of emergent concerns about performance and emissions, ethanol fuel has received significant attention in the past few years. Ethanol has shown promise as a domestically produced alternative to fossil fuels. This is because it is a liquid simplifying distribution and has relatively low toxicity. Ethanol has also shown promise in helping to reduce greenhouse gases, especially if the ethanol is produced by cellulosic or sugarcane feed stocks [10-12]. Ethanol fuel has been shown in copious works to offer some significant benefits over gasoline. Ethanol have high octane rating and it gives the ability to operate at higher compression ratios without pre-ignition [13],

Article available at http://www.matec-conferences.org or http://dx.doi.org/10.1051/matecconf/20164503003

ethanol have greater latent heat of vaporization gives a higher charge density through increased charge cooling [14] and ethanol's higher laminar flame speed and lower flammability limit allows it to be run with leaner, or more dilute, air fuel mixtures [15]. In addition, ethanol fuels generally yield lower criteria pollutant emissions than gasoline [16, 17], lower evaporative emissions due to somewhat lower vapor pressures [18] and when certain renewable feedstock's are used, lower life cycle greenhouse gas emissions [10-12]. Feel strongly that all internal combustion engines control the aggregate of power by controlling the amount of air that is swallowed in the engine, then supplying adequate fuel to efficiently burn. Compression ignition engines allow for an un-throttled operation, and control the aggregate power by varying the amount of fuel metered in the cylinders. Although ultra-lean combustion increases thermal efficiency and decreases emissions, it also typically decreases power output.

1.1 Applications of an Ultra-lean Combustion

A large benefit of lean-burn is operating the engine with the throttle more open, and therefore reduced pumping losses. There is another not so widely known benefit and that air is a better gas for transferring work to the piston than exhaust gas alone (thermodynamically to do with the ratio of specific heats at constant pressure and volume). However a large problem with running lean is that there is an excess of NO_x to CO (carbon monoxide) and HC (hydrocarbons). It is not that the engine produces more NO_x it's just that there is too little CO & HC to reduce the NO_x in the "three-way" catalyst (this is why it's called a 3-way cat because the 3 major pollutants reduce and oxidize each other). The only company I know using lean-burn gasoline engines at the moment is BMW with their efficient dynamics. On the NEDC test (the fuel consumption and CO_2 test quoted by governments and manufacturers) at the very low power points they run lean and then collect the NO_x in very expensive NO_x trap catalysts. They then have to regenerate the trap using an excess of fuel to clear the NO_x, preferably when the engine runs rich at high-load. This is probably the way to keep large multi-cylinder engines in the future with ever tightening fuel consumption regulations, because the alternative for reducing fuel consumption is using downsized and down speeded engines running at stoichiometric. The engines designed for lean burning can employ higher compression ratios and thus provide better performance, efficient fuel use and low exhaust hydrocarbon emissions than those found in conventional petrol engines. Ultra-lean mixtures with very high air-fuel ratios can only be achieved by Direct Injection engines.

1.2 Engine Modification

A single cylinder kirlosker made direct injection diesel engine was chosen to carry out the experiments. This engine was modified to SI engine by incorporating spark plug, ignition system and timing. The experimental study have been accomplished in two stages. In the first stage pure gasoline is used separately in the single cylinder compression ignition modified spark ignition engine. In the second stage, ethanol blends with gasoline are used to determine the performance.

The following modifications are made to the engine for an ultra-lean burn combustion of commercially available gasoline and ethanol with carburetion and spark ignition.

- Thermocouple arrangement is made to measure the temperature of the exhaust gases.
- The squish heights are varied by changing the thickness of the gasket and varying the number of gaskets.
- The fuel injection system is replaced by a carburetor and a spark plug. The spark plug is located in place of the injector with an inclination of 25° to the vertical.
- Provision for a pressure transducer and intake manifold vacuum tapping are made in the cylinder head.
- Exhaust gas sampling points are provided in the exhaust pipe for emission measurement.
- A selected carburetor jet among the available jets of different sizes is used to operate this engine on ethanol blends with gasoline.
- Ignition system is of conventional low energy 12 v magnet and coil contact-less system is fitted to the engine, by mounting the coil and magnet on the crank shaft.

1.3 Approaches for Controlling NOx for Ultra-Lean Combustion

In an ultra-lean burn engine, adjusting the air/fuel ratio towards the ultra-lean operation side increases the volume of air available for the combustion process. This increases the heat capacity of the mixture and lowers the combustion temperature, resulting in lower NO_x formation. EGR (whether internal or external) decreases oxygen attention in the combustion chamber by diluting the entering ambient air with exhaust. During combustion, the lower oxygen content has the effect of dropping flame temperatures, which in turn decreases NO_x production since the NO_x production rate is exponentially proportionate to flame temperature.

1.4 Important Components of the Experimental Setup

The various components of the experimental setup are detailed below. Fig. 1, gives the schematic diagram of the experimental setup and Preparation of ignition coil and magnet on the diesel engine as shown in Fig. 2. Table 1, gives the specifications of the kirlosker engine. The important components of the systemare

- The engine
- Spark plug
- Dynamometer
- Data acquisition
- Computer

The exhaust gas analyzer is switched on quite early so that all its systems will get stabilized before the commencement of the experiment. The data length, frequency range to trigger the data acquisition for computer are carefully selected, based on the approximate cycle time of the engine operation, such that there appeared three TDC signals on the display, with the combustion period occupying the center stage.

Table 1. Specifications of kirlosker engine

No. of cylinders	One
Connecting rod length	230 mm
V_{disp}	552.94 cc
Stroke	110 mm
Bore	80 mm
Rated output	3.68 kW (5 hp)
Inlet valve opens at	527°
Inlet valve closes at	750°
Compression ratio	12
Exhaust valve opens at	340°
Exhaust valve closes at	554°
Speed	1500 rpm
Spark advance	27° BTDC

Figure 1. Experimental Set up for Single cylinder Kirlosker diesel engine

Figure 2. Preparation of ignition coil and magnet on the diesel engine

Ambient condition of pressure and temperature are noted. After the starting of the engine and stabilizing it, air flow, fuel flow, temperature of ambient air, temperature of exhaust gases are noted. The dynamometer readings such as load and speed are also noted. For all the tests the dynamometer is set on constant speed mode. The pressure and TDC signals are recorded in computer and averaged for 100 consecutive cycles.

The experimental study have been performed in two stages. In the first stage normal gasoline are used separately in the catalytic and non-catalytic combustion chamber with a manifold vacuum of 30mm Hg. Then in the second stage, catalytic with a manifold vacuum of 60 mm Hg, to ascertain the performance. For all the tests a cyanide bath coated piston with hemi-spherical combustion chamber which has a squish height of 2.4 mm, with a compression ratio of 12:1 used. High energy electronic contact-less ignition system with extended electrode is used. The electrode is centrally located with a deep penetration of 15mm into the combustion chamber. The plug is water-cooler to avoid pre-ignition. These techniques are employed based on the test results obtained in the earlier research which are presented.

2 Results and Discussion

2.1 Engine Performance

Fig. 3 and Fig. 4 gives the variation of indicated power, brake power and brake thermal efficiency respectively. At a given equivalence ratio when percentage of ethanol in the base fuel increases, the power developed also increases. An increase of 10.1% in the indicated power for 15% blend is noticed compared to gasoline and 10% blend respectively.

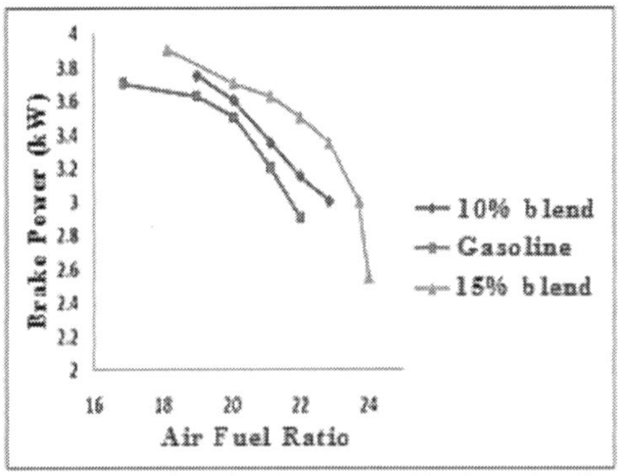

Figure 3. Variation of Brake Power with Air Fuel Ratio

Similarly there is 10.1% increase in brake power for 15% blend compared to gasoline and 10% blend respectively. The brake thermal efficiency is improved by 5.4% for 15% blend compared to gasoline and 10% blend. It is known that, even at the same compression ratio, an engine will have a higher power output and higher efficiency for ethanol blends than operating for gasoline operation. This is due to higher cooling of the charge and faster combustion with ethanol. The variation of brake specific energy consumption (BSEC) in MJ/Kg with equivalence ratio can be noticed that the increase of ethanol in the gasoline increases BSEC. This is expected because of the improvement in the combustion process.

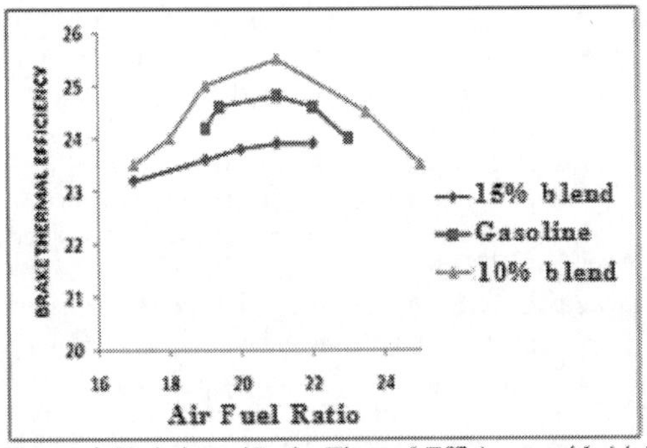

Figure 4. Variation of Brake Thermal Efficiency with Air Fuel Ratio

2.2 Engine Emissions

Figures 5, 6, 7 and 8 explain the variations of MBT timing, exhaust temperature, peak pressure, and combustion duration with equivalence ratio respectively. It is noticed that, as the ethanol blend in gasoline decreases, the MBT timing increases. The percentage decrease in MBT timing for 15% blend are 6.4% compared to gasoline and 10% blend. There is no appreciable change in the exhaust temperature between blends and gasoline. The peak pressure is the highest for gasoline and is decreased as the ethanol proportion in gasoline is reduced. This trend again confirms that addition of ethanol enhances the combustion rate. The combustion duration is decreased as the ethanol content in the blend is increased. This is probably due to the enhanced pre-flame reaction.

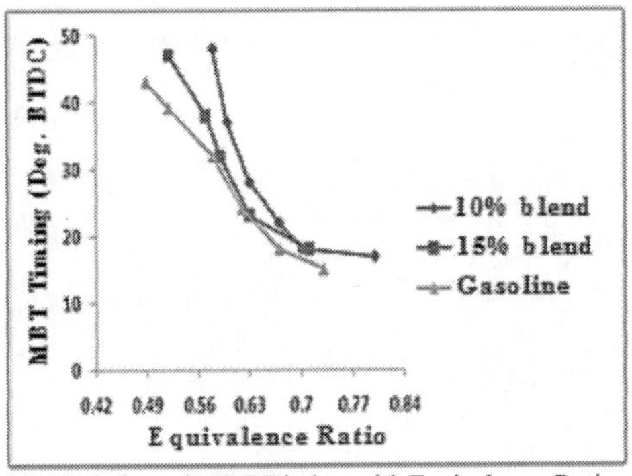

Figure 5. Variation of MBT Timing with Equivalence Ratio

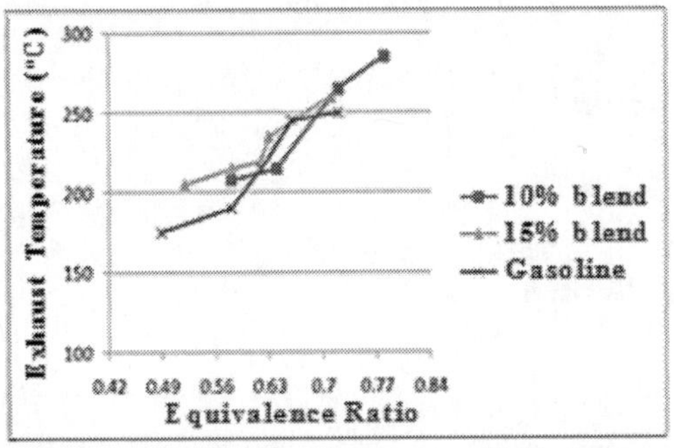

Figure 6. Variation of Exhaust Temparature with Equivalence Ratio

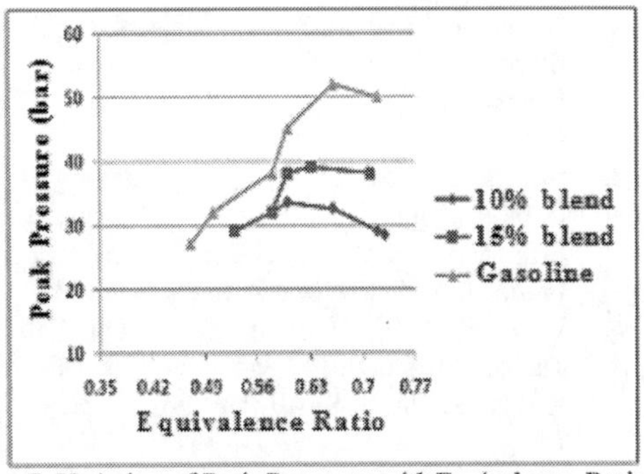

Figure 7. Variation of Peak Pressures with Equivalence Ratio

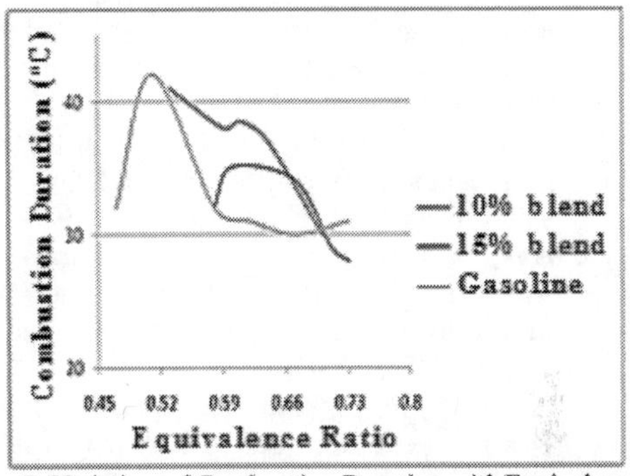

Figure 8. Variation of Combustion Duration with Equivalence Ratio

3 Conclusion

Based on the experimental study of an ultra-lean burn combustion the following conclusions have been made:

- Brake power and brake thermal efficiency are higher for ethanol blends compared to gasoline in all the cases.
- Increase of the ethanol proportion in the blend increases power and brake thermal efficiency.
- CO emission is not influenced by higher percentage of ethanol in the blend. HC emission is found to be reduced with ethanol blends with gasoline.
- Higher swirl increases the heat transfer and hence reduces the efficiency of the engine.
- High energy, high compression ratio with optimum swirl and the use of ethanol blends with gasoline will extend the lean misfire limit (LML) significantly.

References

1. Weinrotter Martin, Herbert Kopecek, Martin Tesch, Ernst Wintner, Maximilian Lackner, Franz Winter, Laser ignition of ultra-lean methane/hydrogen/air mixtures at high temperature and pressure. Experimental thermal and fluid science, 2005. **29**: 569-577, (2005).
2. D. Dunn-Rankin (Ed.), Lean Combustion–Technology and Control, Academic Press, (2008).

3. Henneke, M.R. and J.L. Ellzey, Modeling of filtration combustion in a packed bed. Combustion and flame, **117** (4): p. 832-840, (1999).
4. Swamy K, Harne V, Gunjegaonkar D, and Gopalkrishnan K, Study and Development of Lean Burn Systems on Small 4Stroke Gasoline Engines, SAE 2001 **01** 1801/4222, (2001).
5. Yoji Fukami, Hisashi Yamashita, Shinichi Tamba, Effect of Lean Burn on Emission Reduction in a Small Utility 4-Stroke Spark Ignition Engine, SAE-978486, 1997-10-27, (1997).
6. Yamamoto Hiroshi, Investigation on Relationship between Thermal Efficiency and NOx Formation in Ultra-Lean Combustion, SAE 1999-01-3328, (1999).
7. Bresenham, Damon, John Reisel, the Effect of High Ethanol Blends on Emissions from Small Utility Engines, SAE 1999-01-3345, (1999).
8. Ishii W, Hanajima T, and Tsuzuku H, Application of Air-Fuel Mixture Injection to Lean-Burn Engines for Small Motorcycles, SAE 2004-32-0052, (2004).
9. Pundir, B. P., Emission Reduction in Small SI Engine Generator Sets, SAE 2004-011089 (2004).
10. Pimentel D, Patzek TW. Ethanol production using corn, switchgrass, and wood; biodiesel production using soybean and sunflower. Nat Resour Res 2005; **14** (1):65-76, (2005).
11. Schmer MR, Vogel KP, Mitchell RB, Perrin RK. Net energy of cellulosic ethanol from switch grass. Tech. rep. United States Department of Agriculture, (2007).
12. Wang M, Saricks C, Santini D. Effects of fuel ethanol use on fuel cycle energy and greenhouse gas emissions. Tech. Rep. ANL-CTR. United States Department of Energy; (1999).
13. Radwan MS. Performance and knock limits of ethanol- gasoline blends in spark-ignited engines, SAE Paper 850213.
14. Dodge LG, Shouse K, Grogan J, Leone DM, Whitney KA, Merritt PM. Development of an ethanol-fueled ultra-low emissions vehicle, SAE Paper 981358.
15. Marinov NM. A detailed chemical kinetic model for high temperature ethanol oxidation. Int J Chem Kinet, 1999; **31** (3):183-220, (1999).
16. Kelly KJ, Bailey BK, Coburn T, Clark W, Lissiuk P. Federal test procedure emissions test results from ethanol variable fuel vehicle Chevrolet Luminas. SAE Paper 961092.
17. Sinor JE, Bailey BK, Current and potential future performance of ethanol fuels. SAE Paper 930376.
18. Furey RL. Volatility characteristics of gasoline/alcohol and gasoline/ether fuel blends. SAE Paper 852116, 1985.

MATEC Web of Conferences 45, 03004 (2016)
DOI: 10.1051/matecconf/20164503004

Multi-pulse Orbits and Homoclinic Trees in a Non-autonomous Resonant Hamiltonian System

Sha Zhou[1,2] , Wei Zhang[1,2] and Tian-jun Yu[1,2]
[1] Beijing Key Laboratory of Nonlinear Vibrations and Strength of Mechanical Structures, 100124, China
[2] Beijing University of Technology, College of Mechanical Engineering, 100124, China

Abstract. In this study, we develop the energy-phase method to deal with the high-dimensional non-autonomous nonlinear dynamical systems. Our generalized energy-phase method applies to integrable, two-degree-of freedom non-autonomous resonant Hamiltonian systems. As an example, we investigate the multi-pulse orbits and homoclinic trees for a parametrically excited, simply supported rectangular thin plate of two-mode approximation. In both the Hamiltonian and dissipative case we find homoclinic trees, which describe the repeated bifurcations of multi-pulse solutions, and we present visualizations of these complicated structures.

1 Introduction

The energy-phase method, which was first presented by Haller and Wiggins [1-10], was used to reveal families of multi-pulse solutions and predict the parameter region for chaotic motion may occur.

Although the energy-phase method has been applied widely to engineering problems, it was used to solve autonomous perturbed Hamiltonian systems. It is worth to mention that the energy-phase method has never used in non-autonomous nonlinear dynamical systems. In this paper, we develop the energy-phase method to deal with non-autonomous nonlinear dynamical systems for the first time.

This paper develops the energy-phase method to deal with the high-dimensional non-autonomous nonlinear dynamical systems. In section 2, we formulate the problem and describe the geometrical structure of the phase space of the unperturbed systems, and we study the dynamics in the perturbed system, and derive the explicit formulas of the nth order energy difference function for the non-autonomous systems. In section 3, we apply our developed methods to a specific example: a two-mode truncation of parametrically excited, simply supported rectangular thin plate. And we visualize the homoclinic tree and show by explicit calculations how it breaks up under the effect of dissipation. We make a conclusion in section 4.

2 Generalized of the energy-phase method

Let us consider a two-degree-of -freedom Hamiltonian systemgiven by

$$\dot{x} = JD_x H_0(x,I) + \varepsilon JD_x H_1(x,I,\gamma,\omega t;\varepsilon) + \varepsilon g^x(x,I,\gamma,\phi;\varepsilon) \quad (1a)$$

$$\dot{I} = -\varepsilon D_\gamma H_1(x,I,\gamma,\omega t;\varepsilon) + \varepsilon g^I(x,I,\gamma,\phi;\varepsilon) \quad (1b)$$

$$\dot{\gamma} = D_I H_0(x,I) + \varepsilon D_I H_1(x,I,\gamma,\omega t;\varepsilon) + \varepsilon g^\gamma(x,I,\gamma,\phi;\varepsilon) \quad (1c)$$

$$\dot{\phi} = \omega \quad (1d)$$

$$0 \le \varepsilon \ll 1,\ (x,I,\gamma,\phi) \in \mathbf{R}^2 \times \mathbf{U} \times \mathbf{S}^1 \times \mathbf{S}^1$$

where $\mathbf{U} \subset \mathbf{R}^+$ is an open set, functions H_0 and H_1 are C^{r+1} $(r \ge 2)$ functions, and functions H_1 and $g = (g^x, g^I, g^\gamma)$ are periodic in t with the period $T = 2\pi/\omega$, and

$$J = \begin{pmatrix} 0 & 1 \\ -1 & 0 \end{pmatrix}.$$

Note that when $\varepsilon = 0$ system (1) is an integrable Hamiltonian system on which we make two structural assumptions:

(H1) There exist $I_1, I_2 \in \mathbf{U}, I_1 < I_2$ such that for any $I \in [I_1, I_2]$,

(1) has a hyperbolic fixed point $\bar{x}_0(I)$ and a homoclinic trajectory $x^h(t,I)$, which connects $\bar{x}_0(I)$ to itself.

(H2) (Resonance) There exists $I_r \in (I_1, I_2)$ such that

$$D_I H_0(\bar{x}_0(I_r), I_r) = 0\ ;$$

$$D_I^2 [H_0(\bar{x}_0(I), I)]|_{I=I_r} \neq 0\,.$$

Defining a cross-section in full five-dimensional phase space of equation (2), we have

Article available at http://www.matec-conferences.org or http://dx.doi.org/10.1051/matecconf/20164503004

$$\Sigma^{\phi_0} = \{(x, I, \gamma, \phi) \mid \phi = \phi_0\} \quad (2)$$

Figure 1 gives the geometrical explanation of the cross-section Σ^{ϕ_0}.

Let $q(t) = (x(t), I(t), \gamma(t))$, $(q(t), \phi(t))$ and $(q_\varepsilon(t), \phi(t))$ be trajectories of the unperturbed and perturbed vector fields, respectively.

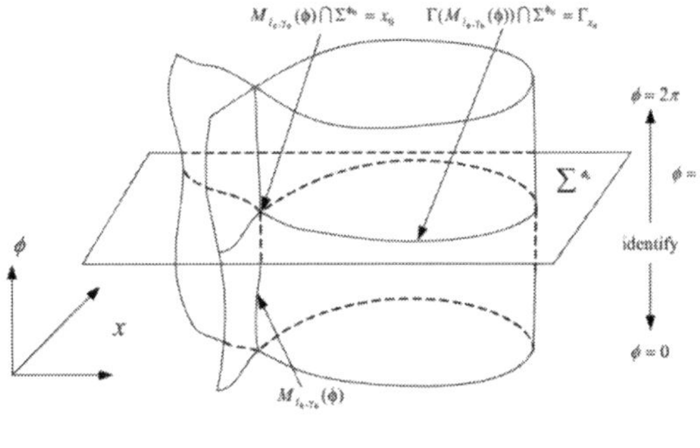

Figure 1 Geometrical structure of the cross-section Σ^{ϕ_0}

Let $(q_\varepsilon^s(0), \phi_0) = s_1$, $(q_\varepsilon^u(0), \phi_0) = p_1$. If p_1 and s_1 coincide, there exists single-pulse homoclinic orbits.

And p_1 and s_1 coincide if

$$\frac{1}{\varepsilon}[H(s_1, \varepsilon) - H(p_1, \varepsilon)] = 0 \quad (3)$$

And the energy difference function can be written as

$$\varepsilon \Delta^1 H(\phi_0) = H(s_1, \varepsilon) - H(p_1, \varepsilon) \quad (4)$$

Therefore

$$\Delta^1 H(t, \phi_0) = H_D(h, \gamma + \Delta\gamma, \phi_0) - H_D(h, \gamma, \phi_0) - \int_{-\infty}^{\infty} \omega D_\phi H_D(h, \gamma, \phi) ds$$

$$- \int_{-\infty}^{\infty} \langle DH_0(x, I), g(x, I, \gamma, \phi) \rangle |_{x^1(t)} \, dt \quad (5)$$

It is known that $g(x, I, \gamma, \phi)$ is bounded for all τ and $DH_0(x, I)$ goes to zero exponentially fast as $\tau \to \pm\infty$ since $x^1(t)$ approaches a hyperbolic fixed point. Therefore, the improper integral $\int_{-\infty}^{\infty} \langle DH_0(x, I), g(x, I, \gamma, \phi) \rangle |_{x^1(t)} \, dt$ converges absolutely. And $\int_{-\infty}^{\infty} \omega D_\phi H(h, \gamma, \phi) dt$ converges at best conditionally. The improper integral makes sense if we approach the limits $\pm\infty$ along the sequences of times $\{\tau_j^s\}, \{\tau_j^u\}, j = 1, 2, \cdots$.

Hence, for ε small enough, if

$$\Delta^1 H(\gamma, \phi_0) = 0, \; D\Delta^1 H(\gamma, \phi_0) \neq 0$$

there exist 1-pulse homoclinic orbits.

To formulate the above algorithm concisely, we introduce the energy difference function

$$\Delta^N H(\phi_0) = H_D(h, \gamma + N\Delta\gamma, \phi_0) - \sum_{j=0}^{N-1} \int_{-\infty}^{\infty} \omega D_\phi H_D(h, \gamma + j\Delta\gamma, \phi) ds$$

$$- H_D(h, \gamma, \phi_0) - \sum_{i=1}^{N} \int_{-\infty}^{\infty} \langle DH_0(x, I), g(x, I, \gamma, \phi) \rangle |_{x^i(t)} \, dt \quad (6)$$

If equation (6) has transverse zero, then there exist N-pulse homoclinic orbit.

3 Equations of motion of the buckled thin plate

We consider the simply supported at the four-edge rectangular thin plate whose edge lengths are a and b and thickness is h. The thin plate is subjected to its plane excitations. We establish a Cartesian coordinate system shown in Figure 2 and the coordinate Oxy is located at the middle surface of the thin plate. It is assumed that u, v and w represent the displacement of a point in the middle plane of the thin plate in the x, y and z directions, respectively. The in-plate excitation of the thin plate may be expressed in the form $p = p_0 - p_1 \cos \Omega t$.

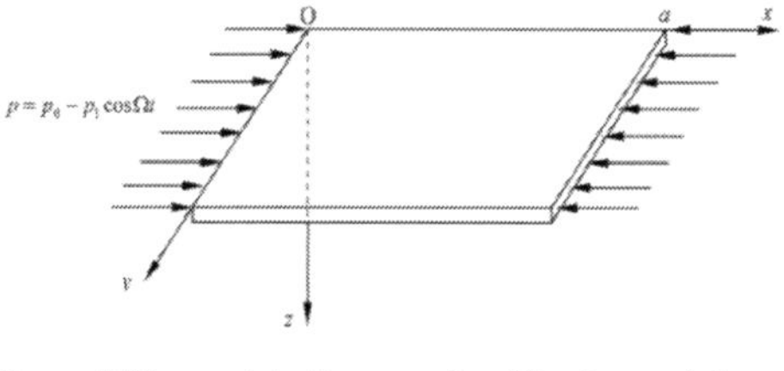

Figure 2 The model of rectangular thin plate and the coordinate system

According to ref [11], the dimensionless equations of motion for the thin plate under parametric excitation are obtained as follows

$$\ddot{x}_1 + \mu\dot{x}_1 - g_1 x_1 + 2x_1 f_1 \cos \Omega t + \alpha_1 x_1^3 + \alpha_2 x_1 x_2^2 = 0 \quad (7a)$$

$$\ddot{x}_2 + \mu\dot{x}_2 - g_2 x_2 + 2x_2 f_2 \cos \Omega t + \beta_1 x_2^3 + \beta_2 x_2 x_1^2 = 0 \quad (7b)$$

Using the normal form theory and the canonial transformation, equation (1) can be transformed to

$$\dot{y}_1 = \frac{\partial H_0}{\partial y_2} + \varepsilon \frac{\partial H_1}{\partial y_2} + \varepsilon g^{y_1} = y_2 \quad (8a)$$

$$\dot{y}_2 = -\frac{\partial H_0}{\partial y_1} - \varepsilon \frac{\partial H_1}{\partial y_1} + \varepsilon g^{y_2}$$
$$= g_2 y_1 - \varepsilon\mu y_2 - \beta y_1^3 - \beta_2 y_1 I - 2\varepsilon y_1 f_2 \cos \phi \quad (8b)$$

$$\dot{I} = \frac{\partial H_0}{\partial \gamma} + \varepsilon \frac{\partial H_1}{\partial \gamma} + \varepsilon g^I$$
$$= -\varepsilon\mu I - 2\varepsilon f I \sin 2\gamma \cos \phi \quad (8c)$$

$$\dot{\gamma} = -\frac{\partial H_0}{\partial I} - \varepsilon \frac{\partial H_1}{\partial I} + \varepsilon g^\gamma$$
$$= -b - \varepsilon f(\cos 2\gamma + 1)\cos \phi - \frac{3}{4}\alpha I - \frac{1}{2}\alpha_2 y_1^2 \quad (8d)$$

$$\dot{\phi} = \Omega \quad (8e)$$

where

$$H_0(y_1, y_2, I) = \frac{y_2^2}{2} - \frac{g_2}{2} y_1^2 + \frac{\beta}{4} y_1^4 + \frac{\beta_2}{2} y_1^2 I + bI + \frac{3}{8}\alpha I^2$$

$$H_1(y_1, y_2, I, \gamma, \phi) = y_1^2 f_2 \cos\phi + fI \cos 2\gamma \cos\phi + fI \cos\phi \quad (9)$$

and g^{y_1} , g^{y_2} , g^I and g^γ are the perturbation terms induced by the dissipative effects

$$g^{y_1} = 0,\ g^{y_2} = -\mu y_2,\ g^I = -\mu I,\ g^\gamma = 0. \quad (10)$$

3.1 The unperturbed dynamics

When $\varepsilon = 0$, it is noted that system (8) is an uncoupled two-degree-of-freedom nonlinear system since $\dot{I} = 0$. Consider the first two decoupled equations

$$\dot{y}_1 = y_2 \quad (11a)$$

$$\dot{y}_2 = g_2 y_1 - \beta y_1^3 - \beta_2 y_1 I \quad (11b)$$

Since $\beta > 0$, system (11) can exhibit the homoclinic bifurcation on the curve defined by $g_2 = \beta_2 I$ .When $g_2 - \beta_2 I < 0$, the trivial zero solution may bifurcate into three solutions through pitchfork bifurcation, which are given by $q_0 = (0,0)$ and $q_\pm(I) = (B,0)$, respectively, where

$$B = \pm\sqrt{\frac{g_2 - \beta_2 I}{\beta}} \quad (12)$$

From the Jacobian matrix evaluated at the non-zero solutions, it is known that the singular point $q_0 = (0,0)$ is the saddle point and the singular points $q_\pm(I) = (B,0)$ are center points.

Thus, for all $I \in [I_1, I_2]$, system (11) has one hyperbolic saddle point q_0 which is connected to a pair of homoclinic orbits $\lim_{T_1 \to \pm\infty} y_\pm^h(T_1, I) = q_0$. Letting $\eta = g_2 - \beta_2 I$, then the homoclinic orbits can be obtained as follows

$$y_1(T_1) = \pm\sqrt{\frac{\eta}{\beta}} \sec h\left(\sqrt{\eta} T_1\right) \quad (13a)$$

$$y_2(T_1) = \mp\frac{\sqrt{2}\eta}{\sqrt{\beta}} \tanh\left(\sqrt{\eta} T_1\right) \sec h\left(\sqrt{\eta} T_1\right) \quad (13b)$$

Therefore, in the full five-dimensional phase space, the set defined by

$$M = \{(y, I, \gamma, \phi) \mid y = q_0 = (0,0), I \in [I_1, I_2], \gamma \in [0, 2\pi), \phi = \Omega t + \phi_0\} \quad (14)$$

is a three dimensional invariant manifold.

Considering the unperturbed system of equation (8) restricted to M , we can calculate the resonant value

$$I_r = \frac{-4b}{3\alpha} \quad (14)$$

and the phase shift $\Delta\gamma$ of the oscillations is defined as

$$\Delta\gamma = \gamma(+\infty, I_r) - \gamma(-\infty, I_r) = -\frac{2\alpha_2}{\beta}\sqrt{\eta} \quad (15)$$

3.2 The perturbed dynamics

It is noticed that the saddle point may persist under small perturbations, in particular, $M \to M_\varepsilon$. Therefore, we obtain

$$M_\varepsilon = \{(y, I, \gamma, \phi) \mid y = q_0 = (0,0), I \in [I_1, I_2], \gamma \in [0, 2\pi), \phi = \Omega t + \phi_0\}.$$

Considering the following cross-section of the phase space

$$\Sigma^{\phi_0} = \{(y, I, \gamma, \phi) \mid \phi = \phi_0\}$$

Let the manifold $M_\varepsilon^{\phi_0}(\phi)$ represent the manifold $M_\varepsilon(\phi)$ on the cross-section Σ^{ϕ_0} . And restricting the system (8) on the manifold $M_\varepsilon^{\phi_0}(\phi)$, we have

$$\dot{I} = -\varepsilon\mu I - 2\varepsilon f I \sin 2\gamma \cos\phi_0 \quad (16a)$$

$$\dot{\gamma} = -b - \frac{3}{4}\alpha I - \frac{1}{2}\alpha_2 y_1^2 - \varepsilon f(\cos 2\gamma + 1)\cos\phi_0 \quad (16b)$$

The n-th order energy-difference function for the dissipative case is given as follows:

$$\Delta^n H(\gamma, \phi_0) = H_D(h, \gamma + n\Delta\gamma, \phi_0) - \sum_{j=0}^{N-1} \int_{-\infty}^{\infty} \Omega D_\phi H_D(h, \gamma + j\Delta\gamma, \phi) ds$$
$$- H_D(h, \gamma, \phi_0) - \sum_{i=1}^{N} \int_{-\infty}^{\infty} \langle DH_0(x, I), g(x, I, \gamma, \phi) \rangle |_{x^i(t)} dt \quad (17)$$

Computing (17) leads to the following expression for the dissipative energy difference function

$$\Delta^n H_D(\gamma, \phi_0) = f[I_r(\cos 2(\gamma + n\Delta\gamma) - \cos 2\gamma)\cos\phi_0 - \frac{4nd\eta\Delta\gamma}{3\alpha_2} + \frac{F}{f}] \quad (18)$$

where

$$F = \Omega f I_r \pi [\frac{\Omega(\Delta\gamma)^3 \sin\phi_0}{24\eta} + \frac{\Delta\gamma \cos\phi_0}{\sqrt{\eta}}] \csc h(\frac{\Omega\pi}{2\sqrt{\eta}}) \cdot \sum_{j=0}^{n-1} \sin 2(\gamma_0 + j\Delta\gamma)$$
$$- \frac{\Omega^2}{2\eta} f I_r \pi (\Delta\gamma)^2 \sin\phi_0 \csc h(\frac{\Omega\pi}{2\sqrt{\eta}}) \cdot \sum_{j=0}^{n-1} \cos 2(\gamma_0 + j\Delta\gamma) \quad (19)$$

Defining a dissipative factor $d = \frac{\mu}{f}$ such that d gives the relative measure of the dissipative effect with respect to the excitation amplitude. Hence the upper bound on the value of the dissipative factor is obtained as follows:

$$|d| < d_{\max} = \left| \frac{6\mu\alpha_2 I_r \cos\phi_0 \sin(n\Delta\gamma)}{3\alpha_2 F - 4n\mu\eta\Delta\gamma} \right| \quad (20)$$

For any small dissipative factor $d < 1$, we obtain an upper bound on the maximum number of pulses

$$|n| < n_{\max} = \left|\frac{3\alpha_2 I_r \cos\phi_0 \sin(n\Delta\gamma)}{2d\eta\Delta\gamma}\right| + \left|\frac{3\alpha_2 F}{4\mu\eta\Delta\gamma}\right| \tag{21}$$

For any n satisfying $n\Delta\gamma \neq 2l\pi\ (l = 0,1,2,\cdots)$, there are two transverse zeroes of the dissipative energy difference function in the interval $\gamma \in [0,\pi]$, that is,

$$\gamma_{-,1}^{n} = \pi - [\frac{n\Delta\gamma}{2} + \frac{\alpha_0}{2}] \bmod \pi \tag{22a}$$

$$\gamma_{-,2}^{n} = \pi - [\frac{\pi}{2} + \frac{n\Delta\gamma}{2} - \frac{\alpha_0}{2}] \bmod \pi \tag{22b}$$

where $\alpha_0 = \arcsin[\frac{F}{2fI_r \cos\phi_0 \sin(n\Delta\gamma)} - \frac{2n\eta d\Delta\gamma}{3\alpha_2 I_r \cos\phi_0 \sin(n\Delta\gamma)}]$.

We now define the set of transveral zeros of $\Delta^n H_D(\gamma,\phi_0)$ as follows

$$Z_-^n = \{(h,\gamma) \mid \Delta^n H_D(\gamma,\phi_0) = 0, D_\gamma \Delta^n H_D(\gamma,\phi_0) \neq 0\}.$$

Let us start by defining the energy sequence

$$h_0 = H_D(0,\gamma_s),\ h_n = \max[H_D(0,\gamma_{-,1}^n), H_D(0,\gamma_{-,2}^n)] \tag{23}$$

and the sequence of sets

$$A_0 = \phi,\ A_n = \{(h,\gamma) \in S_0 \mid H_D(h,\gamma,\phi_0) < h_n\},\quad n \geq 1 \tag{24}$$

Next we define the pulse sequence

$$N_1 = 1,\ N_k = \min\{n \in Z \mid n > N_{k-1}, h_n > h_{N_{k-1}}\},\quad k \geq 2 \tag{25}$$

Further, we define the layer sequence

$$L_{N_k} = Int(A_{N_k} \setminus A_{N_{k-1}}) \tag{26}$$

where $Int(\cdot)$ refers to the interior of a set. The construction of the layer sequence is shown in Figure 3. And the corresponding inner angular radii of the layers in the layer sequence is defined as

$$r_{N_k} = \min(|\gamma_c - \gamma_{-,1}^{N_k}|, |\gamma_c - \gamma_{-,2}^{N_k}|) \tag{27}$$

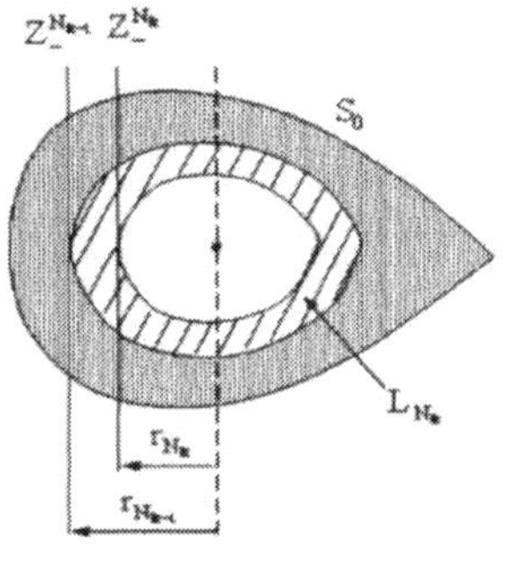

Figure 3 The construction of the layer sequence

However, all these sequence is finite due to (20). For any periodic orbit $\gamma \subset L_{N_k}$, the pulse number is $N(\gamma) = N_k$.

Figure 4 gives the pulse sequence and layer radius sequence as a function of the phase shift in the Hamiltonian case when $N = 100$ at the cross-section $\phi = 0$. In the left diagram, the horizontal line segments at each level N indicate that an infinity of N-pulse orbits exist for all values of the phase shift in the interval below that line. And this diagram shows a fairly stable pulse distribution for lower pulses and increasing sensitivity to small changes in the parameters for higher pulses. In the right diagram, $\Delta\gamma$ can be regarded as an independent bifurcation parameter, and the resulting bifurcation diagram is an infinite binary tree, which can be called the homoclinic tree. The diagram for the layer radii also has a secondary meaning: for fixed $\Delta\gamma$, $\{r_{N_k}\}$ gives the angular distance of the take-off curves of multi-pulse orbits from the nearest center on the manifold.

Figure 5 gives the pulse sequence and layer radius sequence as a function of the phase shift in the dissipative case when $N = 100$ at the cross-section $\phi = 0$. The layer radius diagrams now do not refer to layers of periodic orbits, but their secondary meaning remains valid for the full dissipative system: they show the approximate angular distance of the take-off curves of multi-pulse orbits from the nearest sinks on the manifold.

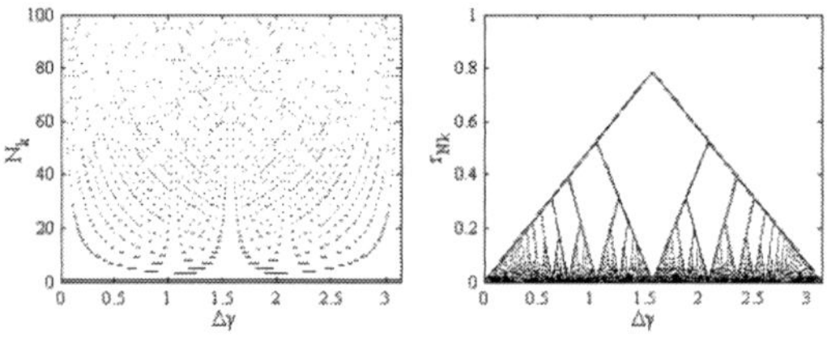

Figure 4 The pulse sequence and the layer sequence as a function of a function of the phase shift in the Hamiltonian case

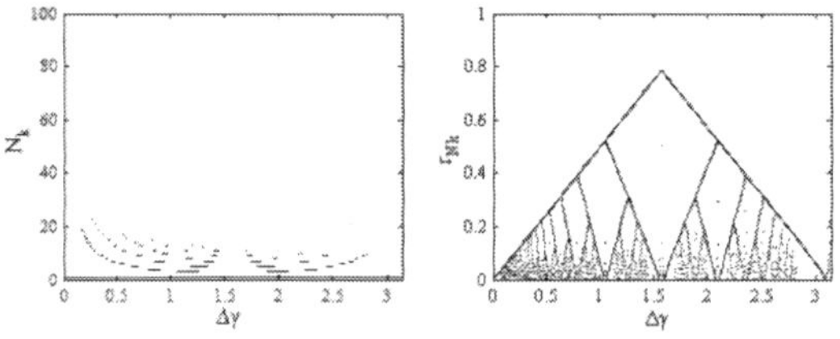

$d = 0.001$

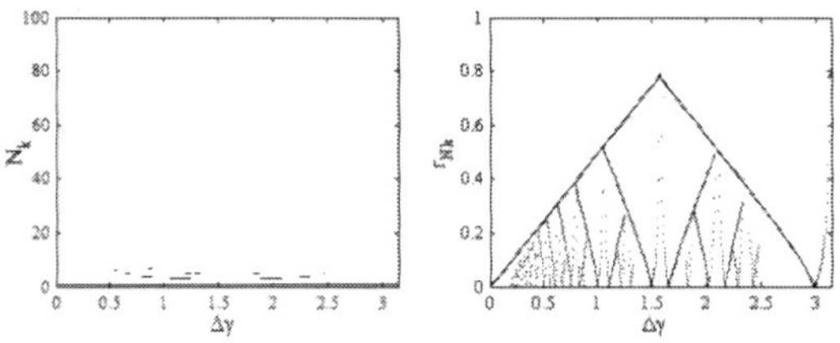

$d = 0.01$

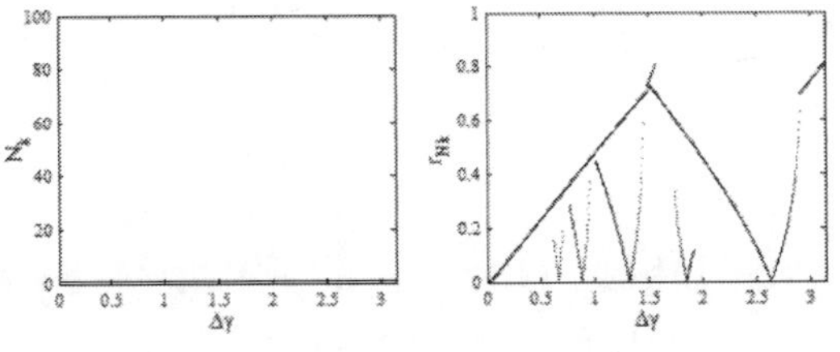

$d = 0.1$

Figure 5 The pulse sequence and the layer sequence as a function of a function of the phase shift in the dissipative case

4 Conclusion

In this paper, we have generalized the energy-phase method to deal with the non-autonomous nonlinear system and derive the explicit formulas of the energy-different function. Applying our method to a two-mode approximation of parametrically excited, simply supported rectangular thin plate, we find the homoclinic tree both in the Hamiltonian case and the dissipative case are shown in Figure 4 and Figure 5. The pulse number diagrams show a variety of ways in which N can change. And the layer radius diagrams gives the layers of periodic orbits, and the angular distance of the take-off curves of multi-pulse orbits from the nearest center on the manifold in the Hamiltonian case. But the layer radius diagrams in the dissipative case just show the approximate angular distance of the take-off curves of multi-pulse orbits from the nearest sinks on the manifold.

Acknowledgments

The authors gratefully acknowledge the support of National Natural Science Foundation of China (NNSFC) through grant Nos.11290152 and 11072008, the Funding Project for Academic Human Resources Development in Institutions of Higher Learning under the Jurisdiction of Beijing Municipality (PHRIHLB).

References

1. G. Haller, S. Wiggins. Phys D, **66**(1993).
2. G. Haller. Phys. Lett. A, **200**(1995).
3. G. Haller, S. Wiggins. Arch. Ration. Mech. and Anal, **130**(1995).
4. G. Haller, S. Wiggins. Phys D, **85**(1995)
5. G. Haller, S. Wiggins. Phys D, **90**(1996).
6. G. Haller. J. S. Phys, **86**(1997).
7. G. Haller. Commun. Math. Phys, **193**(1998).
8. G. Haller. Commun. Appl. Math, **LII**(1999).
9. G. Haller, G. Menon. Phys. Lett. A, **263**(1993).
10. G. Haller. *Chaos near resonance*. Springer-Verlag(1999).
11. W. Zhang. J. Sound Vib, **239**(2001).

MATEC Web of Conferences **45**, 03005 (2016)
DOI: 10.1051/matecconf/20164503005

Chaotic Dynamics Analysis for a Class of Delay Nonlinear Finance Systems

Ge Kai[1,2], Wei Zhang[1]

[1] College of Mechanical Engineering Beijing University of Technology, Beijing, 100124, PR China
[2] College of Statistics and Mathematics Inner Mongolia University of Finance and Economics, Hohhot, 010070, PR China

Abstract. This article focuses on a class of nonlinear chaotic finance model with feedback control problem. The dynamic responses of the delayed finance system were analyzed and the chaos control problems were considerd. The main work consists of three steps: (i) for a financial system model with the delayed feedback control, the fixed point was obtained, and a new system was obtained by shifting the fixed point to the coordinate origin; (ii) the delayed term was added to the new system, the characteristic equation of the new system was solved, and the distribution of the characteristic equation roots was analyzed. Since the system with time delay undergoes Hopf bifurcation at the equilibrium point under certain conditions, and the fixed point exists stability switching phenomenon, then the intervals of the stable and unstable fixed point were specifically given;(iii) the stable periodic solution and the stable fixed point were simulated under a set of specific parameters, therefore the previous theoretical results obtained by numerical simulation were verified.

Introduction

Financial system is composed of many elements,which is an open and extremely complex nonlinear systems. In this non-linear system, delay is an important factor. In the financial system, we can not fully describe some economic phenomenon with the general differential equation because of many uncertain factors. Some scholars added the time delay factor in differential equations when they studied the economic dynamics [1-5]. In these studies, time delay was focused on a fixed point. However, in practical applications, time delay is often distributed over a range, so the introduction of continuous distributed delays can better describe the actual economic problems [6].

This paper studies a class of nonlinear chaotic financial system. Since the system appears chaotic response with certain parameters, the delayed feedback control method is introduced into the financial system. We analyse the dynamic response of the finance system with time delay, and the chaos control problem is considered.

1 The proposal of the model

Since the 1980s, economists found that the chaos phenomenon exists in economic system [7-9] and it has a tremendous impact on the western mainstream economics , because the economic system appear chaotic phenomenon means macroeconomic movement itself is inherently instability. Literature [10-12] developed a chaotic financial system differential equations model composed of the manufacturer sub-blocks, sub-block currencies, securities sub-block and sub-block. The equations can be expressed as follows

$$\begin{cases} \dot{X} = Z + (Y - a)X \\ \dot{Y} = 1 - bY - X^2 \\ \dot{Z} = -X - cZ \end{cases} \tag{1-1}$$

In finance, the above parameter variables have practical significance. X represents the interest rate, Y represents the investment demand, Z represents the price index, a represents the amount of storage, b represents the investment growth, c represents the supply and demand factor. These parameters play an important part in the market supply and demand.

By linear transformation [13]:

$$x(t) = X(t), y(t) = Y(t) - 1/b, z(t) = Z(t),$$

the new system is obtained as follow:

$$\begin{cases} \dot{x} = (1/b - a)x + z + xy \\ \dot{y} = -by - x^2 \\ \dot{z} = -x - cz \end{cases} \tag{1-2}$$

System (1-1) and (1-2) are topologically equivalent, then we can reflect dynamical properties of the system (1-1) by studying the kinetic properties of the system (1-2). Cai *et al.* analyzed the fixed point stability of the system (1-2) in detail and a new class of chaotic attractors are discovered under the particular parameter value. In this paper, we try to control this system to be stable at the fixed points by introducing delayed feedback control system.

2 Stability analysis of equilibrium

2.1 Delay finance chaos feedback control system

First, we add delay to the first system of equations (1-2)and consider the delay feedback control problem of the system. Thus, the finance model can be simplified into the following dynamic system model:

$$\begin{cases} \dot{x} = (1/b - a)x + z + xy + k(x(t-\tau) - x(t)) \\ \dot{y} = -by - x^2 \\ \dot{z} = -x - cz \end{cases} \quad (2\text{-}1)$$

In the following discussions, we analyze the fixed points of the equation (1-1) in three cases.

(1) (0,0,0) is a fixed point of the system (1-1).

(2) If $c-b-abc<0$, the system has only one fixed point $(0,0,0)$.

(3) If $c-b-abc>0$, the system (1-1) has two fixed points:

$$(x_1, y_1, z_1) = (-\sqrt{\frac{c-b-abc}{c}}, \frac{abc+b-c}{bc}, \frac{1}{c}\sqrt{\frac{c-b-abc}{c}})$$

$$(x_2, y_2, z_2) = (\sqrt{\frac{c-b-abc}{c}}, \frac{abc+b-c}{bc}, -\frac{1}{c}\sqrt{\frac{c-b-abc}{c}})$$

This article achieves the purpose of chaos control by introducing delay feedback. we will not consider fixed point (0,0,0) and only analyze the dynamic stability of the other two point when $c-b-abc>0$.

We define

$w_1(t) = x(t) - x_i, w_2(t) = y(t) - y_i, w_3(t) = z(t) - z_i, (i = 1,2)$,

the equations are obtained as follows:

$$\begin{cases} \frac{dw_1(t)}{dt} = (1/b - a + y_i)w_1(t) + x_i w_2(t) \\ + w_3(t) + w_1(t)w_2(t) + k(w_1(t-\tau) - w_1(t)) \\ \frac{dw_2(t)}{dt} == -2x_i w_1(t) - bw_2(t) - w_1(t)^2 \\ \frac{dw_3(t)}{dt} == -w_1(t) - cw_3(t) \end{cases} \quad (2\text{-}2)$$

The characteristic equation of the equation (2-2) at the origin of coordinates is as follow.

$$\lambda^3 + R_2^*\lambda^2 + R_1^*\lambda + R_0^* - k(\lambda^2 + S_1^*\lambda + S_0^*)e^{-\lambda\tau} = 0 \quad (2\text{-}3)$$

where,

$$\begin{aligned} R_0^* &= kbc + 2(c - b - abc) \\ R_1^* &= bc - 3b/c + k(b+c) + 2 - 2ab \\ R_2^* &= b + c - 1/c + k \\ S_0^* &= bc, S_1^* = b + c \end{aligned} \quad (2\text{-}4)$$

We found that the models are same for two non-trivial fixed points of the characteristic equation. And the method to discuss this two fixed point is similar. So we just need to discuss one fixed point. When $\tau = 0$ the parameters are taken into equation (2-3), and equation (2-4) is obtained.

$$\begin{aligned} &\lambda^3 + (b + c - \frac{1}{c})\lambda^2 + (bc - \frac{3b}{c} + 2 - 2ab)\lambda \\ &+ 2(c - b - abc) = 0 \end{aligned} \quad (2\text{-}5)$$

According to the Routh-Hurwitz criterion, we know that if the following equations are observed,

$$\begin{cases} (b + c - \frac{1}{c}) > 0 \\ (b + c - \frac{1}{c})(bc - \frac{3b}{c} + 2 - 2ab) - 2(c - b - abc) > 0 \end{cases},$$

the real part of the solution of the equation (2-3) are located in the left half plane of the axis. In this case, the fixed point $(x_i, y_i, z_i)(i = 1,2)$ of chaotic system (2-1) remains stable.

The purpose of this paper is to control the chaos, which means that the unstable fixed points are controlled to be stable. Therefore, the fixed point is unstable when we discuss the situation of $\tau = 0$. So the parameters need to meet the following conditions:

$$(b + c - \frac{1}{c}) \le 0 \quad \text{or}$$

$$(b + c - \frac{1}{c})(bc - \frac{3b}{c} + 2 - 2ab) - 2(c - b - abc) \le 0.$$

2. 2 Analysis of the characteristic root

Here, we analyze the characteristic root of the system. For equation (2-3), we set $\tau = 0$. The characteristic equation of the original system (2-3) has the following form:

$$\lambda^3 + (R_2^* - k)\lambda^2 + (R_1^* - kS_1^*)\lambda + R_0^* - kS_0^* = 0 \quad (2\text{-}6)$$

When $\tau = 0$, equations (2-3) and (2-5) are equivalent. When $\tau \ne 0$, the characteristic equation is as follow:

$$\lambda^3 + R_2^*\lambda^2 + R_1^*\lambda + R_0^* - k(\lambda^2 + S_1^*\lambda + S_0^*)e^{-\lambda\tau} = 0$$

We assume the characteristic root of the above equation is $\lambda(\tau) = \alpha(\tau) + \beta(\tau)i$. When $\tau = \tau_*$, if $\alpha(\tau_*) = 0, \beta(\tau_*) \ne 0$, the system (2-1) experiences a Hopf bifurcation at (0,0,0) with $\tau = \tau_*$. Then, we assume that the equation has a pure imaginary root $\lambda = i\omega$ $(\omega > 0)$, and take it into the above equation, the following equations are obtained.

$$\begin{aligned} &-R_2^*\omega^2 + R_0^* + (K\omega^2 + KS_0^*)\cos\omega\tau - KS_1^* w\sin\omega\tau \\ &-(\omega^3 - R_1^*\omega + (K\omega^2 - KS_0^*)\sin\omega\tau + KS_1^* w\cos\omega\tau)i = 0 \end{aligned}$$

$$\begin{cases} -\omega^3 + R_1^*\omega = (K\omega^2 - KS_0^*)\sin\omega\tau + KS_1^* w\cos\omega\tau \\ -R_2^*\omega^2 + R_0^* = -(K\omega^2 - KS_0^*)\cos\omega\tau + KS_1^* w\sin\omega\tau \end{cases} \quad (2\text{-}7)$$

After calculation, the equation (2-7) is given.

$$\begin{aligned} &\omega^6 + (R_2^{*2} - 2R_0^* - K^2)w^4 + (R_1^{*2} - 2R_0^*R_2^* \\ &+ 2K^2S_0^* - K^2S_1^{*2})\omega^2 + R_0^{*2} - K^2S_0^{*2} = 0 \end{aligned} \quad (2\text{-}8)$$

Then, we have to discuss a six-order equation. After observation, we found that we can simplify this equation to a cubic equation. When we assume that $p = \omega^2$, the equation (2-8) is obtained:

$$p^3 + l_2p^2 + l_1p + l_0 = 0 \quad (2\text{-}9)$$

Where,

$$l_0 = R_0^{*2} - K^2 S_0^{*2},$$

$$l_1 = (R_1^{*2} - 2R_0^* R_2^* + 2K^2 S_0^* - K^2 S_1^{*2}), \qquad (2\text{-}10)$$

$$l_2 = (R_2^{*2} - 2R_0^* - K^2)$$

If equation $p^3 + l_2 p^2 + l_1 p + l_0 = 0$ has two or three positive roots, the equilibrium point may has stable range. So we just need to discuss the existence of positive root of equation (2-9). With the help of the Matlab software, it is easy to achieve in numerical results.

3 Numerical Simulation

3.1. Numerical simulation of the non-Delay Systems

The main purpose of this subsection is to select the appropriate control intensity k and time delay τ . Thus,we can control the fixed-point into a stable, or branched stable periodic solution in the vicinity of the fixed point.

Considering the system (1-2), Let $a = 0.00001, b = 0.1, c = 1$, we have:

$$\begin{cases} \dot{x} = 9.99999x + z + xy \\ \dot{y} = -0.1y - x^2 \\ \dot{z} = -x - cz \end{cases} \qquad (3\text{-}1)$$

The system (3-1) is chaotic and has three equilibria:
(0,0,0),
$(x_1, y_1, z_1) = (0.9487, -9.0000, -0.9487)$,
$(x_2, y_2, z_2) = (-0.9487, -9.0000, 0.9487)$

By the time $\tau = 0, k = 0$, the system (1-2) and the system given in [13] are the same. In this case, the numerical simulations of nonlinear chaotic systems are as follows:

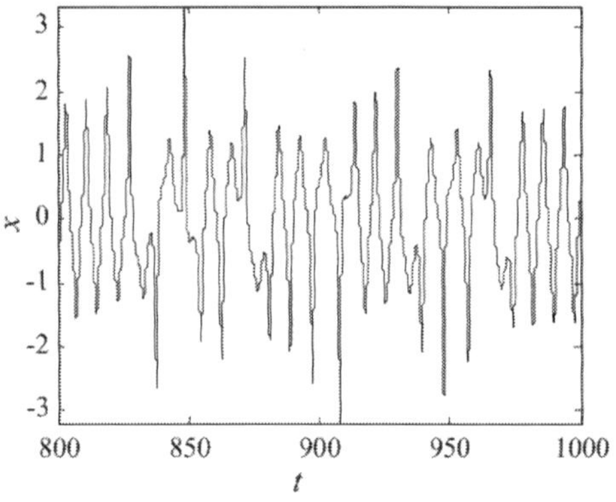

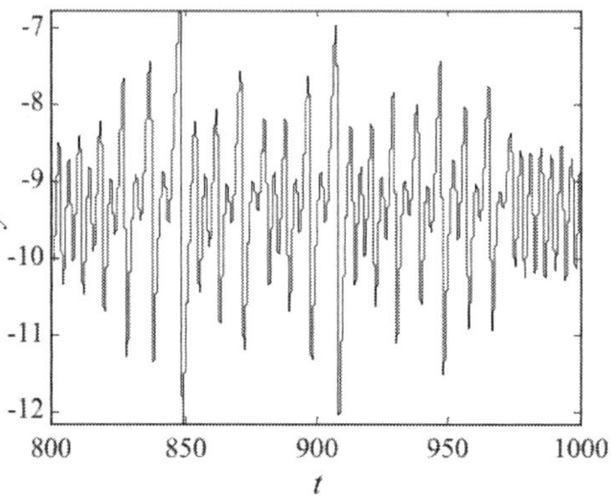

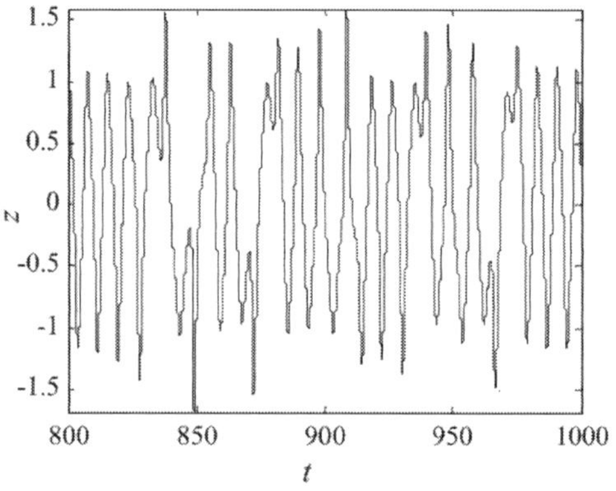

Figure1.when $\tau = 0$,waveforms for (x_1, y_1, z_1) , respectively

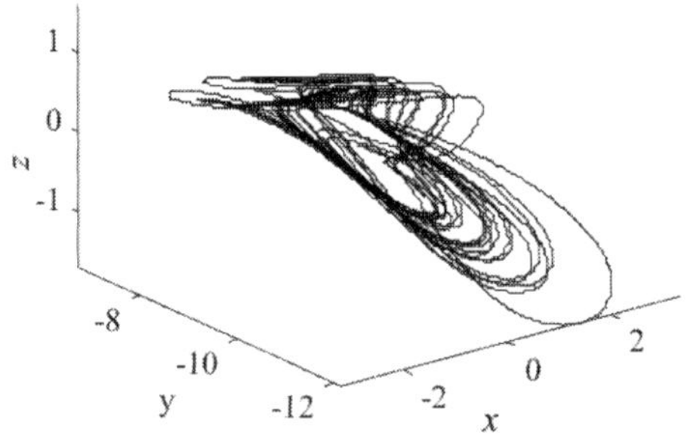

Figure 2.when $\tau = 0$, chaotic motion

It is found that the system is unstable under this condition. According to the discussion in section II, the range of k is obtained when (2-9) has a positive root. Under this situation, the system will produce Hopf bifurcation.

Let $k = 0.33$, we have Substituting $a = 0.00001, b = 0.1, c = 1, k = 0.33$ into （2-4） and（2-10） yields

$$R_0^* = 1.8330, R_1^* = 2.1630, R_2^* = 0.4300,$$

$$S_0^* = 0.1000, S_1^* = 1.1000, l_0 = 3.3588$$

$$l_1 = 2.9922, l_2 = 3.3588$$

Hence, we obtain the positive roots of the equation: $h(p)=p^3+l_2p^2+l_1p+l_0=0$.

In this case, this equation has two positive roots and one negative root:

$p_1=2.2171, p_2=2.6127, p_3=-0.5798$

Discuss the stability of the equilibria:

$(x_1,y_1,z_1)=(-0.9487,-9.0000,0.9487)$

At this time:

$p_1=2.2171, h(p_1)=-1.1065<0, \omega_1=\sqrt{p_1}=1.4890,$

$p_2=2.6127, h(p_2)=1.6164>0, \omega_2=\sqrt{p_2}=1.6164,$

$\tau_0^{(0)}=1.8664<\tau_1^{(0)}=2.1700<\tau_1^{(1)}=6.0572<\cdots$

Next, we will discuss the Hopf bifurcation and stability.

When $\tau=\tau_0^{(0)}=1.8664$, we find:

$\mathrm{Re}(C_1(0))=-1.5759<0, \mu_2=-3.1518<0,$

$\beta_2=-3.1518<0, T_2=51.0390>0$

Since there exists the periodic solutions of Hopf bifurcation when $\tau<\tau_0^{(0)}$, then bifurcation periodic solution is stable in the center manifold and the period of the bifurcation periodic solution increases.

When $\tau=\tau_0^{(0)}=2.1700$, we find:

$\mathrm{Re}(C_1(0))=-1.2672<0, \mu_2=5.4593>0,$

$\beta_2=-2.5344<0, T_2=-44.8958<0$

Hopf bifurcation periodic solution exists when $\tau>\tau_1^{(0)}$; the bifurcation periodic solution is stable in the center manifold andthe period of the bifurcation periodic solution is reduced.

3.2 Numerical simulation of systems with time delay feedback

Next, numerical simulation is performed for the parameter of system (2-1) to verify the results discussed above. Simulation results provide a strong basis for the correctness of the conclusions.

We take $a=0.00001, b=0.1, c=1, k=0.33$, and the initial value is $(0.8,-8,-0.8)$.

Notice that $\tau=0$ has been discussed before, and shown in Figure 1 and Figure 2. In the following part, numerical simulation for (x_1,y_1,z_1) are made for the case of the fixed point $\tau=1.4<\tau_0^{(0)}=1.8664$

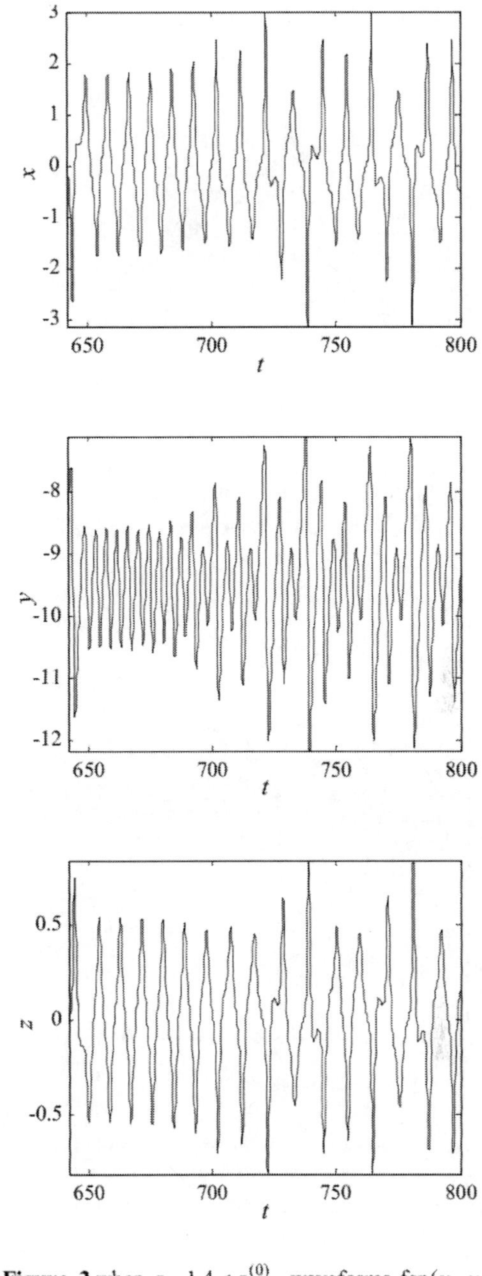

Figure 3.when $\tau=1.4<\tau_0^{(0)}$, waveforms for (x_1,y_1,z_1), respectively

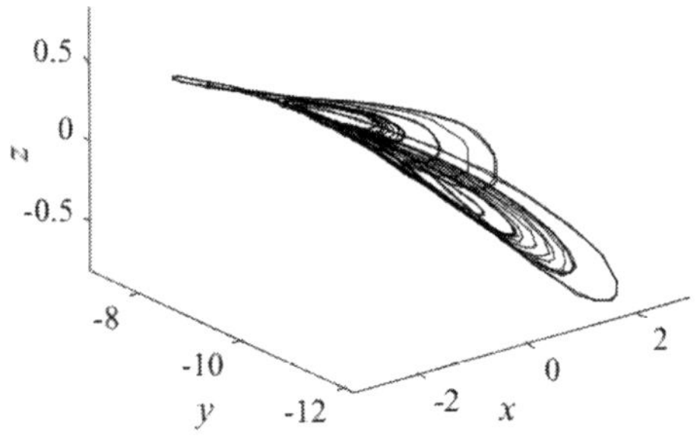

Figure 4. when $\tau = 1.4 < \tau_0^{(0)}$, the chaotic motion

According to Figure 3and Figure 4 , when $0 < \tau = 1.4 < \tau_0^{(0)}$, we can clearly see that the system is still chaotic at a fixed point (x_1, y_1, z_1) .

Next, we will analyse the situation of (x_1, y_1, z_1) when $\tau_0^{(0)} < \tau < \tau_1^{(0)}$

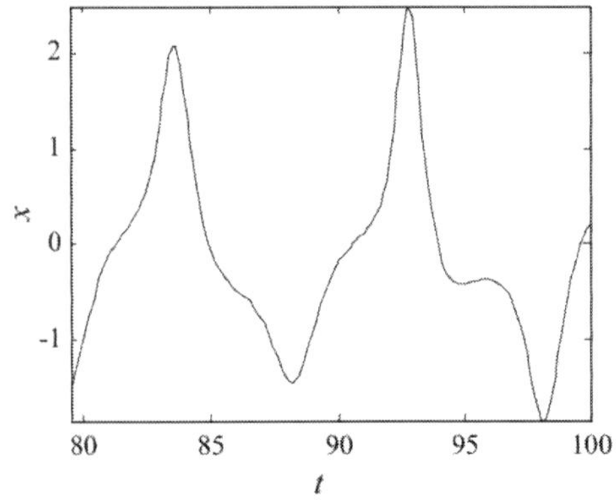

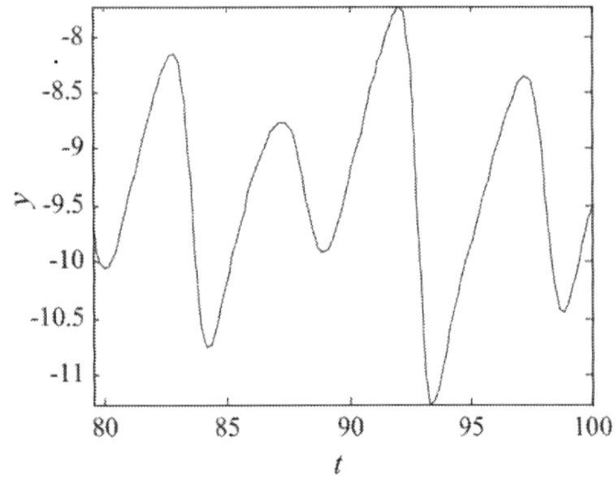

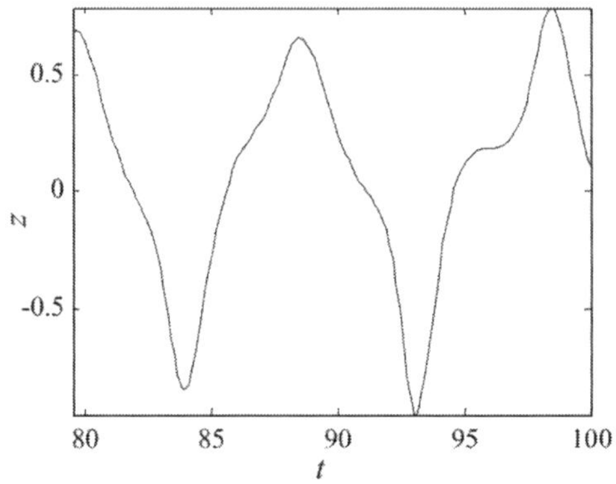

Figure 5. when $\tau_0^{(0)} < \tau = 1.9 < \tau_1^{(0)}$,waveforms for (x_1, y_1, z_1) , respectively

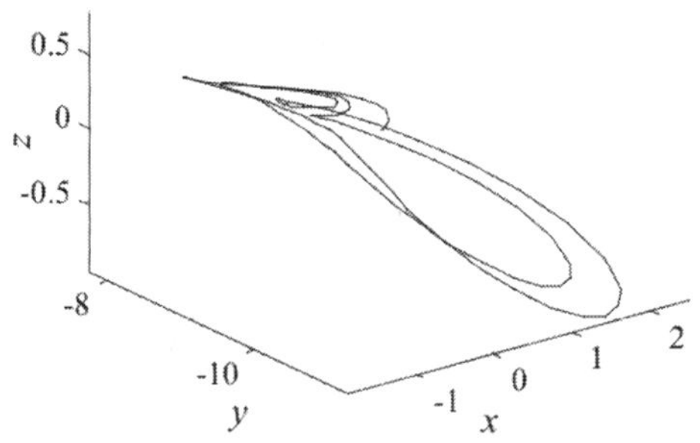

Figure 6. when $\tau_0^{(0)} < \tau = 1.9 < \tau_1^{(0)}$,the fixed point (x_1, y_1, z_1) is stable

When the fixed point near the phase diagram is as follows:

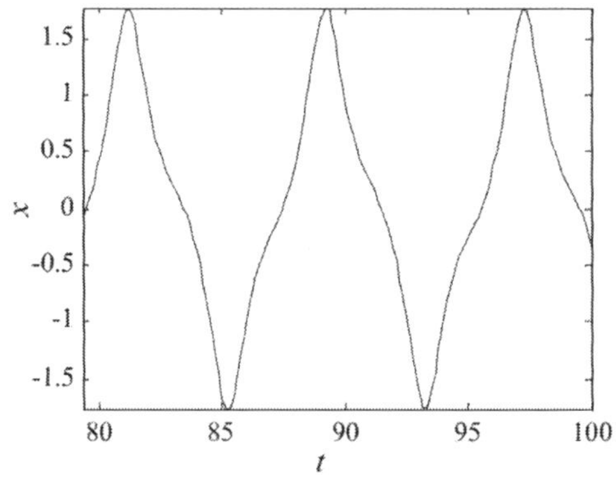

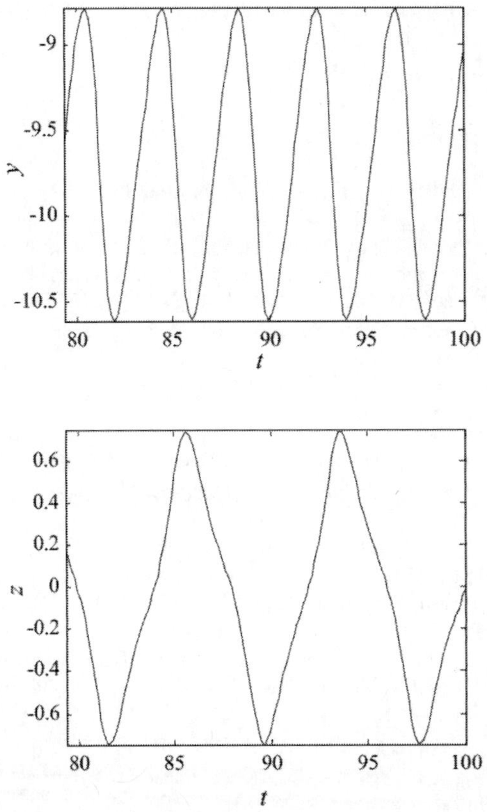

Figure 7. when $\tau = 2.8000 > \tau_1^{(0)}$, waveforms for (x_1, y_1, z_1) respectively

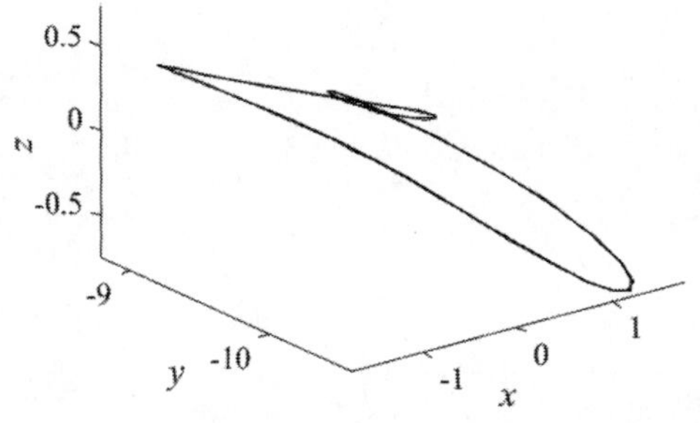

Figure 8. when $\tau = 2.8000 > \tau_1^{(0)}$, period solutions of Fixed Points (x_1, y_1, z_1)

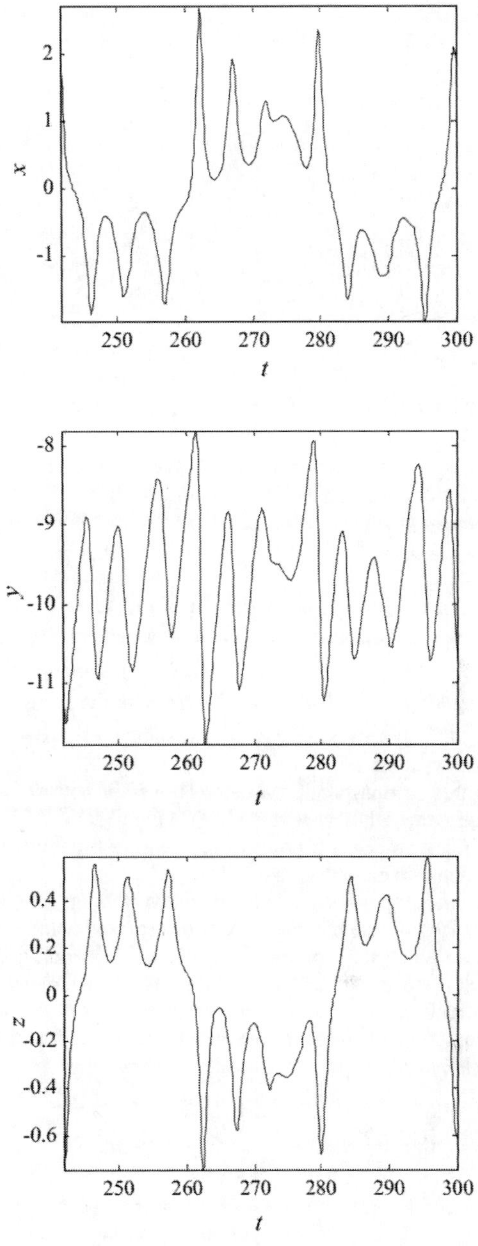

Figure 9. when $\tau = 10 > \tau_1^{(0)}$, waveforms for (x_1, y_1, z_1) respectively

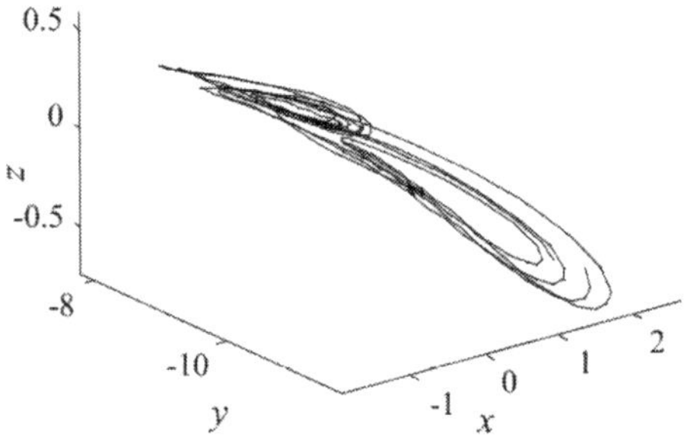

Figure 10. when $\tau = 10 > \tau_1^{(0)}$, the chaotic motion

We can see that there will be no large-scale solutions near the point of the cycle. We can not explain this phenomenon in theory, but from the numerical simulation, we can clearly find that there exists a similar periodic solutions.

It can be found that under the same set a,b,c of parameters and the same initial value, if we take the value k as the positive root of equation $h(p) = p^3 + l_2 p^2 + l_1 p + l_0 = 0$ With the increase of τ, the system will produce a stable or unstable periodic solution.

In this section, numerical simulations are performed by using the DDE application of Matlab toolkit. The Matlab software has a powerful computing capabilities and graphics capabilities. Using the data discussed in the finance model provided in [13], we successfully make the chaotic system stable in the vicinity of the fixed point. At the same time, we discovered the Hopf phenomenon and verified the correctness of the theoretical derivationbyusing the powerful DDE45 toolkit of Matlab, . Thereby, through combining theory and practice, we achieve the purpose of illustrating this problem.

4 Conclusions

This paper discusses a class of finance models with chaotic characteristics and the delay feedback control problem for these models. According to the finance chaos model which was discussed in the literature [13] by Cai, when the delay feedback was added into the equation of the system, the unstable fixed point of chaos system was controlled into a stable fixed point, or the stable periodic solution was branched near the fixed point. The numerical simulation are performed with the assistance of the computer. It is found that the unsuitable parameter combination is the root of the appearance of chaos for the system.Therefore, the study of the stability of this control system as well as the analysis of the characteristics of the internal dynamics for the system are very necessary. We also found that the financial system with nonlinear chaotic characteristic and delay feedback has rich dynamic responses. It is indicated that the delay feedback item is a sensitive factor in the financial system. This study has an important practical value which can provide a theoretical reference for the relevant departments to regulate the economic activity.

References

1. L. Fanti, P. Manfredi. Chaos, Soliton. Fract. **32**, 736-744 (2007)
2. K. Pyragas. Phys. Lett. A. **206,** 323-330 (1995)
3. K. Pyragas. Phys. Rev. Lett. **86,** 2265-2268 (2001).
4. Q. Gao, J.H. Ma. Nonlinear Dyn. **58,** 209-216 (2009).
5. W.C. Chen. Chaos Soliton. Fract. **36**, 1305- 1314 (2008)
6. Y. Wang, Y.H. Zhai Int. J. of Appl. Math. and Mech. **6**, 1-13 (2010)
7. M.J. Stutzer. J. Econ. Dyn. Control **2**, 353-376 (1980)
8. M.J. Stutzer. J. South. Econ. **6**, 55-62 (1982)
9. M.J. Stutzer. Public Finance Quarterly **77**, 79-95 (1984)
10. J.H. Ma, Y.S. Chen. Appl. Math. Mech. **22**, 1240-1251 (2001)
11. J.H. Ma, Y.S. Chen. Appl. Math. Mech. **22**, 1375-1382 (2001)
12. J.H. Ma, H.L. Tu. Nonlinear Dyn. **9**, 497-508 (2014)
13. G.L.Cai. Int. J. of Nonlinear Sci. **17**, 213-220 (2007)

MATEC Web of Conferences **45**, 03006 (2016)
DOI: 10.1051/matecconf/20164503006

A power recirculating test rig for ball screw endurance tests

Hermes Giberti and Andrea Collina

Department of Mechanical Engineering, Politecnico di Milano, Campus Bovisa Sud - via La Masa 1, 20156 Milano, Italy

Abstract. A conceptual design of an innovative test rig for endurance tests of ball screws is presented in this paper. The test rig layout is based on the power recirculating principle and it also allows to overtake the main critical issues of the ball screw endurance tests. Among these there are the high power required to make the test, the lengthy duration of the same and the high loads between the screw and the frame that holds it. The article describes the test rig designed scheme, the kinematic expedients to be adopted in order to obtain the required performance and functionality and the sizing procedure to choose the actuation system.

1 Introduction

The mechanical screw transmissions concern the transforming of a rotational movement into a linear one and vice versa. Usually the input motion into the transmission is rotational and driven by an electric motor. According to the required application the movement is applied to the screw or to the nut. The two most important features of these mechanisms are the heightened precision in positioning and increased load capacity. For these reasons it is widely used in a large number of applications: from automatic machines to robotic systems and more generally in applications in which high precision linear movements are required.

Ball screw mechanisms are closed systems and it is difficult to experimentally investigate what happens between balls, screw and nut during the movement. Several mathematical models are developed to describe the behaviour of this mechanical transmission. The most important studies have been done by Levit [1][2] and in general these approaches are based on the settled theories developed for ball bearings [3]. More recently a revision of Levit's work has been proposed and more generally to the classical approach to the study of ball screw kinematic [4-7] introducing pseudosliding in nut-screw contact. Indeed the analysis of the movement of the balls during the working cycle has shown microslipping in the contact areas. This phenomenon has great effects both in the mechanical system design phase and in its use. In particular it causes a reduction of the maximum theoretical mechanical efficiency that is possible to obtain from the transmission [5] and links efficiency to the load operating conditions. The non-traditional approach to the study of recirculating ball screw mechanisms is not supported by a wide experimental analysis available in scientific literature. Thus it is difficult to characterize the parameters on which this approach is based and the behaviour of the system related to their variation. The object of this work is to propose a new layout of endurance test rig capable of performing endurance tests under real load conditions and measure physics values useful in developing the mathematical models set out above.

In particular this paper shows the conceptual design of this device. The proposed layout is well suited for tests on large size ball screws (diameter to 50 mm) but for description of the system and the analysis of its performance a 25 mm diameter maximum ball screw has been taken into account. The originality of the system consists in the recirculating power principle applied in the ball screw test. This article sets out the functional groups constituting the test rig, with reference to the design choices carried out. The principal forces acting on the mechanical organs and on the drive system are evaluated by means of a simple mathematical model in order to define all the elements necessary for the design of the system.

2 Test rig description

The test rig uses the recirculating power principle borrowed from gear test benches concept [8]. To simplify it is possible to schematise a gear test rig, called "close cycle system" in two identical pairs of gears held at the end of two shafts. The gears are preloaded twisting the two shafts. This torsion is obtained by giving relative rotation to the two elements constituting of a joint by means of a predetermined torque. A motor moves a shaft introducing only the necessary power to overcome the friction while strong forces are transmitted among the gears due to the preload.

The same idea is duplicate for recirculating ball screws mounting two nuts on the same screw which are

Article available at http://www.matec-conferences.org or http://dx.doi.org/10.1051/matecconf/20164503006

pushed or pulled against each other by means of spring pack, as shown in figure 1.

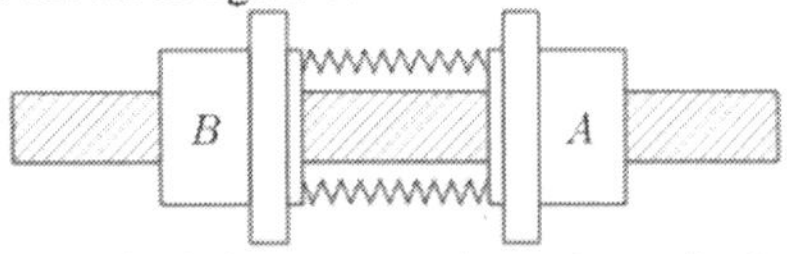

Figure 1. Recirculating power scheme for recirculating ball screws.

In this way the same force acts on the two nuts and the mechanical power resulting from these forces is nullified because they are equal in modules but opposite in direction. The inertia forces, the friction of the nut-screw contact, the friction of the slider-guide contact, the friction of the bearings that hold the screw are the only loads in action. It is necessary to point out that this layout, by nature, is particularly suitable for testing large size ball screws.

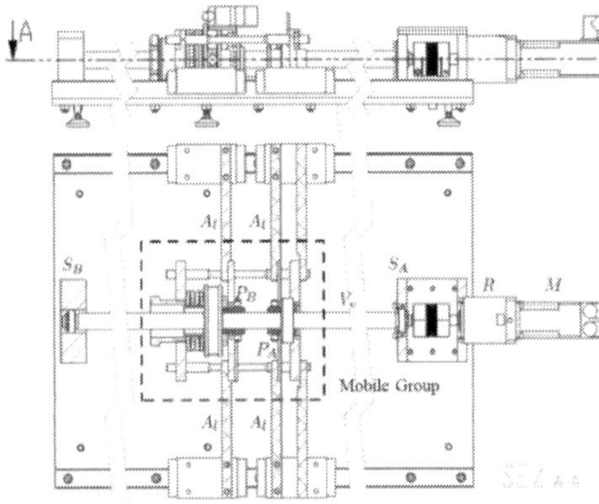

Figure 2. Test rig cross section.

Figure 2 shows a cross section of the design test rig. The screw (V_v), the supports that hold the end of the screw (S_A, S_B), the electric motor (M) and the reducer (R) that provide the movement of the screw, the mobile equipment and the two nuts are all shown. The screw is held by two SKF 6003 bearings: the one nearest to the motor is axially fixed while the other is free to move in such direction. Two steel plate (P_A, P_B) are joined to the two nuts. Two holes are made at the ends of the plates in order to fix four aluminium beams (Al_1, Al_2, Al_3, Al_4). These elements, hereafter referred to as wings, rest on four ball bearing slides (HIWIN HG-H-30-CA2R-1160-ZA-HII) capable to blocking the rotation of the nut and ensuring its translation.

Figure 3 shows a cross section of the mobile equipment. The two nuts mounted on the same screw, among which the load generate by the springs M_K, is shown. The spring block has been placed outside the two nuts in order to install the load system without disassembling the ball screw. We have chosen the belleville springs to generate the load because they are able to produce high force with small displacement. On the one hand the springs transmit to the plate (P_B) joined with nut B and on the other hand to plate P_A connected to nut A via four beams (Tr_1, Tr_2, Tr_3, Tr_4). The spring force passes through to SKF 811109 TN bearing before discharging on plate P_B while the other passes through plate P_C, the beams, plate P_D and bearing SKF GAC 40 F before discharging on plate P_A. In this way the two nuts are pushed against each other and the screw between them is compressed.

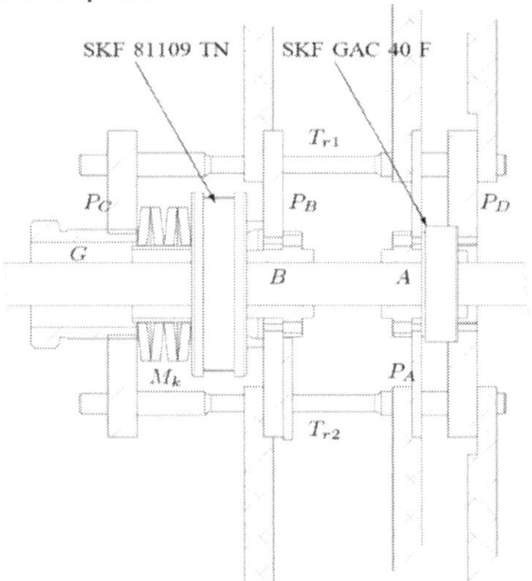

Figure 3. Mobile group cross section.

In order to have an axial load along the screw without load components in any other direction, the following expedients were taken into account: 1) insertion of an axial bearing (SKF 81109 TN) between the belleville springs and the plate joined to the nut B. 2) insertion of an orientable bearing (SKF GAC 40 F) between the nut A and the plate P_D.

In this way the two nuts are free to rotate one respect to the other along the screw axis and the system that generates the load is able to auto-align with the two nuts.

By means of the screwing or unscrewing of the ring nut G it is possible to modify the springs deflection and consequently to adjust the load that acts on the two nuts. Note that the plate P_D is joined to the beams and transmits the load directly to the orientable bearing and the latter does the same with nut A. This plate is equipped with two tiny wings. These are used only to keep the position of the mobile equipment during the setup phase of the machine in order to facilitate this task. Finally the screw is actuated by means of a motor-reducer group adequately sized and joined to the screw by a coupling. When the mobile group moves in steady state conditions, the only power required by the electric motor is that to overcome the friction forces within the screw-nut contact, within the sliders that block the rotation of the nuts and within the ball bearings that hold the screw. A device with adequate instrumentation as set out before, allows one to carry out the endurance test on the screws with pay load requiring the least possible mechanical power and thus making the trial intrinsically faster and cheaper.

3 Mathematical model

A mathematical model of the system is necessary to evaluate the required motor torque during the endurance test and the reaction forces on the anti-spin system. This model has to be as simple as possible and must accurately estimate these forces, to be use in the design phase of the test rig. It will be noted that the system is not simple to analyse because it is hyperstatic. Every nut is fixed to the ground both by the screw, which blocks the translations in all directions except along the screw axis, and by two tiny wings, which block the spinning around the screw. The motor ``sees" the translating block, in which the recirculating power takes place, as a black box dissipating mechanical power due to the non ideal transmission efficiency. To evaluate the power dissipation it is necessary to enter into this black box and to analyse what occurs internally.

3.1 Model simplifications

The mechanical system is shown in a simplified way in figure 3. In particular the following components are taken into account: the screw, the two nuts and the tiny wings. All the masses of the components not taken into consideration in the scheme are adequately distributed among the previously components, in order to maintain their inertial contribution. Note that by reason of this simplification the position of the center of mass of each component is not retained, thus the moments of the weight forces on the constraints are neglected.

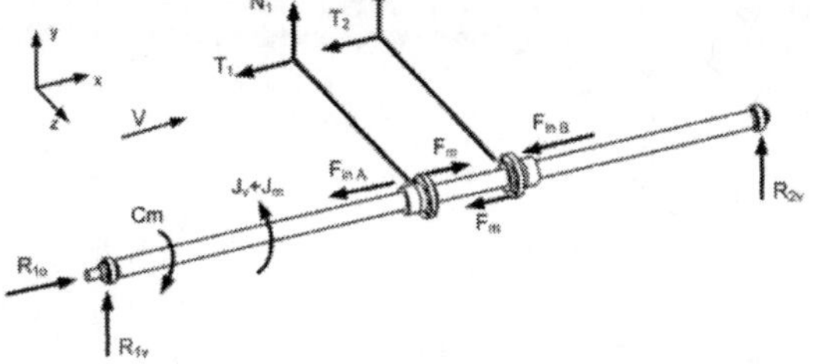

Figure 4. Simplified test rig scheme.

Figure 3 shows the external forces acting on the system, the constraint reaction forces on the slider that block the rotation of the tiny wings, the constraint reaction forces on the bearings, the motor torque and the inertia forces. Further simplifications have been effected to simplify the analysis of the system: 1)the weight force is neglected. This simplification is acceptable because the weight of the mobile pack is applied almost entirely on the sliders that hold the rotation of the nuts, hence it does not significantly modify the efficiency in the screw-nut contact; 2) a tiny wing of the anti-spin system for each nut has not been taken into consideration so as to make this constraint isostatic. The value obtained has to halved to consider the missing wing.

For the purposes of this mathematical model a complicated description of the thread-ball contact is not used. A recirculating ball screw has been studied as a normal screw where the friction force in the screw-nut contact is evaluated by the value of the efficiency supplied by the ball screw maker. The coupling of screw and nut can be represented by two wedges each with a single degree of freedom: The geometry of each wedge has to respect the tread characteristics (diameter and pitch), and the relationship between the pitch and the slant of the thread is: $\tan(\alpha)=p/(2\pi\, r_v)$ where α is the slant of the thread, p is the lead of the screw and $r_v = D_p/2$ is half pitch diameter (D_p).

Introducing the friction coefficient in the contact between the wedges it is possible to get the transmission efficiency (direct and indirect) by means the following equations:

$$\eta_{dir} = \frac{\tan\alpha}{\tan(\alpha+\varphi)}, \eta_{inv} = \frac{\tan(\alpha-\varphi)}{\tan\alpha} \tag{1}$$

It is possible to obtain the values of the screw efficiency from the manufacture's catalogue. In effect these values depend on several factors, for example, the load and the temperature. The friction coefficient and the friction angle φ are tied by the following equation:

$$\varphi = \arctan\left(\frac{f_G}{r_b \sin\alpha}\right) \tag{2}$$

From the literature [2] and from experimental tests [9] the size of the friction coefficient is F_G=0.001 cm.

3.2 Non recursive system model

On the hypothesis that the friction force on the slider is omitted, it is possible to simplify the equations that allow one to obtain the required motor torque during an endurance test. In this manner the resolution method is not recursive. The inertia forces are defined completely when the mass of each component and their motion law are chosen. The test rig is divided into two subsystems named A e B. Each subsystem contains a nut and a part of screw. Inertia forces are applied to these and called F_{inA}, F_{inB} while the screw inertia torque $J_v\ \omega'_v$ is divided into equal parts on each subsystems. The spring force (F_m) acting on the two screws is an external force taking into account the view of each subsystem. Due to this force the two nuts, that are fixed rotationally by means of the tiny wings, try to rotate the piece of screw between them in opposite directions. So the screw in this zone is subject to torsion. In the equivalent scheme based on the wedges this torsion is described as a resistant force opposite to the translation of the wedge-screw, as shown in figure 5. The module of this force is equal to C_v/r_v where C_v is the resistant torque and r_v is the pitch radius of the screw. Note that nut A is close to the motor while the other nut is called B. In the equivalent scheme based on wedge the nut A ``sees", in addition to the resistant force, another horizontal force due to the action of the motor (C_m/r_v).

The variables of this subsystem are two: the screw resistant torque C_v and the torque required by the motor C_m. Figure 5 shows a summarized scheme of the two subsystem A e B with all the forces applied. Note that between the two wedges there is a contact force S. Without friction between nut and screw this force would be perpendicular to the contact surface otherwise a friction force T is present and opposite to the motion of the two wedges. The component of S perpendicular to the surface, called N, has a different direction in the two subsystems A and B as is the force (F_m) generated by the spring on them. Subsystem A, closes to the motor, has two variable C_v and C_m at the same time, while the subsystem B has only the variable C_v (resistant moment). Firstly we can resolve the equilibrium equations of subsystem B, in order to obtain the common variable, and then those of the subsystem A to obtain the motor torque C_m required to move the mobile equipment.

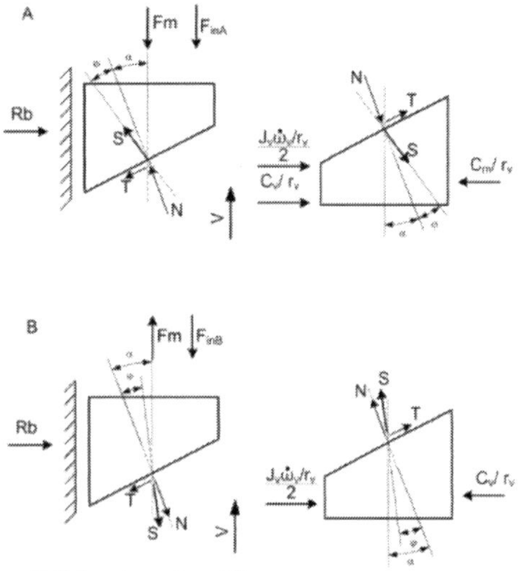

Figure 5. Subsystems A and B.

3.2.1 Solving subsystem B

Solving the equilibrium equations of subsystem B we obtain the variable C_v in the function of the spring force F_m applied on the two nuts. In particular we have:

$$\begin{cases} \dfrac{C_v}{r_v} = -N[\sin\alpha - \tan\varphi\cos\alpha] + \dfrac{J_v\dot{\omega}_v}{2r_v} \\ F_m - F_{inB} = N[\cos\alpha + \tan\varphi\sin\alpha] \end{cases} \quad (3)$$

From the second equation of 3 it is possible to calculate N while from the first equation we get C_v in function of the spring force F_m acting on the two nuts:

$$N = \left(\frac{F_m - F_{inB}}{\cos\alpha + \tan\varphi\sin\alpha} \right) \quad (4)$$

$$C_v = \frac{J_v\dot{\omega}_v}{2} - (F_m - F_{inB}) r_v \tan(\alpha - \varphi) \quad (5)$$

As shown in figure 6, the wing fixed on subsystem B has to sustain the constraint reaction N_b equal to the ratio between the torque C_v and the gap l between the guide and the screw axis:

$$N_b = \frac{2C_v - J_v\dot{\omega}_v}{2l} \quad (6)$$

Obviously this value has to be divided in two in order to consider the presence of the two wings.

3.2.2 Solving subsystem A

By virtue of the scheme shown in figure 5 and knowing the value of C_v in respect to the spring force F_m, it is possible to resolve subsystem A and to obtain the value of C_m.The equilibrium equations are similar to those of subsystem B. The variable C_m, in function to the spring force F_m is stated as follows:

$$C_m = J_v\dot{\omega}_v - (F_m - F_{inB}) r_v \tan(\alpha - \varphi) + (F_m - F_{inA}) r_v \tan(\alpha + \varphi) \quad (7)$$

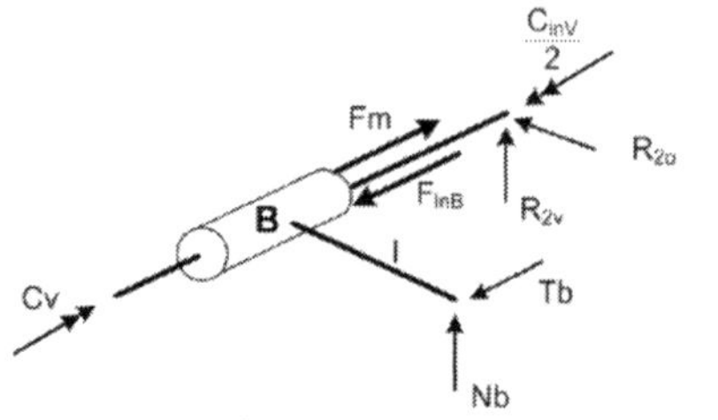

Figure 6. Subsystem B.

As shown in figure 7, the wing fixed on the subsystem A must sustain the constraint reaction N_a equal to the ratio between the resultant torque on the nut and the gap l between the guide and the screw axis:

$$N_a = \frac{(F_m + F_{inA}) r_v \tan(\alpha + \varphi)}{l} \quad (8)$$

It can be noted that the two constraint reactions N_a and N_b, have different modulus and opposite direction according to the spring force that tends to rotate the piece of the screw between the two nuts in opposite directions.

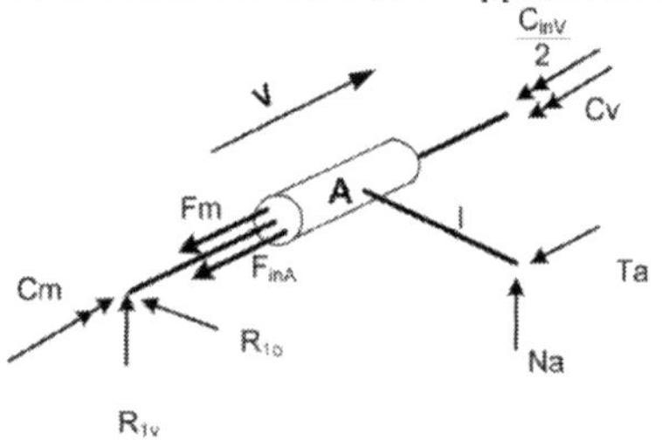

Figure 7. Subsystem A.

3.2.3 Inversion of the rotation of the screw

The developed mathematical model takes into consideration a certain sense of rotation and so a translation direction of the mobile equipment as shown with letter v in figures. The reversal of the sense of rotation implicates a different reply from the system because of the friction forces changing direction. Indeed friction forces are always opposite to the direction of the velocity. Revisiting the mathematical steps presented in previous sections it is possible to obtain the equations necessary to calculate the motor torque and the value of the slider constraint reaction in such working conditions.

4 Motor reducer sizing

4.1 Motion law definition

The nominal life of a ball screw is defined as the number of revolutions that it can reach. The lifetime is closely linked to the payload and accordingly two different load conditions are defined: 1) Dynamic Load C_a, the constant unidirectional axial load that a certain number of equal ball screws are able to bear without problems for 1 million of revolutions 2) Static load C_{0a}, the constant unidirectional axial load applied in static condition that generates, in the maximum stress area, a permanent deformation of 0.0001 times the diameter of the rolling ball. If we apply to the screw the load C_a its lifetime is 10^6 revolutions. When loads are smaller the lifetime is longer and it follows the equation: $n=(C_d/C_e)^3\ 10^6$. Where the variable C_e represents an equivalent load applied on the screw during the working cycle.

Having chosen the applied load, lifetime of the screw only depends on the motion law or in other words the rise h and the time t_a for the rise h. In particular for the case study the value of the rise h is the movement of the nuts mounted on the screw and it must be at least equal to the use length of the nut and no greater than the distance between the two nuts. The limit to the maximum rise is due to the fact that every nut must work over its piece of screw without involving the screw used by the other nut. It can be noted that during the design of the test rig the distance between the two nuts is an important parameter and it must be at least 1.25 times the length of a nut subject to test. The time t_a for the rise depends on the maximum velocity of the motor, the reducer and the screw, values available on the manufacture's catalogues.

4.2 Motor reducer sizing

Generally the choice of the motor and the reducer has to be made at the same time. The continuum operation area in the speed-torque performance curve of a brushless motor can be schematized as a rectangle limited by the torque C_n whereas the maximum torque curve $C_{m,max}$ defines the short term operation area [10].

Usually C_n is constant up to the maximum allowed velocity $\omega_{m,max}$ whereas $C_{m,max}$ decreases starting from a certain value of the velocity ω_m. The value of the torque C_n is obtained by thermal evaluation. Supposing that the dissipated power of the actuator and transformed into heat, is proportional to the square of motor torque C it is necessary that root mean square of motor torque $C_{m,q}$.be inferior or equal to the value of C_n. For a proper sizing of the motor not only should $C_{m,q} <= C_n$ but also the time t_a must be less than the motor thermal constant τ_{th}. The optimal choice of motor and reducer is a complex task [11], because the following conditions must be verified together: 1) Maximum torque limit: $C_{max} <= C_{m,max}$, . 2) Root mean square torque limit: $C_{m,q} <= C_n$, 3) Maximum velocity limit: $\omega_{max} <= \omega_{m,max}$

A sizing procedure capable of considering all the aspects before described is presented in [11][12], applied in [13] and it has been used for choosing motor and reducer of this test rig.

5 Conclusions

The paper presents a conceptual design of an innovative test rig for endurance tests on recirculating ball screws. The originality of the layout is the use of the recirculating power principal by which the test rig can perform tests with a minimum required power, in a short time and by means of a simple machine. Due to these qualities the proposed solution can be used for large size screw. The paper shows the key elements of the test rig analysing the main construction aspects and through the development of a simple mathematical model regarding the sizing of actuation system. The conceptual design of this new layout has been developed and for design convenience a 25 mm diameter ball screw has been taken into account, but this layout, by nature, is particularly suitable for testing large size ball screws.

References

1. Levit, G. A., *Recirculating Ball Screw and Nut Units*, Machines and Tooling, Vol. XXXIV, No. 4, pp. 3-8, (1963).
2. Levit, G. A., *Calculations of Recirculating Ball Screw and Nut Transmissions*, Machines and Tooling, Vol. XXXIV, No. 5, pp. 9-16, (1963)
3. Harris, T. A., *Rolling Bearing Analysis*, 1984, John Wiley and Sons, New York, 2nd edition.
4. Lin, M. C., Ravani, B., Velinsky, S. A., *Kinematics of the ball screw mechanism*, Journal of mechanical design, vol. 116, No. 3, pp. 849-855, (1994).
5. Lin, M. C., Ravani, B., Velinsky, S. A., *Design of the ball screw mechanism for optimal efficiency*, Journal of mechanical design, vol. 116, No. 3, pp. 856-861, (1994).
6. Huang, H. T., Ravani, B., *Contact Stress Analysis in Ball Screw Mechanism Using the Tubular Medial Axis Representation of Contacting Surfaces*, Journal of mechanical design, 119, pp. 8-14, (1997).
7. Wei, C.C., Lin, J.F., *Kinematic Analysis of the Ball Screw Mechanism Considering Variable Contact Angles and Elastic Deformations*, Journal of Mechanical Design, Vol. 125, pp. 717-733, (2003).
8. AA.VV, *Dynamic Loads on Gear Teeth*, Report of the A.S.M.E. Special Research Committee on the Strength of Gear Teeth, The American Society of Mechanical Engineers, New York, 1931
9. Silvestri, M., Prati, E., Tasora, A., *Frictional behavior of radial lip seals under dynamic operating conditions for different elastomer rings*, (2007) Sealing for Pollution Prevention and Control - 19th International Conference on Fluid Sealing, pp. 121-134.
10. Giberti, H., Clerici, A., Cinquemani, S., *Specific accelerating factor: One more tool in motor sizing projects*, (2014) Mechatronics, 24 (7), pp. 898-905.
11. Giberti, H., Cinquemani, S., Legnani, G., 2011, *A practical approach to the selection of the motor-reducer unit in electric drive systems*, Mechanics Based Design of Structures and Machines, Volume 39, Issue 3, July 2011, Pages 303-319
12. Giberti, H., Cinquemani, S., Legnani, G., 2010, *Effects of transmission mechanical characteristics on the choice of a motor-reducer*, Mechatronics, Volume 20, Issue 5, August 2010, Pages 604-610.
13. Giberti, H., Pagani, *A. Flexibility oriented design of a horizontal wrapping machine* (2015) Mechanical Sciences, 6 (2), pp. 109-118.

MATEC Web of Conferences **45**, 04001 (2016)
DOI: 10.1051/matecconf/20164504001

A development of the method of the control signal formation for the hot plate mill automation systems to improve the flatness of the finish plate

Stanislav S. Voronin , Vadim R. Gasiyarov , Andrey A. Radionov

South Ural State University (National Research University), 454080, Lenin ave. – 76, Chelyabinsk, Russian Federation

Abstract. This article describes how to control of the hot plate mill automation system to improve the quality metrics if the final strip. Based on the data of the modern hot rolling mills the classification of the cage equipment was designed. Depending on the degree of influence on the magnitude of reduction, the equipment was divided into categories. The functioning of every system including the main and the vertical cages was described. The conditions of electrical and hydraulic mechanisms was marked. The developed algorithm allows to improve defects based on the finite number of the thickness measurements given by special non-contact sensors. The example of regulators signals calculating was shown. The result of the algorithm operating was illustrated.

1 Introduction

The requirements for geometric parameters of the finished rolling strip is growing year per year, not depending on product assortment of the hot plate mills. To satisfy all specifications of the plate flatness concerning the four-high mill stand "quarto" a number of special control systems were designed and constructed by engineers and scientists around the world. This systems help to hold the roll gap in a given range during all the rolling process.

The main goal of the rolling stand automation system is signal formation for local regulators of the every flatness control system. In other case the displacement of the thickness and width on the final rolled strip may occur. To escape this problem it is necessary to develop a new algorithm of the formation of control signal for automation systems of the hot plate mill.

2 Classification of flatness control systems of the hot plate mill

All the flatness control systems can be divided into two groups: the main group and the auxiliary group [1]. The main group includes equipment to form an initial roll gap, including rough tuning and gap correction at the rolling moment. Equipment of the auxiliary group systems may be different, depends on the type of rolling mill [2]. This group is responsible for the accurate thickness adjustment along the strip. Thickness deviation arise from the two basic reasons.

The first reason is a number of external factors that make an additional value to previously exposed roll gap [3]. The work roll bending under the metal pressure, the rolls wear and thermal expansion, the backlashes and deformations of the roll stand - this are some of the external factors. It is necessary to take into consideration all the factors to offset their negative impact. Some values could be determined by computer modelling before rolling. Based on the model results the settings of the auxiliary control systems are performed.

The second reason is a thickness deviation in every measurement point along the slab before rolling process. Moreover, the slab often has internal defects, obtained at the casting stage, which can make a bad impact on the final strip. In this case, some auxiliary control systems put into work during rolling. All systems conditions are determined by sensors, which help to know about metal thickness every rolling moment. Some systems also perform an active role in a case of the rolling with different thickness along the strip. Typical classification of flatness control systems of the hot plate mill is shown at figure 1.

Every flatness control system must provide both separately mode and simultaneously mode with other systems. Electromechanical gap control system (EGCS) moves only between passes, that is when there is no a plate in the rolling stand [4]. Roll system can move in a vertical plane via rotating of screw-down mechanism. EGCS allows changing the gap in wide range (several hundred millimeters). At the modern rolling mills, EGCS consist of two pressure screws to change the roll gap at the edges of the plate. Every pressure screw is driven by own electric motor. To provide stable feedback it is necessary to use screw-setting transducers.

Hydraulic gap control system (HGCS) is designed to accurate correction of the roll gap. HGCS activates during the rolling process. Hydraulic system includes two cylinders that are moved by own servo valve, which changes amount and moving direction of the hydraulic fluid [5]. Actual position is calculated by averaging feedback signals from two position-sensitive transducers. Rolling pressure is determined via special load cells.

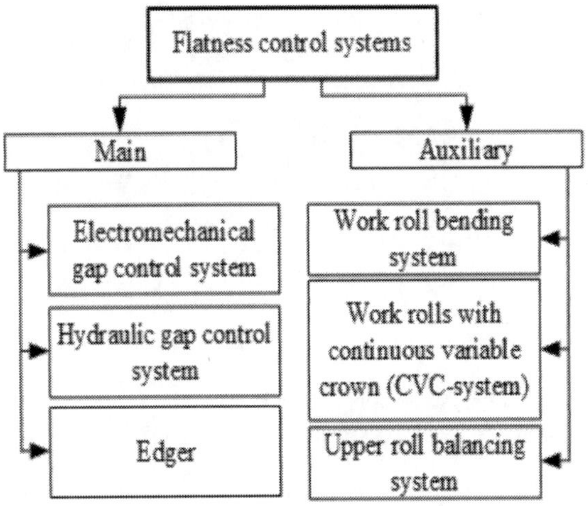

Figure 1. Flatness control system classification

Vertical cage (edger) has own control subsystems like EGCS, HGCS, width control system (WCS), roll balancing system [6]. Operation of these subsystems is similar with main cage.

Work roll bending system (WRBS) is needed to compensate rolls bending cases by high-pressure level at reducing moments [7]. The bending value depends on the rolling pressure, plate width, type of the rolls. WRBS helps to minimize thickness variation along the plate width by compensation the bending forces between metal plate and work rolls. Bending system uses four hydraulic cylinders (eight cylinders total) by each side (figure 2).

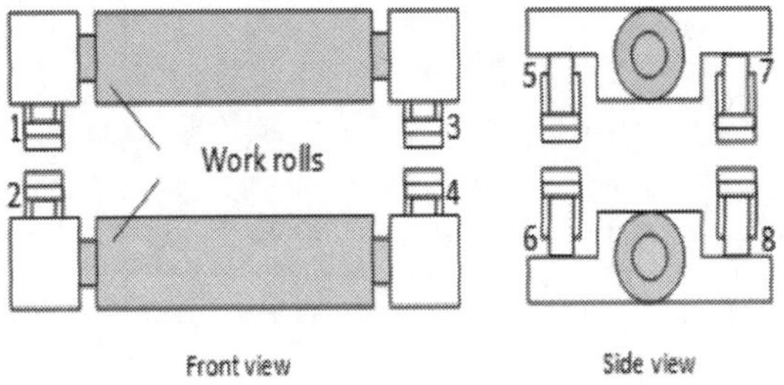

Figure 2. Arrangement of the hydraulic cylinders

Another auxiliary system to improve cross thickness is CVC-system. Work rolls have S-type polish with same contour and moving in horizontal plane between rolling passes. The value of the rolls shift influences on the roll force in the center and at the edges of the plate. Four hydraulic cylinders are used to shift upper and lower work roll. The shift value is limited by cylinder stroke and controlled by position sensors built-in each cylinder.

Upper roll balancing system (RBS) is used to minimize backlashes between rolls and stand. RBS has hydraulic and mechanical blocking device to hold upper roll at given level.

3 Flatness of the hot plate as a mathematical function

It is difficult to determine the final value of the roll gap, taking into accounts all internal and external rolling factors. In some cases, for example, it is necessary to calculate the distance between the rolls at a particular point or to determine the variation of the gap over the entire length of the roll. The most convenient solution of this problem is to consider all the main cage systems, affecting the final roll gap, as a kind of mathematical function depending on the distance between the work rolls and the rolls length (or the width of the rolled metal). Simple mathematical manipulations with obtained functions allow to obtain the values of the roll gap at any point. Based on the same approach is convenient to calculate the position of the screw-down mechanism of the cage and the hydraulic cylinders position for auxiliary control systems for next pass of the metal depending on the size and compression requirements of shape and flatness of the workpiece.

To view the principle of constructing mathematical functions, put the rolling mill quarto plate mill in the coordinate system (figure 3).The best place for a starting point is a point of intersection of the centerline of rolling (the horizontal axis) and the middle of the roll length (the vertical axis). The distance between the x-axis and the work rolls is the same. When the roll gap equals zero, the line of contact between the gap of the work rolls coincides with the line of rolling. This reference system is the most convenient for further calculations, as most of the rolling stand control systems operate symmetrically with respect to the axes, which greatly simplifies the derivation of formulas.

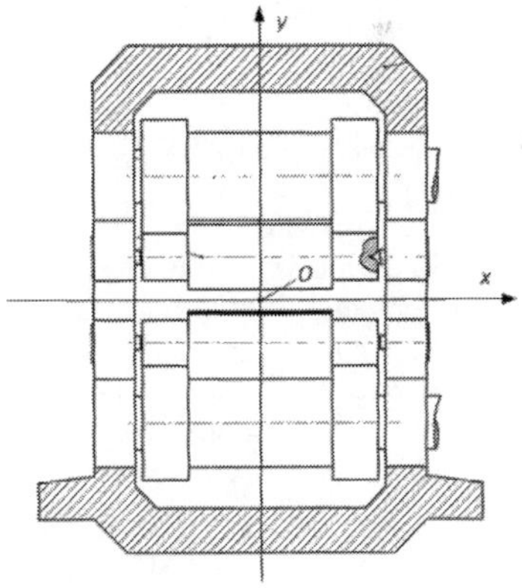

Figure 3. Quarto stand in the coordinate system

Figure 4 shows in more detail the space between the rolls and defines the basic parameters and constraints that may occur in the mathematical description. The dependence of h(l) is called a "roll gap function". The distance between the rolls (h_r) at any point is defined as the sum of the distances between the x-axis and the rolls (h_{ur}, h_{br}). In this domain of the function is limited to the length of the rolls (l_r). The domain of the function also is limited to the width of rolled metal, but in this case, the accuracy decreases due to elastic and plastic deformation of the whole roll.

A similar method is applicable to the rolled sheet. Before the rolling a sheet don't have a perfect rectangle shape - the thickness of the sheet in the middle and at the edges is different. A convenient way to know how to act on the work rolls during the rolling process is to pre-calculate the dependence of the sheet width and the its thickness. In this case, the key factor is the correct definition of the thickness of the sheet between passes. The system of indirect measurement of the sheet thickness allows partially compensate the flatness at the rolling moment. To accurately determine the thickness after each pass the thickness gauge is installed at the outlet of the cage that measure thickness at several points of the sheet.

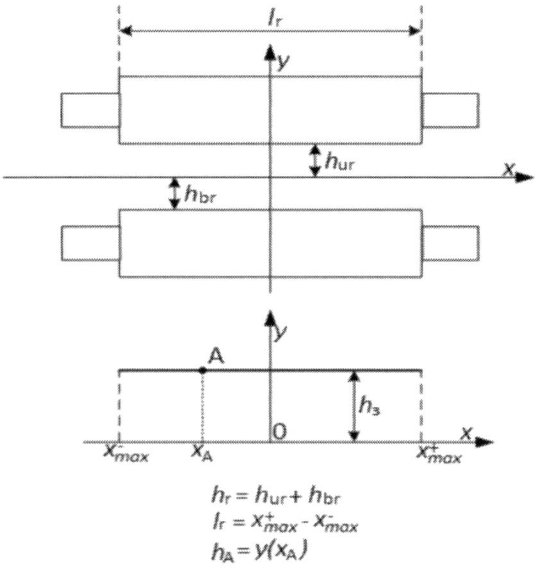

Figure 4. The roll gap as a mathematical function

Figure 5 shows an example of a local measurement of thickness in a single plane of the sheet. The thickness variation is most pronounced at the edges of the sheet, but also has a convexity in the central part of the sheet. In fact, the difference between the highest and the lowest value is usually less than a few millimeters, but in modern conditions such accuracy is insufficient to the end product.

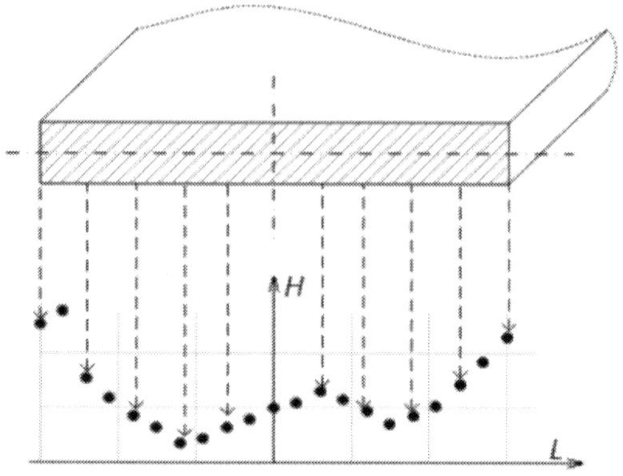

Figure 5. A measurement of the thickness at several points

To compensate for the thickness deviation the given measurements are converted into "sheet thickness function» H(L). Then the function compared with the "roll gap function" h(l) (see figure 4). Optimal conditions for compression are selected using the adjustment of the "roll gap function".

Finally, it is necessary to restore a functional relationship between the thickness and width of the sheet. To solve this problem the methods of approximation functions are used.

4 The foundation of optimum control values for flatness systems

Most of the hot plate mills have a thickness measurement system besides the main cage. The most useful sensors are calipers and non-contact thickness sensors. Usually after the thickness measurement and secondary signal treatment, the following array of data are possible to use:

$$H[i,j] = \begin{pmatrix} h_{11} & h_{12} & \dots & h_{1j} \\ h_{21} & h_{22} & \dots & h_{2j} \\ \dots & \dots & \dots & \dots \\ h_{i1} & h_{i2} & \dots & h_{ij} \end{pmatrix}, \qquad (1)$$

$H[i,j]$ is array of thickness data, [mm]; h_{ij} is plate thickness at a specific point, [mm].

From the other side the same array could be received from the flatness control systems. The difference between two data arrays is a thickness deviation. Thereby the plate flatness could be minimized by changing the systems parameters. It is worth nothing that each system affects only on the part of whole plate and cannot reduce all thickness defects. Thus, it is necessary to simultaneous control of all flatness systems. The result of the system functioning is formation of the signals for local regulators of the electric and hydraulic drives at every rolling moment. It was realized by special program algorithm operating between rolling passes.

The algorithm based on the consequent calculation all the signals of each flatness system. This is caused by that some systems would operate only between rolling passes, whereas other systems work at the pass time. Moreover, the force value of main and auxiliary systems cause different impact on the metal. So there are three steps of algorithm can be distinguished: the first step includes the calculation of the thickness variation in each plate point. The second step contains the foundation an optimal signal value of selected flatness system. On the third step, the calculation of the new thickness array and its transfer to the next system is occurred. The algorithm repeats until all the regulator's signals will be obtained.

EGCS is the fundamental system to change the gap between the rolls. The system also used for compensate all deformations of stand and rolls.

To minimize thickness variation along the strip, the HGCS is used. The most important parameters of the system are speed of the rolls moving [mm/s], the system performance, the reaction time, the moving value between the two adjacent thickness measurement points. The last parameter shows the value of the thickness deviation that HGCS could fully compensate (Eq. 2).

$$\Delta l_{HGCS} = \frac{(\Delta m - V \cdot R) \cdot v}{V}, \quad (2)$$

Δm is the distance between the two adjacent thickness measurement points, [mm]; V is rolling speed, [mm/s]; R is HGCS reaction time, [s]; v is max HGCS speed, [mm/s].

In case of big defect's values exceeding the maximum HGCS moving value, the system could remove all deviation only after some rolling passes.

To minimize thickness variation across the strip, the CVC-system is used [8-14]. Figure 6 shows the changing of the gap across the rolls. The reducing pressure across the strip is different, depending on the rolls offset.

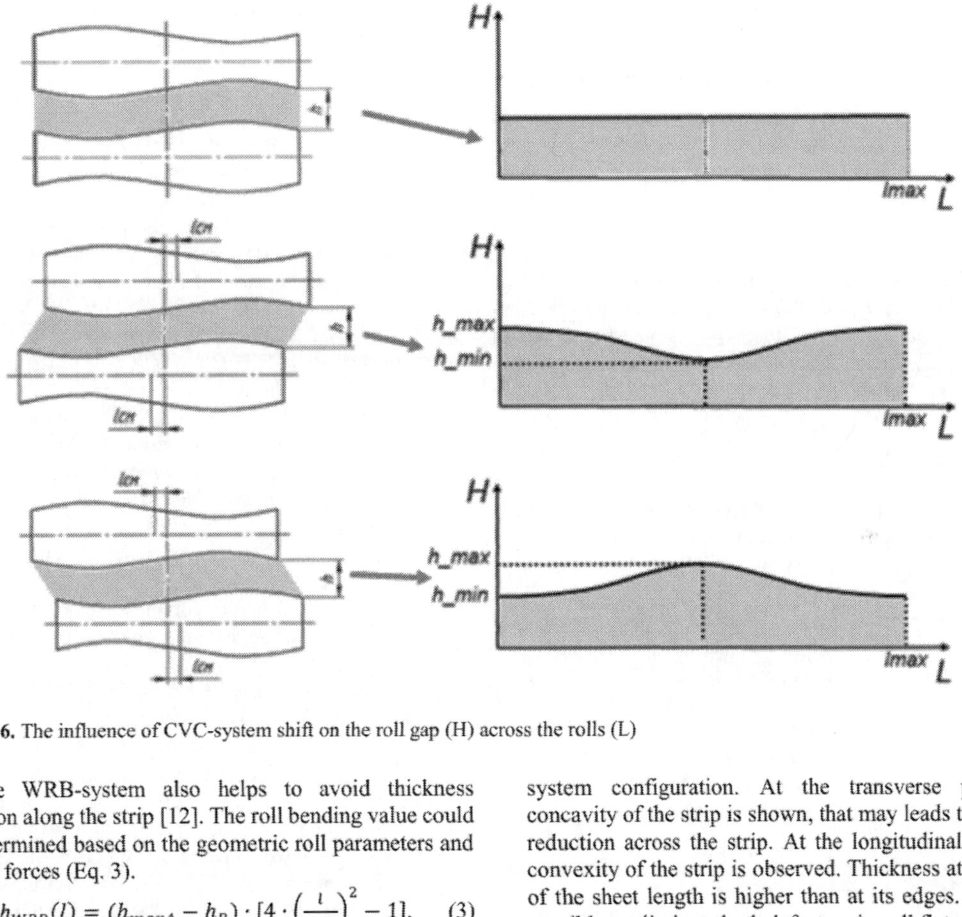

Figure 6. The influence of CVC-system shift on the roll gap (H) across the rolls (L)

The WRB-system also helps to avoid thickness variation along the strip [12]. The roll bending value could be determined based on the geometric roll parameters and rolling forces (Eq. 3).

$$h_{WRB}(l) = (h_{maxA} - h_B) \cdot [4 \cdot \left(\frac{l}{l_{roll}}\right)^2 - 1], \quad (3)$$

$h_{WRB}(l)$ is work roll bending function, [mm]; h_{maxA} is maximum roll bending, [mm]; h_B is WRB reaction, [mm]; l is calculated point, [mm]; l_{roll} is roll's length, [mm].

Parameter h_{maxA} proportional to the force created by hydraulic cylinders and could be determined by summing all the forces of each WRB cylinders. Thickness deviation is minimized by foundation the optimal value on every section along the metal plate.

Finally, the result is presented as data massive including all system shifts and moving. Figure 7 shows how the thickness variation changes before and after system configuration. At the transverse plane the concavity of the strip is shown, that may leads to different reduction across the strip. At the longitudinal plane the convexity of the strip is observed. Thickness at the center of the sheet length is higher than at its edges. It may be possible to eliminate both defects using all flatness control systems simultaneously.

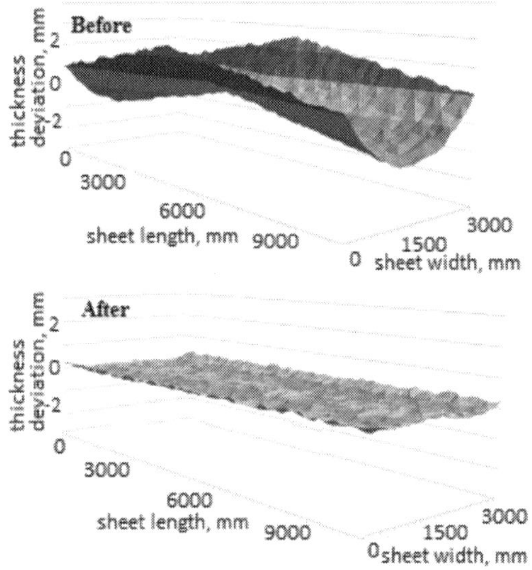

Figure 7. Three-dimensional view of thickness deviation of the hot sheet before and after rolling pass

The developed algorithm allows to eliminate defects of the sheet at the first and the last passes equally. The rough thickness irregularities are removed at the first several passes by EGCS and HGCS systems. Then at the finish passes the final flatness of the strip are formed via CVC and WRB systems. The algorithm's precision depends on the number of thickness measurements obtained from non-contact caliper.

5 Summary

The equipment of the rolling mills has complex structure. During development an automation system, it is necessary to examine each cage system individually and its impact on the rolling process. The designed equipment classification permit more detailed to explore the dependence between geometric dimensions of the sheet and flatness control system operating. The investigations show the flatness forming algorithm including calculations of the regulator's signals may be used for development of technological rolling schemes in the hot plate mill conditions. This will achieve the required thickness of the sheet with fewer reduction passes.

References

1. V. R. Gasiyarov, O. A. Zalogin, A. A. Radionov, "The strip thickness, profile and flatness control systems of the Magnitogorsk steel works hot plate mill 5000," *Urals science and industry*. pp. 107-112, (2010)
2. N. N. Druzhinin, *Continuous rolling mills as an automatiozation objects*. (1975)
3. V. R. Gasiyarov, D. U. Usatiy, "The thickness control system of the hot plate mill 5000," *Technology and production automation processes in metallurgy*. pp. 126-130, (2012)
4. S. S. Voronin, D. U. Usatiy, "A variable compression slab technology in the vertical and horizontal rolling cages of the hot plate mills," *Actual problems of modern science, technology and education*. **71**, (2013)
5. D. N. Popov, *Dynamics and regulation of hydraulic and pneumatic systems*. (1976)
6. V. R. Gasiyarov, A. A. Radionov, L. V. Radionova, D. U. Usatiy, "Automatic control of the geometry of the hot plate rolling mill 5000," *All-Russian multiconference on control problems*. **4**, pp. 307-309, (2011)
7. S. S. Voronin, D. U. Usatiy, "Work roll bending as a parameter that affects the lateral variation in thickness of the sheet and work roll bending control system of the hot rolling mill," *Russian Internet Journal of Industrial Engineering*. **1**, pp. 51-55, (2013). URL: http://www.indust-engineering.ru
8. S. S. Voronin, D. U. Usatiy, V. R. Gasiyarov, A. A. Radionov, The using of the CVC system to improve the roll gap on the hot rolling plate mills, *Russian Internet Journal of Industrial Engineering*. **1**, pp. 45-49, (2015). URL: http://www.indust-engineering.ru
9. Radionov, A.A., Gasiyarov, V.R., Gasiyarova, O.A., Automatic gap control of Plan View Pattern Control Mechatronics System, *Source of the Document 2015 International Siberian Conference on Control and Communications, SIBCON 2015 – Proceedings*, pp. 1-4.
10. Radionov, A.A., Maklakov, A.S., Gasiyarov, V.R., Smart Grid for main electric drive of plate mill rolling stand, *Proceedings of 2014 International Conference on Mechanical Engineering, Automation and Control Systems, MEACS 2014*, pp. 1-4.
11. Karandaev, A.S., Radionov, A.A., Khramshin, V.R., Andryushin, I.Y., Shubin, A.G., Automatic gauge control system with combined control of the screw-down arrangement position, *2014 12th International Conference on Actual Problems of Electronic Instrument Engineering, APEIE 2014 – Proceedings*, pp. 88-94.
12. Shilyaev, P.V., Andryushin, I.Yu., Golovin, V.V., Radionov, A.A., Karandaev, A.S., Khramshin, V.R., Algorithms of a digital automatic system for tension and loop control in a wide-strip hot-rolling mill, *Russian Electrical Engineering*. **84 (10)**, pp. 533-541
13. A.S. Maklakov, A.A. Radionov, *Proceedings of the 2015 IEEE North West Russia Section Young Researchers in Electrical and Electronic Engineering Conference, ElConRusNW 2015, 236-240 (2015)*
14. A.A. Radionov, A.S. Maklakov, E.A. Karyakina, *Proceedings 2015 International Siberian Conference on Control and Communications, SIBCON 2015, (2015)*.

MATEC Web of Conferences 45, 04002 (2016)
DOI: 10.1051/matecconf/20164504002

Surface Defects on Forged Nut of Stainless Steel

Tomonori Sasa[1], Zhigang Wang[1], Takashi Soga[2]

[1]*Dept. of Mechanical Engineering, Gifu University, 1-1 Yanagido, Gifu, 501-1193, Japan*
[2]*Ihara Science Corporation, Ihara Takanawa Building, 3-11-3 Takanawa Minato-ku, Tokyo, 108-0074 Japan*

Abstract. Surface quality of nuts used in gas pipe joints is an important issue because the surface state affects the fatigue characteristics. Surface defects of nuts produced by warm forging from a stainless steel are observed and the cause of defect generation is discussed based on the numerical results of FEM simulation. It is revealed that there exists the lap defect on the drawn wire surface with an opened end and the surface defects on the forged nut are generated by the tensile stress during forging since there exists a moment at almost the whole area of the billet surface that a tensile stress over 200MPa is generated during nut forging.

1 Introduction

Nuts under investigation in the present paper are used in gas pipe joints (**Figure 1**) for experimental devices and buildings. These gas pipes serve mainly for poisonous gases and flammable gases. Serious accidents occur if the gas pipe joint in the service had been broken. Therefore the demand for nuts is severe and the surface quality of the nut is considered as an important issue because the surface state affects the fatigue characteristics [1, 2]. However, the nuts produced by warm forging sometimes have surface defects as shown in **Figure 2**. The nut is produced from a drawn wire of SUS316 by four stages forging; upsetting, 1st extrusion, 2nd extrusion and piercing. Figure 2(d) shows a check mark on the nut surface, Figure 2(e) gives a cave-in. The cave-in appears near the bottom of the nut. A few check marks exist almost on a straight line.

These defects on nut surface perhaps become the start point of fatigue crack and cause failure finally under repeated stress [3]. Therefore the defects on the surface are regarded as a critical item and are inspected distinctly. The generation of surface defects reduces yield of material and brings cost up.

The purpose of the present work is to clarify the generation cause of surface defects on forged nuts. Surface defects are observed by SEM and then FEM simulation is carried out to discuss the stress state in warm forging.

Figure 1. Nuts used in gas pipe joint.

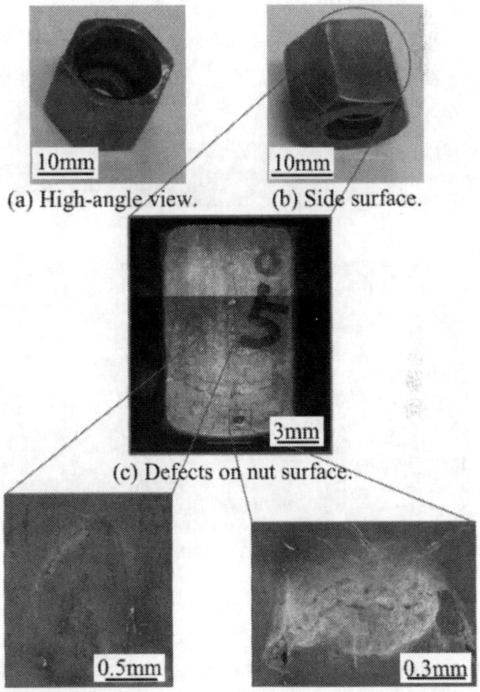

Figure 2. Defects on nut surface.

2 Observation of defects on forged nut

First, the cause of defects on the forged nut shown in Figure 2 is considered to be peeling off of hard particles.

As shown in **Figure 3**, some particulate zones are seen and their surfaces are covered by lead. As shown in **Table 1**, the chemical composition in the particulate zones is distinctly different with that of the stainless steel shown in **Table 2**. The main compositions, Pb, C, S in the particulate zones perhaps come from ingredients of the lubricant used in the forging process [4, 5]. No hard particles can be found on the surface of forged nuts.

Table 1. Chemical compositions in particulate zone (wt.%).

K	C	O	Si	S	Ca	Cr	Fe	Pb
0.5	33.4	7.1	1.4	7.6	1.1	2.0	8.1	38.8

Table 2. Chemical compositions of stainless steel (wt. %).

C	Si	Mn	P	S	Ni	Cr	Mo
0.04	0.48	1.6	0.027	0.02	10.09	16.56	2.04

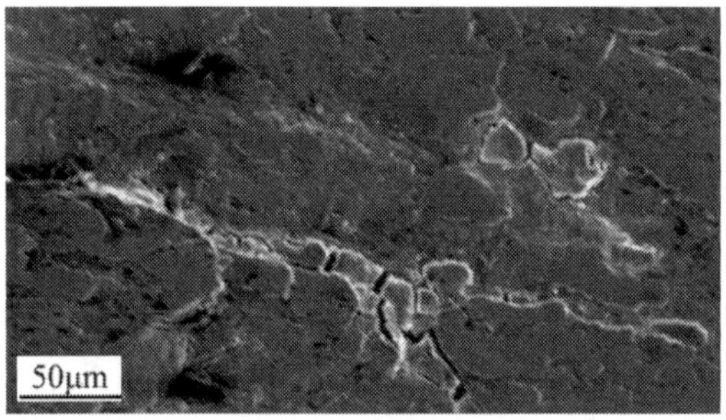

(a) Nut surface (SEM)

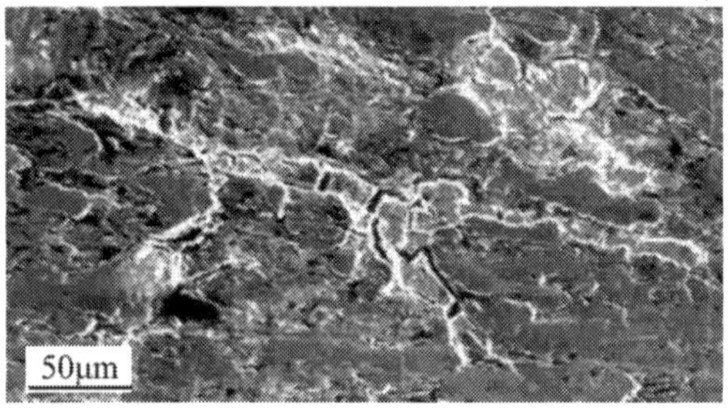

(b) Distribution state of Pb

Figure 3. Photograph of nut surface.

3 Observation of as-drawn wire surface

The billet in warm forging is cut off from a coil of drawn wire. First, the billet surface is observed with a microscope. **Figure 4** shows the observation results. The lap defects perhaps become the cause of the surface cracks of nuts.

The cross-sectional structure of the lap defect is observed after electrolytic etching. In electrolytic etching, the test piece with the observation plane burnished by buffing is anode; an austenitic stainless steel plate is cathode. They are connected to a power supply device and dipped into 10% oxalate solution and etched for 90 seconds under the condition of electric current 1A for etching area 1cm^2. After etching, the test piece is observed by 3D laser microscope (OLS4100).

Three cross sections as shown in Figure 4 are observed. Observation procedure is shown in **Figure 5**. First, the billet is cut parallel with the drawing direction. 1st cross section is observed (observation 1) after etching. Then the billet is cut vertically to the drawing direction and the 2nd and 3rd cross section are observed (observation 2, 3).

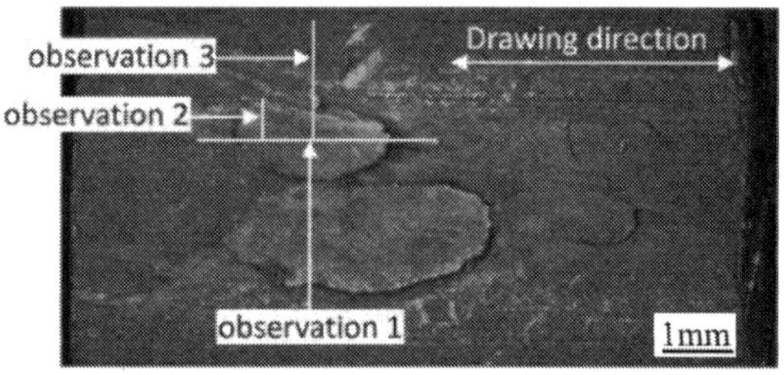

Figure 4. Defects on wire surface.

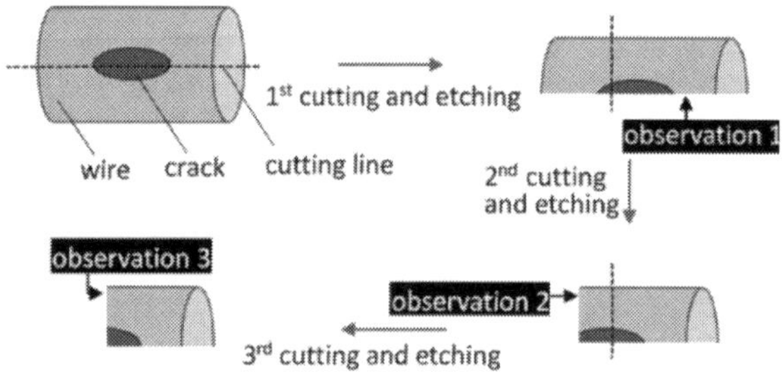

Figure 5. Process images of cross-sectional observation of defects on wire surface.

Cross-sectional structure of the defects at observation 1 is shown in **Figure 6**. The defect has an open end and connects perfectly at the other end. Observation 2 and 3 are made to check connecting situations in the hoop direction of billet. As shown in **Figures. 7, 8**, observation 2 shows almost perfect connecting although there exist caves at the ends of the defects. Observation 3 shows a crack throughout the length of the defect. It is found that the defects on the wire surface in the present study have the connecting part with the substrate material and the cracking part. It is difficult to clarify the cause of defect generation, but the possibility is extremely high that the cause of defect generation is due to unsuitable forming conditions of the rolling and wire-drawing process [6-8].

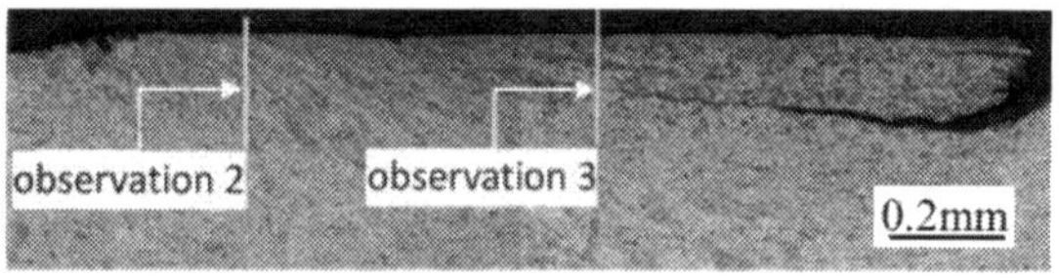

Figure 6. Surface crack at the plane of observation 1.

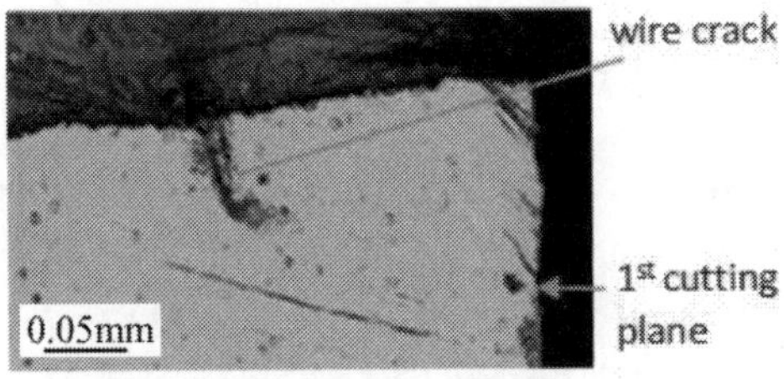

Figure 7. Crack at observation 2.

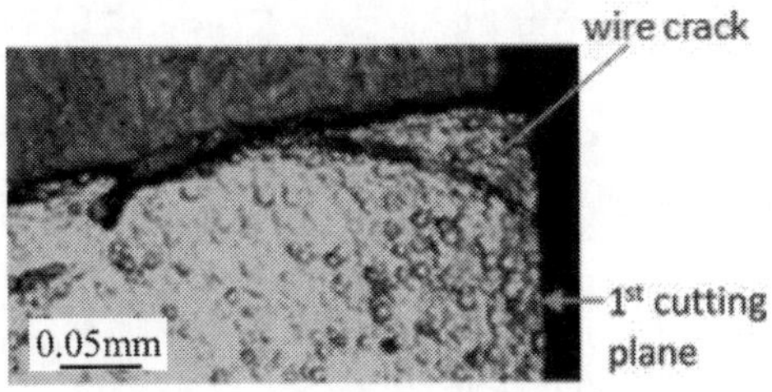

Figure 8. Crack at observation 3.

4 FEM simulations

FEM simulations are carried out to clarify the stress state at the billet surface in the forging process of nut. A commercial FEM code "DEFORM-3D" is used. The simulation model is shown in **Figure 9**. The simulation is performed by using 1/6 model to reduce the time of calculation. Friction coefficient is 0.1. The strain-stress curve of SUS316 in the software is used. The billet temperature is 300 Celsius degree that is the same as the operation in forging shop.

The nut is formed by four stages; upsetting, 1st extrusion, 2nd extrusion and piercing. Because the piercing process does not give any change of the nut surface, it is ignored. In upsetting, the billet is compressed by the compressing punch to close the hexagonal die and is formed to the shape of a hexagonal cylinder. 1st extrusion is operated to make a shallow shape and then 2nd extrusion is carried out to give a deep hexagonal hole.

Because the billet shape becomes to hexagonal at upsetting stage, the deformation state during upsetting is shown in **Figure 10**. The billet shape varies drastically to the hexagonal shape at the punch stroke of 3.5-5.0mm. The effective stain at the surface layer is about 0.6.

Figure 11 shows the stress state of the surface in the radial direction when the point 1 comes to contact with the die surface. It can be found that the surrounding area of the contact area with the die surface is in a tensile stress state.

Relation between the radial stress and punch stroke is shown in **Figure 12**. Horizontal axis is the punch stroke divided by the initial height of billet. A series of optional points are chosen and numbered in the order of entering contact with the die. The billet is compressed to become a shape with barrel because of friction. The die shape is hexagonal cylinder, therefore contact timing with the die

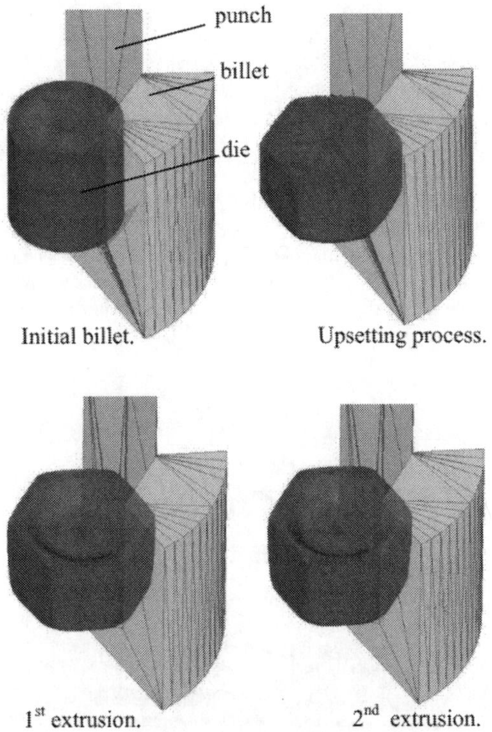

Figure 9. Forging processes of nut.

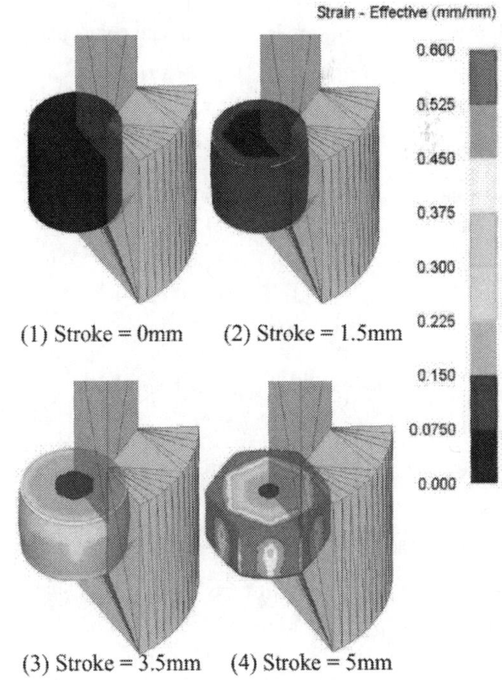

Figure 10. Change of billet shape in upsetting

varies from point to point. When a point on the billet surface comes to contact with varies from point to point. When a point on the billet surface comes to contact with the die, a compressive stress is caused on that point, the stress value rapidly shifts to minus in Figure 12. Before the stress shifts to minus, a tensile stress near 200MPa is generated on the surrounding area. The tensile stress is perhaps caused by bulging as a reaction when the compressive stress is caused on the material beneath the die.

In the forging process of the nut, it is found that there exists a moment for each point on the billet surface that a tensile stress is generated. The tensile stress is larger enough to make the lap defect shown in Figure 2 separate from the billet surface and generate a cave on the forged nut surface [9, 10].

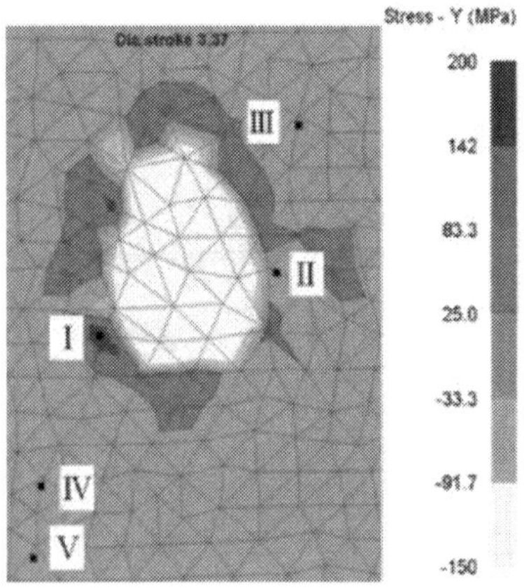

Figure 11. Image of simulation result and points position.

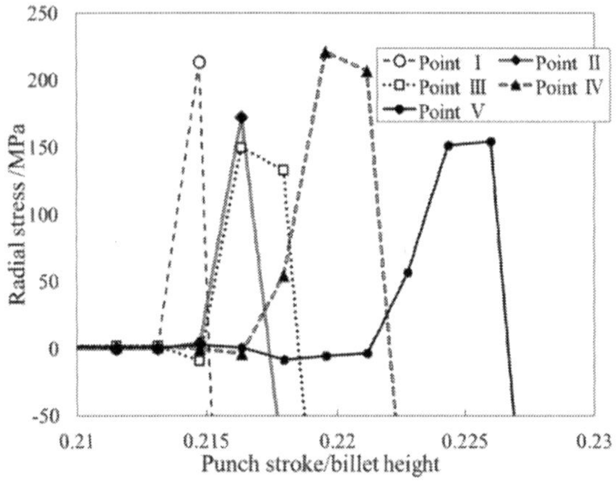

Figure 12. Relation between radial stress and punch stroke at simulated process.

5 Conclusions

Surface defects on the forged nut of a stainless steel are observed and the cause of defect generation is discussed based on the numerical results of FEM simulation. It is revealed that the cause of the surface defect is the lap defect with an opened end on the drawn surface and the surface defect on the forged nut are generated by the tensile stress during forging since there exists a moment at almost the whole area of billet surface that a tensile stress over 200MPa is generated during nut forging.

References

1. E.Sumami, Y.Sagae, *Tetsu-to-Hagane*, **68**, 415 (1982)
2. H.Satoh, Y.Kawaguchi, M.Nakamura, M.Terashita, T.Kobayashi, Y.Isono, *Wire J Int*, **3**, 120 (1995)
3. T.Shinohara, K.Yoshida, *Tetsu-to-Hagane*, **90**, 1010 (2004)
4. K.Yuasa, Y.Suzuki, *Tetsu-to-Hagane*, **72**, 1488 (1986)
5. K.Hirai, *Journal of JSTP*, **25**, 878 (1984)
6. T.Ishikawa, N.Yukawa, Y.Yoshida, Y.Tonohara, *Tetsu-to-Hagane*, **89**, 1142 (2003)
7. N. Yukawa, T. Ishikawa, Y. Yoshida, A. Koyachi, *Tetsu-to-Hagane*, **91**, 861 (2005)
8. N. Yukawa, Y.Yoshida, T. Ishikawa, *Tetsu-to-Hagane*, **92**,661 (2006)
9. Y. Murakami, S. Kodama, S. Komuma, *Journal of JSME, Series A*, **54**, 688 (1988)
10. M.Kumagai, S.Takagi, F.Yakabe, Y. Jinbo, Y. Hasegawa, *Surface technology*, **57**, 428 (2006)

MATEC Web of Conferences **45**, 04003 (2016)
DOI: 10.1051/matecconf/20164504003

Revamping Grooving Process for Sustainability using Fuzzy Expert System

Asif Iqbal

Department of Mechanical & Aerospace Engineering, Air University, Islamabad, Pakistan

Abstract. The article presents an application of a fuzzy expert system for renovating a metal cutting process to cope with the sustainability requirements. The work seeks a sustainable balance between energy consumption, productivity and tool damage. Cylindrical grooving experiments were performed to generate data related to quantification of the effects of material hardness, cutting speed, width of cut and feed rate on the aforementioned sustainability measures. A fuzzy knowledge-base was developed that suggests the most suitable adjustments of the controlled variables that would lead to achievement of various combinations of the objectives.

1 Introduction

For last 15 years or so, manufacturing processes are being readjusted in accordance with the environmental benignity requirements. The quest of the current research activities is to reduce negative environmental impacts by finding ways to lift productivity and curb consumption of energy and other natural resources. The metal cutting processes, thus, are under intense study for embracing sustainability norms.

An analytical method for minimizing process cost and energy consumption for machining process was presented [1]. The method searches for a cutting speed value that guarantees longest tool life and minimum energy cost. Diaz et al have compared demand of power in micro-machining of steel with those of polycarbonate and aluminum at various productivity levels [2]. The authors in another work have presented a multi-objective analysis of energy consumption, process cost, and work surface roughness of a machining process based on optimization of cutting speed, feed rate, and depth of cut [3]. In a recent work, a comparison between tool life, process cost, productivity, and specific energy consumption has been provided [4]. Another work presents a model that optimizes cutting parameters for maximization of profit per unit energy consumption in a single pass turning process [5]. A mathematical model was put forward that quantifies energy requirements in three different states of machining for various machines [6]. In [7], the authors have presented an experimental investigation for energy cost modeling of turning hard steel. The authors have used Box-Behnken Design and artificial neural networks for the modeling purpose.

With regard to application of intelligent systems, Hashmi *et al* have developed a fuzzy knowledge-based system for optimization of cutting speed as a function of work hardness based on various combinations of work material, depth of cut and tool material [8]. A prediction model for surface roughness and thrust force for a drilling process was developed using radial basis function network [9]. A fusion of fuzzy logic, neural network, and genetic algorithm was presented for optimizing parameters of a turning process [10]. A self-developing fuzzy expert system capable of modeling and optimizing manufacturing process without requirement of manually developing fuzzy sets or rules was presented [11].

The current work presents a fuzzy expert system developed from experimental data generated through cylindrical grooving experiments performed on a high-strength low alloy steel. The aim of the expert system is to revamp the grooving process in accordance with the requirements of manufacturing sustainability, such as minimize specific energy consumed, maximize productivity (in terms of material removal rate), and maximize tool life.

2 Experimental work

Four predictor variables were controlled in a full-factorial, mixed level design of experiments. The process of cylindrical grooving was studied on a high-strength low alloy steel, AISI 4340. Cylindrical grooving was selected because it is a metal cutting process that most closely matches an orthogonal cutting process. Following is the detail of the input parameters (predictor variables):

1. Work material hardness, *H*. The material rods were heat-treated differently to obtain two levels of hardness: 31 and 38 HRc.
2. Cutting speed, V_c. Three levels were tested: 28, 57, and 89 m/min

Article available at http://www.matec-conferences.org or http://dx.doi.org/10.1051/matecconf/20164504003

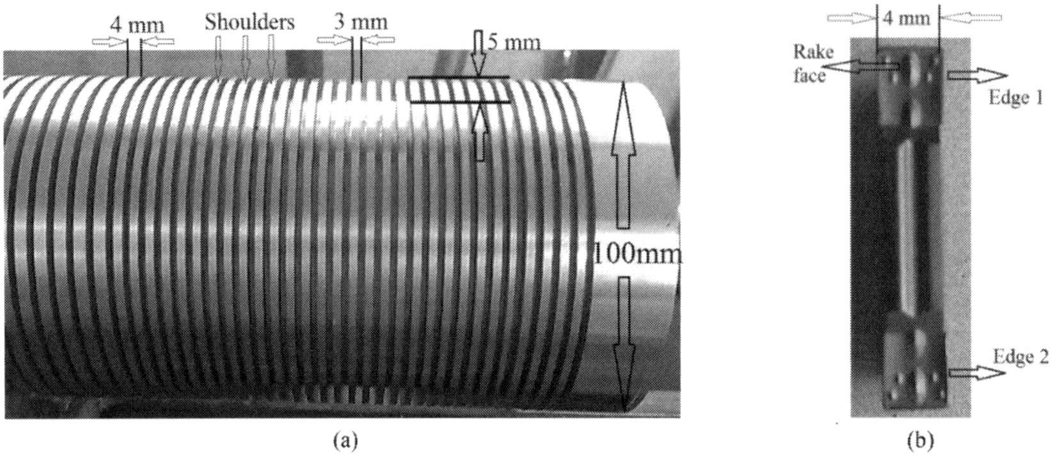

Figure 1. Experimental setup: (a) an AISI 4340 rod with shoulders cut into its surface; (b) a grooving insert

3. Feed rate, *f*. Two levels were used: 0.07 and 0.17 mm/rev.
4. Width of cut, *w* (mm). *w* was controlled by the following three widths of cutting inserts used: 2, 3, and 4 mm.

The number of the levels of the four predictors suggest that 36 (= 2 × 3 × 2 × 3) experiments were conducted in total. All the experimental runs were conducted on a CNC lathe with a maximum spindle speed 3500 rpm and maximum motor power 11 kW. The AISI 4340 material was used in the form of rods having length 500mm and diameter 100mm. Before start of the experiments, 108 (= 36 experimental runs × 3) cylindrical shoulders of 5mm depth and 2, 3, and 4mm widths were prepared on the bars. The widths of the shoulders correspond to the widths of the inserts chosen. Figure 1 gives the details. Uncoated tungsten carbide cutting inserts having flank angle 12° and rake angle 1.5° were used. Each of the 18 inserts (each insert has 2 cutting edges) was required to cut 3 relevant shoulders. The area of material removed (AMR) in cutting each shoulder is 1492.3mm² [= $\pi\{(50)^2 - (50-5)^2\}$] and for three shoulders is 4477mm². The total volume of material removed by each insert of 2, 3, and 4mm width is 8954 (= 2 × 4477), 13431, and 17908 mm³, respectively. The flank wear was measured, using 10× toolmaker's microscope, after cutting each of the three assigned shoulders. The three wear levels are denoted by *VB1*, *VB2*, and *VB3*. Material removal rate *MRR* (mm³/sec) was calculated by using the relationship: $MRR = w \times f \times (V_c/60 \times 1000)$.

The following responses were assessed for each run:

1. Width (maximum) of wear land on flank face of the tool after cutting each of the three shoulders, *VB1*, *VB2*, and *VB3* (mm).
2. Total energy consumed in cutting of three shoulders, *E* (kJ).
3. Specific cutting energy, *SE* (J/mm³). The *SE* is evaluated by dividing average cutting power by *MRR*.

Electric power drawn by the machine tool was measured by applying Fluke 345 on its power cable. The non-cutting electric power used up by the lathe, for each of the 6 (= 3 × 2) combinations of feed and cutting speed, was found out by revolving a rod and advancing the tool inwards at the given combination of the cutting parameters, but without touching the rod. The actual cutting power was evaluated by lessening the total power measured during the cutting process from the relevant non-cutting power. The cutting power was then averaged and multiplied with the relevant cutting time to obtain the value of *E* for each experiment.

2.1 Experimental results

Table 1 shows the results for the 36 runs. Few observations can directly be extracted from the table: (1) the higher levels of cutting speed, width of cut and feed rate render higher magnitude of tool damage; (2) cutting of the harder temper causes higher magnitude of tool damage; (3) Higher levels of speed, feed, and width of cut cause slash in specific cutting energy; and (4) the harder temper of material causes rapid tool damage as well as more consumption of specific energy.

3 Development of the fuzzy expert system

3.1 Data fuzzification

Triangular fuzzy sets were worked out for the predictors and responses. As the predictors are user controlled, their membership functions are evenly distributed. On the other hand, the experimental values of the responses are not distributed evenly or even randomly; their distributions show existence of two to three dense zones. For the sets related to the response variables, the dense zones are assigned more membership functions than the rare areas.

Table 1. The experimental results.

S/ No.	*H* (HRc)	V_c (m/min)	*f* (mm/r)	*w* (mm)	*VB1* (mm)	*VB2* (mm)	*VB3* (mm)	*E* (kJ)	*MRR* (mm^3/sec)	*SE* (J/mm^3)
1	31	28	0.07	2	0.02	0.049	0.057	282.9	65.97	6.32
2	31	28	0.07	3	0.028	0.051	0.065	278.7	98.96	4.15
3	31	28	0.07	4	0.036	0.058	0.078	360.8	131.95	4.03
4	31	28	0.17	2	0.053	0.059	0.069	125.8	160.22	2.81
5	31	28	0.17	3	0.043	0.061	0.092	165.2	240.33	2.460
6	31	28	0.17	4	0.045	0.071	0.097	163	320.44	1.820
7	31	57	0.07	2	0.032	0.042	0.066	353.2	131.95	7.890
8	31	57	0.07	3	0.032	0.055	0.072	325	197.92	4.840
9	31	57	0.07	4	0.041	0.062	0.082	357.2	263.9	3.990
10	31	57	0.17	2	0.047	0.055	0.08	156.2	320.44	3.490
11	31	57	0.17	3	0.039	0.055	0.0457	187.4	480.66	2.790
12	31	57	0.17	4	0.055	0.076	0.112	146.8	640.89	1.640
13	31	89	0.07	2	0.041	0.067	0.079	233.2	207.8	5.210
14	31	89	0.07	3	0.041	0.067	0.084	296.8	311.73	4.420
15	31	89	0.07	4	0.044	0.072	0.095	368	415.63	4.110
16	31	89	0.17	2	0.045	0.053	0.088	110.1	504.7	2.460
17	31	89	0.17	3	0.028	0.044	0.0437	165.2	757.05	2.460
18	31	89	0.17	4	0.06	0.084	0.132	135.2	1009.4	1.510
19	38	28	0.07	2	0.024	0.054	0.078	346.1	65.97	7.730
20	38	28	0.07	3	0.034	0.066	0.092	331.7	98.96	4.940
21	38	28	0.07	4	0.043	0.075	0.108	410.1	131.95	4.580
22	38	28	0.17	2	0.062	0.076	0.099	166.1	160.22	3.710
23	38	28	0.17	3	0.052	0.082	0.127	216.2	240.33	3.220
24	38	28	0.17	4	0.062	0.092	0.128	252.5	320.44	2.820
25	38	57	0.07	2	0.038	0.057	0.088	367.1	131.95	8.200
26	38	57	0.07	3	0.041	0.075	0.099	392.2	197.92	5.840
27	38	57	0.07	4	0.049	0.080	0.115	442.3	263.9	4.940
28	38	57	0.17	2	0.056	0.071	0.110	186.2	320.44	4.160
29	38	57	0.17	3	0.047	0.071	0.088	235.7	480.66	3.510
30	38	57	0.17	4	0.065	0.098	0.155	180.0	640.89	2.010
31	38	89	0.07	2	0.048	0.0864	0.104	291.4	207.8	6.510
32	38	89	0.07	3	0.049	0.083	0.116	347.2	311.73	5.170
33	38	89	0.07	4	0.053	0.091	0.128	445.9	415.63	4.980
34	38	89	0.17	2	0.054	0.072	0.118	137.9	504.7	3.080
35	38	89	0.17	3	0.034	0.057	0.093	208.2	757.05	3.100
36	38	89	0.17	4	0.069	0.108	0.182	165.6	1009.4	1.850

A fuzzy version of CLIPS software was used for formulating the expert system. Following is presentation of the triangular fuzzy sets of the variables (input and output) in CLIPS format:

H: **So** {*Soft*} [(27, 1) (31, 1) (38, 0)]; **Ha** {*Hard*} [(31, 0) (38, 1) (42, 1)]

V_c: **Sl** {*Slow*} [(21, 1) (28, 1) (57, 0)]; **Me** {*Medium*} [(28, 0) (57, 1) (89, 0)]; **Fa** {*Fast*} [(57, 0) (89, 1) (96, 1)]

f: **Sm** {*Small*} [(0.05, 1) (0.07, 1) (0.17, 0)]; **La** {Large} [(0.07, 0) (0.17, 1) (0.19, 1)]

w: **Na** {*Narrow*} [(1, 1) (2, 1) (3, 0)]; **Me** {*Medium*} [(2, 0) (3, 1) (4, 0)]; **Wi** {*Wide*} [(3, 0) (4, 1) (5, 1)]

MRR: **EVL** {*Extremely very low*} [(50, 1) (65, 1) (120, 0)]; **VL** {*Very low*} [(65, 0) (120, 1) (175, 0)]; **L** {*Low*} [(120, 0) (200, 1) (250, 0)]; **SL** {*Slightly low*} [(200, 0) (250, 1) (300, 0)]; **SH** {*Slightly high*} [(250, 0) (310, 1) (370, 0)]; **H** {*High*} [(310, 0) (410, 1) (500, 0)]; **H** {*High*} [(310, 0) (410, 1) (500, 0)]; **VH** {*Very high*} [(410, 0) (570, 1) (730, 0)]; **EVH** {*Extremely very high*} [(520, 0) (960, 1) (1025, 1)]

VB3: **EEL** {*Extremely extremely low*} [(0.035, 1) (0.05, 1) (0.065, 0)]; **EVL** {*Extremely very low*} [(0.05, 0) (0.064, 1) (0.078, 0)]; **VL** [(0.064, 0) (0.074, 1) (0.084, 0)]; **L** [(0.074, 0) (0.082, 1) (0.09, 0)]; **SL** [(0.085, 0) (0.09, 1) (0.095, 0)]; **SH** [(0.093, 0) (0.098, 1) (0.103, 0)]; **H** [(0.1, 0) (0.109, 1) (0.118, 0)]; **VH** [(0.11, 0) (0.122, 1) (0.134, 0)]; **EVH** [(0.125, 0) (0.143, 1) (0.161, 0)]; **EEH** {*Extremely extremely high*} [(0.145, 0) (0.165, 1) (0.185, 1)]

E: **EEL** [(70, 1) (100, 1) (140, 0)]; **EVL** [(100, 0) (130, 1) (160, 0)]; **VL** [(135, 0) (155, 1) (175, 0)]; **L** [(160, 0) (175, 1) (190, 0)]; **SL** [(175, 0) (200, 1) (225, 0)]; **SH** [(200, 0) (240, 1) (270, 0)]; **H** [(240, 0) (270, 1) (300, 0)]; **VH** [(270, 0) (310, 1) (350, 0)]; **EVH** [(310, 0) (355, 1) (400, 0)]; **EEH** [(360, 0) (425, 1) (450, 1)]

SE: **EEL** [(0.5, 1) (1, 1) (1.8, 0)]; **EVL** [(1, 0) (1.8, 1) (2.6, 0)]; **VL** [(1.8, 0) (2.4, 1) (3, 0)]; **L** [(2.4, 0) (3, 1) (3.6, 0)]; **SL** [(3, 0) (3.5, 1) (4, 0)]; **SH** [(3.5, 0) (4, 1) (4.5, 0)]; **H** [(4, 0) (4.4, 1) (4.8, 0)]; **VH** [(4.4, 0) (5, 1) (5.6, 0)]; **EVH** [(4.8, 0) (6.4, 1) (8, 0)]; **EEH** [(6.4, 0) (8.1, 1) (9.5, 1)]

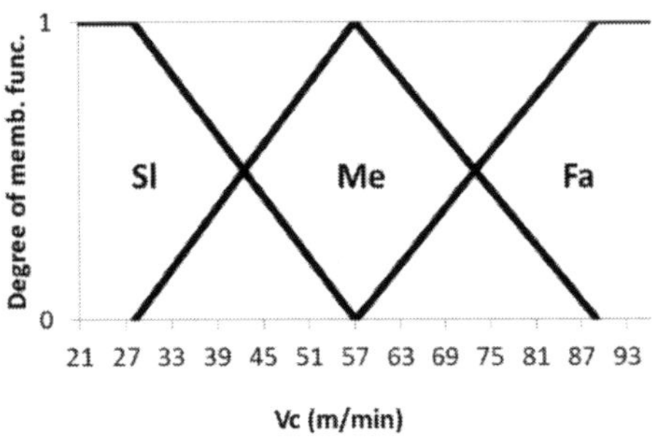

Figure 2. The triangular fuzzy set of V_c.

For graphical description, the fuzzy set of cutting speed is presented in Figure 2.

3.2 The optimization approach

The aim of the expert system is to suggest the ideal values of the unfixed predictors (input variables) according to a given combination of the objectives. The process kicks off with the user's input of the desired objective(s) and values of the input parameters. It is compulsory to select an objective or a combination of objectives, while the four input parameters may or may not be fixed according to the problem on hand. Any of the following eight combinations can be chosen as the objective:

1. Minimize *E*
2. Minimize *VB3*
3. Minimize *SE*
4. Min. *E* and max. *MRR*, simultaneously
5. Min. *E* and min. *VB3*, simultaneously
6. Min. *VB3* and max. *MRR*, simultaneously
7. Min. *VB3*, minimize *E*, and max. *MRR*, simultaneously
8. Min. *VB3* and min. *SE*, simultaneously.

Next comes fuzzification of the numeric values of the input parameters fixed by the user. Fuzzification is a process of converting format in which numeric data are changed into a lingual form, which is well understandable by human minds. Fuzzification of the experimental values is done according to the relevant fuzzy sets. The sequence of parameters optimization starts with fixing of *H* and follows with that of *f*, V_c, and *w*. The knowledge-base of the expert system is as follows:

1. IF ***H*** is unfixed THEN ***H*** is **So**.
2. IF ***f*** is unfixed and Objective is Minimize *VB3* THEN ***f*** is **Sm** ELSE ***f*** is **La**.
3. IF V_c is unfixed and Objective is Minimize *VB3* or "Minimize *VB3* and Minimize *E*" THEN V_c is **Sl** ELSE V_c is **Fa**.
4. IF ***w*** is unfixed and Objective is "Minimize *VB3* and Maximize *MRR*" or "Minimize *VB3* and Minimize *E* and Maximize *MRR*" THEN ***w*** is **Me&Wi**.
5. IF ***w*** is unfixed and Objective is Minimize *VB3* or Minimize *E* or "Minimize *VB3* and Minimize *E*" THEN ***w*** is **Na**.
6. IF ***w*** is unfixed and Objective is "Minimize *E* and Maximize *MRR*" THEN ***w*** is **Me**.

"&" is a union operator (used in rule 4), which produces an amalgamated fuzzy value of two successive membership functions of a set.

The aforementioned rules are applicable to the objectives in which energy consumed is dealt in its absolute form, i.e., *E*. Following is the list of the rules in which energy consumed is presented in its specific form, i.e., *SE*.

1. IF ***H*** is unfixed THEN ***H*** is **So**.
2. IF ***f*** is unfixed THEN ***f*** is **La**.
3. IF V_c is unfixed and Objective is Minimize *SE* THEN V_c is **Fa** ELSE V_c is **Sl**.
4. IF ***w*** is unfixed and Objective is "Minimize *SE* and Minimize *VB3*" THEN ***w*** is **Me** ELSE ***w*** is **Wi**.

The output of the optimization step is fuzzified values of the optimized input parameters in form of membership functions of the relevant fuzzy sets. In the last step, the fuzzy values of the optimized predictors are defuzzified to get numeric and distinct values. The process of defuzzification transforms a fuzzy value to a number that is most central to the fuzzy distribution.

4 Application examples

Two optimization examples, related to grooving of AISI 4340, are provided. Further experiments were performed at the optimized settings of the input parameters to find the values of the maximized/minimized output parameters.

In the first example, it is desired to find ideal values of cutting speed and width of cut for the following combinations of objectives: (a) minimize *SE* & minimize *VB3*, simultaneously and (b) minimize *SE* & maximize *MRR*, simultaneously. The values of material hardness and feed rate are fixed as 33HRc and 0.14mm/rev, respectively.

In the second example, it is desired to optimize cutting speed and feed rate for the following objectives: (a) minimize *VB3* and (b) minimize *SE* and maximize *MRR*, simultaneously. The values of material hardness and width of cut are fixed as 36HRc and 3.5mm, respectively. Tables 2 and 3 present the optimization results and the measured values of the responses against the optimized values of the predictors for example 1 and 2, respectively.

The results suggest that energy consumption can be minimized by operating the machining process at the maximum levels of the cutting parameters, especially feed rate and cutting speed. Such a combination would also guarantee high level of productivity.

Table 2. The optimization results and their comparisons with the experimental values for example 1

Parameter	(a) Min. *SE* & min *VB3*	(b) Min. *SE* & max MRR
V_c (m/min) [*optimized*]	28	89
w (mm) [*optimized*]	2.8	3
SE (J/mm^3) [*measured*]	3.29	2.96
*VB*3 (mm) [*measured*]	0.073	0.089
MRR (mm^3/s) [*calculated*]	184.7	623.5

Table 3. The optimization results and their comparisons with the experimental values for example 2

Parameter	(a) Min. *VB3*	(b) Min. *SE* & max *MRR*
V_c (m/min) [*optimized*]	28	89
f (mm/rev) [*optimized*]	0.07	0.17
SE (J/mm^3) [*measured*]	4.69	2.23
*VB*3 (mm) [*measured*]	0.101	0.14
MRR (mm^3/s) [*calculated*]	115.5	883

Conclusions

The article puts forward a fuzzy expert system used to upgrading a machining process for reduced energy consumption and tooling cost and enhanced productivity. The optimization objectives target the realistic needs of manufacturing industry and are in accordance with the sustainability principles.

Following pertinent conclusions can be drawn from the presented work:

1. Expert system, employing fuzzy reasoning, is a good tool to model highly complex manufacturing processes such as machining.
2. In order to slash energy demand of a cutting process, it is important to operate the process at high cutting speed and feed rate. Increase in width of cut is also helpful for this cause but with comparatively lower influence. This approach will also cause increase in productivity. Oppositely, high levels of the three cutting parameters cause rapid tool damage and increase in tooling cost.

References

1. M.F. Rajemi, P.T. Mativenga, A. Aramcharoen, J. Clean. Prod. **18**, 1059-1065 (2010)
2. N. Diaz, E. Redelsheimer, D. Dornfeld, In: *Globalized Solutions for Sustainability in Manufacturing*. Springer Berlin Heidelberg. 263-267 (2011)
3. Q. Wang, F. Liu, X. Wang, Int. J. Adv. Manuf. Technol. **71**, 1133-1142 (2014)
4. K.A. Al-Ghamdi, A. Iqbal, J. Clean. Prod. **In press**, doi:10.1016/j.jclepro.2015.05.132 (2015)
5. M. Arif, I.A.Stroud, O. Akten, Proc. Inst. Mech. Eng. (B): J. Eng. Manuf., **229**, 266-274 (2014)
6. V.A. Balogun, P.T. Mativenga, J. Clean. Prod., **41**, 179-186 (2013)
7. M.H.F. Al-Hazza, E.Y.T. Adesta, A.M. Ali, D. Agsman, M.Y. Suprianto, J. Appl. Sci., **11**, 2578-2584 (2011)
8. K. Hashmi, I.D. Graham, B. Mills, J. Mat. Proc. Technol., **135**, 44-58 (2003)
9. C.C. Tsao, H. Hocheng, J. Mat. Proc. Technol, **203**, 342-348 (2008)
10. D.H. Ojha, U.S. Dixit, J.P. Davim, Int. J. Mat. & Prod. Technol., **35**, 145-166 (2009)
11. A. Iqbal, N.U. Dar, N. He, M.M.I. Hammouda, L. Li. J. Intel. Manuf. **21**, 761-776 (2010)

MATEC Web of Conferences **45**, 04004 (2016)
DOI: 10.1051/matecconf/20164504004

Prototype Design and Experimental Test of a Rotorcraft Capable of Adhering to and Moving on the Ceiling

Masaya Yasunaga , Jae Hoon Lee and Shingo Okamoto

Graduate School of Science and Engineering, Ehime University, Japan

Abstract. This paper presents the prototype design and evaluation of a novel rotorcraft that can adhere to and move on the ceiling. It has four propellers for flying like conventional rotorcraft. In addition, a specialized guard frame and two active wheels are attached on its body frame for adhering to and moving on the ceiling. The propellers create negative pressure in the guard frame that acts as a suction cup for generating adhesion force. The active wheels are used for moving on the ceiling face. Experiments are carried out to show the feasibility of flying in the air, adhering to, and moving on the ceiling.

1 Introduction

There are about 700,000 bridges in Japan. In 2010, 50 years had elapsed since about 8% of them had been constructed. Moreover, it is expected that about 26 % and 53%, in 2020 and in 2030 respectively, will be considered deteriorated bridges with ages of more than 50 years [1]. Likewise, rapid deterioration of physical infrastructure, including bridges, is recognized as a big problem in Japan. Old infrastructures requires a huge maintenance cost. The safety of the worker and the worker's accessibility to the inspection area are the main difficulties in inspection tasks. Therefore, robotics technology has been attracting attention from researchers because it is recognized as a solution for these problems. Particularly, several types of multi-rotor crafts have been developed for effective inspection processes because they show high performance in mobility to access target positions [2-4]. Recently, experimental tests with unmanned aerial vehicles (UAVs) for bridge inspection have been carried out field situations [5].

The most important capability of a multi-rotor craft for inspection applications is stable hovering. Namely, the rotorcraft should have the capability of stable flight in a high place near the bridge for inspection tasks such as taking pictures. However, the fatal drawback while using a multi-rotor craft is the fact that it is sensitive to wind. Generally, it shows good performance in navigating to the goal position through free space because of excellent motion control technology with GPS sensor system. However, in high places near the structure, sometimes there are complex and strong winds, so the rotorcraft cannot keep its posture and may collide with the structure. In addition, the fundamental limitation of a UAV is that, it cannot access the ceiling surface of bridge structure or a building. For this reason, robots having the capability of adhering to the ceiling or wall has got an attention from the robotics research society [6-10]. So far, their fundamental limitation is that they can move only on the face of the wall, and cannot fly in the air. Thus, it is impossible to move directly to the task position.

To cope with the above-mentioned problems, a novel rotorcraft is presented in this research. In addition to the basic flying functions of a conventional rotorcraft, a special device to provide the capabilities for adhering to and moving on the ceiling was include in the proposed system. For this purpose, a specialized frame with active wheels was designed. Additionally, the prototype was constructed and experimental tests were carried out.

2 Method for adhering to and moving on the ceiling

2.1 Adhering to the ceiling

For a rotorcraft's adhesion function on the ceiling, several methods, such as magnetics or suction cups could be applied. However, most rotorcraft cannot carry additional heavy devices because their loading capacity is so small.

The innovative idea of this paper is to design a specialized guard frame for utilizing propellers in both the flying and adhering objectives as shown in Fig. 1. Using lightweight material for the guard is very effective in reducing the weight of the device, while providing additional functionality. Propellers are used for flying in the air in the same way as other rotorcraft. At the same time, by installing a special guard frame around the propeller, an enclosed space is formed between the ceiling surface and the propeller while adhering. Moreover, the propeller expels the air in the enclosed space, creating a negative pressure in the space. Resultantly, the propeller guard helps form a vacuum.

2.2 Moving on the ceiling

While adhering on the ceiling, the propellers cannot generate moving force in the horizontal direction because the lift forces of the propellers act vertically. In order to move the rotorcraft, two active wheels that can be controlled independently were installed on the platform. By changing the velocities of both wheels, translational and rotational motions of the system can be achieved. A set of more than three omni-directional wheels could be used for generating moving force. In that case, the rotorcraft could move on the ceiling surface omni-directionally; i.e., both translational and rotational motions could be achieved independently. However, it would require more than three wheels and motors, with their corresponding extra weight. Therefore, two active wheels were employed in this research.

The forces acting on the rotorcraft while it moves on the ceiling are depicted in Fig. 2. The rotation of the propeller generates the lift force, F_{thrust}, which compensates for the rotorcraft's weight, $F_{gravity}$, and lifts up the system during both flying and attaching. The suction force, F_{vacuum}, makes the system attach to the ceiling face, and becomes zero during flying motion. By rotating the wheels while attaching, the moving force $F_{traction}$ is generated. When it is larger than the friction force between the ceiling face and the guard, $F_{friction}$, the system can move.

3 Development of rotorcraft prototype

In order to examine the feasibility of the proposed design, a four-rotor craft prototype has been constructed.

3.1 Platform design

The structure of a well-known quad-copter with four propellers was employed as the basic platform design. A ready-made carbon frame of IRON MAN 650 made by TAROT Co. was used for the system. Motors and propellers were selected that can provide sufficient payload capacity for additional devices including the special guard and inspection instrument. Two servomotors were used to move the system by rotating the wheels on the ceiling. A 6-cell lithium-polymer battery was embedded in the platform as the power source. The total mass of the aircraft prototype, including the battery is about 3.3 kg. The four-rotor craft prototype is displayed in Fig. 3.

3.2 Special guard frame for generating suction force

The proposed special guard was designed to enable the functions of adhering to and moving on the ceiling by creating a vacuum. It was made with lightweight Styrofoam and attached to the platform. The radical gap between each propeller edge and its guard was made as small as possible to maximize the effects of negative pressure inside the guard itself. In the prototype, this gap was set to about 1 cm. Figure 3 shows the guard, made of white Styrofoam board, constructed for this study. Additionally, a flexible suction plate, with a semi-circular cross section, was installed on the upper surface of the guard to increase sealing performance and reduce the friction with the ceiling surface. It was made with PET plate in this research. Moreover, it helps maintain the suction force on uneven surfaces. The prototype is 940 mm wide, 940 mm of long, and 300 mm of high.

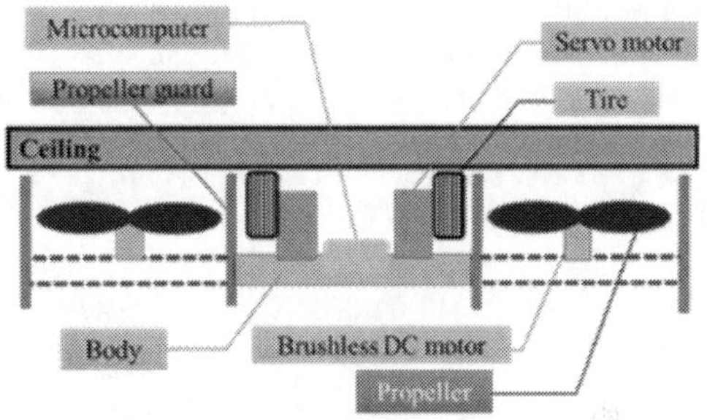

Figure 1. Schematic diagram of the rotorcraft prototype

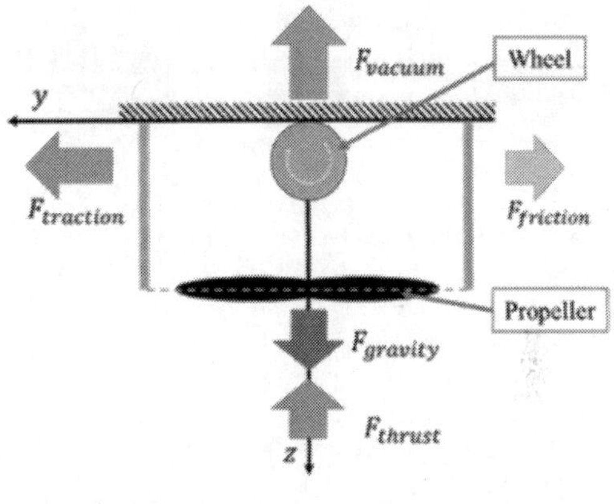

Figure 2. Schematic diagram of the forces acting on the rotorcraft while moving on the ceiling

Figure 3. The rotorcraft prototype developed in this research

3.3 Controller configuration in the developed prototype

Figure 4 shows the schematic diagram of the controller configuration in the developed four-rotor craft. It consists of an embedded controller with a receiver, four propeller-driver sets, two servo motors for active wheels, a battery, a computer for data monitoring, and a transmitter for user interface. The embedded controller is connected to the computer and the transmitter with Bluetooth wireless communications. Operating commands are sent from the transmitter to the controller, and include reference velocities for rotational and vertical motion. Then, the controller translates them into rotational velocities for each propeller, and sends them to each motor driver. In addition, basic attitude control was included in the embedded controller; thus, it is possible to stably hover even without input from the transmitter. The computer is used to record information such as posture angles and operating commands.

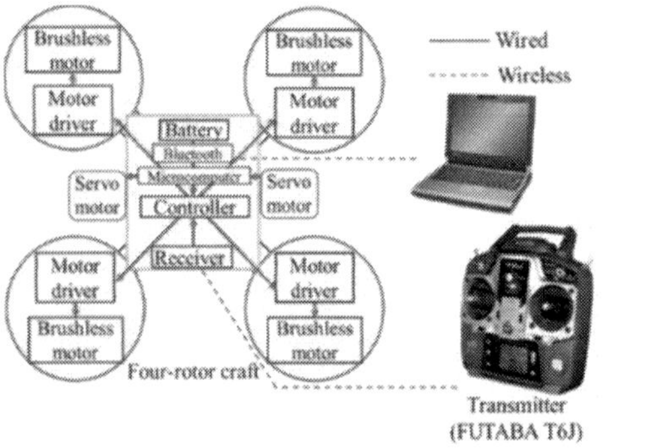

Figure 4. Configuration of the control system of the developed rotorcraft

4 Experimental tests

Experimental studies were carried out to verify the characteristics of the rotorcraft prototype developed in this research.

4.1 Method for experiments

The basic function, flying in the air, and the additional functions, adhering to and moving on the ceiling, were demonstrated with the developed rotorcraft prototype. All motion was generated by manual operation of the transmitter. The experimental motion scenario consists of taking off, hovering in the air, moving to the ceiling, moving on the ceiling, and landing. The experiments were conducted in an indoor environment. The ceiling surface has a decorative of board whose irregularity of about 1 cm in height.

Two designs, with the propeller inside of or below the guard, were constructed, in order to check their suction performance. In the A-type prototype, each propeller is installed inside of the suction plate as shown in Fig. 5.

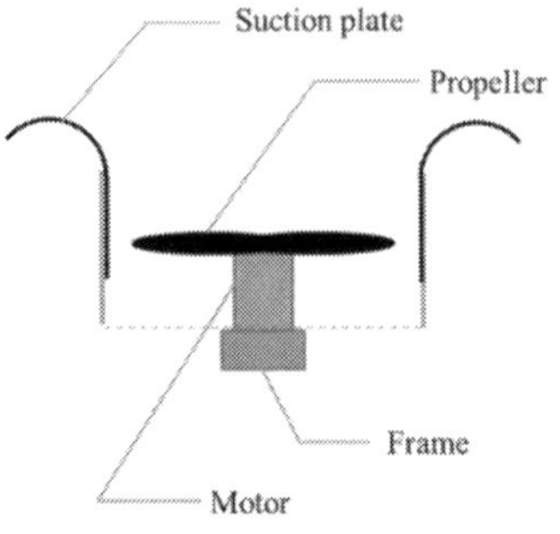

(a) Schematic diagram of A-type prototype

(b) Side view of A-type prototype

Figure 5. Prototype rotorcraft of A-type: Propellers are installed inside of suction plate

On the other hand, the propellers are installed below the suction plate in the B-type as shown in Fig. 6. Both prototypes were subjected to the same experimental scenario and results were compared.

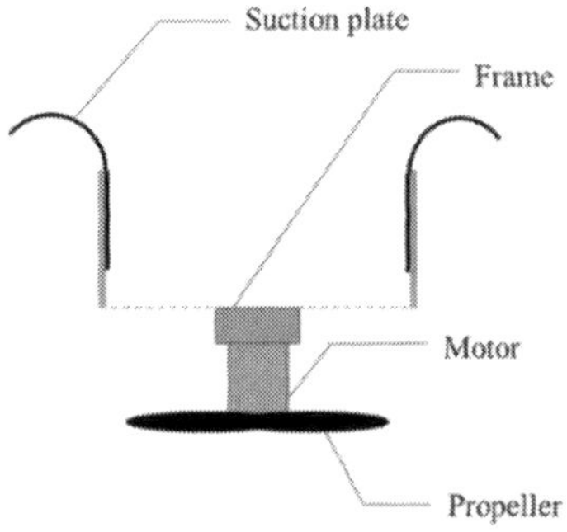

(a) Schematic diagram of B-type prototype

(b) Side view of B-type prototype

Figure 6. Prototype rotorcraft of B-type: Propellers are attached below the suction plate

Figure 7. Experimental result by using the A-type rotorcraft

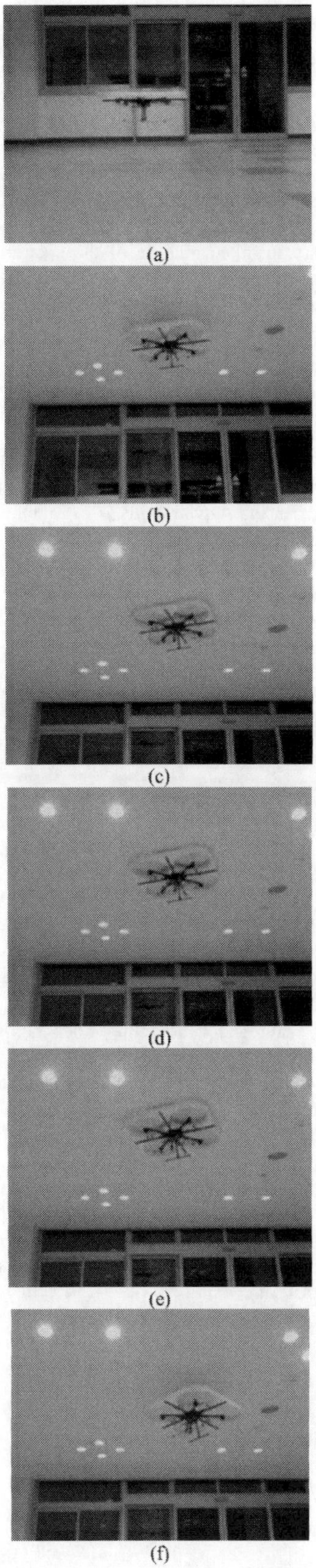

Figure 8. Experimental result by using the B-type rotorcraft

4.2 Experimental results

Figures 7 and 8 show the scenes of experiments with A-type and B-type prototypes. It was possible to fly in the air for both cases. In addition, both prototypes could move on the ceiling while adhering.

It was observed that the performance of each motion type was different four each prototype. The B-type has better flying capability than the A-type because its propellers are installed outside of the suction plate and can generate a lager lift force. Conversely, for adhering motion, the A-type rotorcraft shows better performance than the B-type. The required power for both kinds of motion, i.e., hovering and adhering to the ceiling, was measured experimentally in this research. The differences between the prototype are outlined in Table 1. For a hovering motion, the required motor power for the B-type body is 7% smaller than for the A-type. In contrast to this, when the prototype is adhering to the ceiling, the required motor power for the A-type body is 11% smaller than for the B-type.

Table 1. Motor power required for each motion with two type of rotorcraft prototype

Type of rotor craft	Motion of the rotorcraft	
	Hovering	Adhering to the ceiling
A-Type	47%	22%
B-Type	40%	33%

5 Conclusion and future works

A novel rotorcraft design for inspecting high-rise infrastructure was developed in this study. Two prototypes were proposed and developed. The creative idea to utilize the propeller for both flying and attaching on the ceiling was realized by manufacturing a special guard. Through experiments with the developed system, it was verified that the proposed design is very effective and energy saving for attaching and moving on the ceiling surface.

In this study, we carried out experiments where a four-rotor craft adhered not only to a smooth surface but also a surface with irregularities in width and height of about 1 cm. However, the bottom surface of an old bridge is likely to be a more complex surface. Therefore, future challenges include coping with irregular surfaces, and designing the suction plate to be more flexible to ensure the suction force. Currently, research on autonomous flight control and outdoor field inspection trials are ongoing.

Acknowledgements

This research was supported by Okumuragumi Research Fund.

References

1. Ministry of Land, Infrastructure, Transport and Tourism, The maintenance of national road network in Japan, (2009) (http://www.mlit.go.jp/road/road_e/03key_challenges/1-2-2.pdf)
2. P. Liu, A.Y. Chen, Y.-N. Huang, J.-Y. Han, J.-S. Lai, S.-C. Kang, T.-H. Wu, M.-C. Wen, and M.-H. Tsai, A Review of Rotorcraft Unmanned Aerial Vehicle Developments and Applications in Civil Engineering, Smart Structures and Systems, **13**, 6 (2014)
3. L. Wallace, A. Lucieer, C. Watson, and D. Turner, Development of a UAV-LiDAR System with Application to Forest Inventory, Remote Sensing, **4**, 6(2012)
4. C. Deng, S. Wang, Z. Huang, Z. Tan, and J. Liu, Unmanned Aerial Vehicles for Power Line Inspection: A cooperative Way in Platforms and Communications, Journal of Communications, **9**, 9 (2014).
5. Barritte Lovelace, Principal Investigator Collins Engineers Inc., Unmanned Aerial Vehicle Bridge Inspection Demonstration Project, Research Project Final Report (2015).
6. I.-G. Koo, T.-D. Trong, Y.-H. Lee, H. Moon, J. Koo, S.-K. Park, H.-R. Choi, Development of Wall Climbing Robot System by Using Impeller Type Adhesion Mechanism, journal of intelligent and Robotic System, 57-72, (2013).
7. Mahmoud Tavakoli, C. Viegas, L. Marques, J. N. Pires, A. T. de Almedia, OmniClimbers: Omni-directional magnetic wheeled climbing robots for inspection of ferromagnetic structures, Robotics ad Autonomous System, **61**,997-1007,(2013).
8. D. Schmidt, C. Hillenbrand and K. Berns, Omnidirectional locomotion and traction control of the wheel-driven, wallclimbing robot, Robotica, **29**,991-1003,(2011).
9. Kiyoshi Tsuru, S. Hirose, Development of Vmax Ⅲ: Magnetic Wall Climbing Robot with Holonomic and Omni-directional Mobility, Journal of the Robotics Society of Japan ,**30**, 639-647,(2012).
10. J.C. Grieco, M. Prieto, M. Armada, P. Gonzales de Santos, A Six-Legged Climbing Robot for High Payloads, Proceeding of the 1998 IEEE International Conference on Control Applications, **1**, 446-450,(1998).

MATEC Web of Conferences **45**, 04005 (2016)
DOI: 10.1051/matecconf/20164504005

Research on Ontology Modeling of Steel Manufacturing Process Based on Big Data Analysis

Qing Bao, Jian Wang and Jin Cheng

CIMS research center, Tongji University, 201804 Shanghai, China

Abstract. As an important method that steel industries ride the Indutrie 4.0 wave, knowledge management is expected to be versatile, effective and intelligent. Mechanism modeling difficulties, numerous influencing factors and complex industrial chains hinder the development of knowledge and information integration. Using data potentials, big data analysis can be an effective way to deal with knowledge acquisition as it solves the inaccuracy and imperfection mechanism modeling may lead to. This paper proposes a big data knowledge management system(BDAKMS) adhering to data driven, intelligent analysis, service publication, dynamic update principle which can effectively extracts knowledge from mass data. Then, ontology modeling gives the knowledge unified descriptions as well as inference details combined with semantic web techniques.

1 Introduction

Indutrie 4.0 was put forward in Hannover Messe and brings prospects to hasten the strategic transformation of steel industries as well. One of the aspects that Industrie 4.0 propels manufacturing revolution is comprehensive, unified and domain-oriented knowledge collection to instruct business operation. Ref.[1] proposes fast matrix-based approaches to knowledge acquisition in covering information systems and reduce the computational time compared with classical rough set, especially large-scale data sets. However, this method is limited to static information and discretization is not universal. Ref.[2] builds a generic EUEDE contextual architecture for knowledge acquisition, compatible with current DSS. Ontology is used for knowledge formalization, so that the relationship between the knowledge items contribute to the rules for decision making, but EUEDE excludes external factors which are important to decision knowledge extraction. Ref.[3] combines knowledge graph with case analysis to minimize the gap among distinct levels of expertise in access to knowledge. More than the difference or diversity as it has addressed, the inconsistency should not be neglected in multi-source tacit knowledge.

Ontology makes it possible to knowledge interchange and reusability which is essential to knowledge modeling. Manufacturing taxonomy helps ontology modeling and results in ontology based knowledge hierarchy[4]. Top-down development approach can satisfy diversified levels of ontology granularity. Lifecycle[5] management is proposed and brings an impetus to ontology based knowledge update. Ontology has not only been applied in the manufacturing process[6] but also product design and cost budgeting[7]. Ontology is the core of semantic web. While ontology enriches semantic description, knowledge is expressed more comprehensively and integrated from distributed systems. Evolved from RDF Schema, ontology web language(OWL) has a richer vocabulary to define relations for inference[8] and is developed to describe ontologies.

Big data are produced in every segment in manufacturing: For enterprise performance, big data help identify the product demands, productivity and performance via diverse business objectives. For production, big data make it possible to detect the right facility causing the product flaw. Currently, steel making industries have applied big data analysis to quality management, ore blending solution optimization, supply chain management, equipment fault diagnosis and financial, etc. For example, BAO Steel finds better ore blending schemes with analysis of historical data. Anyang steel industry employ big data analysis to find out the key factors that cause the steel cracks[9].

Focusing on the difficulties in knowledge acquisition of steel industries, this paper presents a big data analysis based knowledge management structure (BDAKMS). Ontology is used to model the domain knowledge and finally enhance the usability and interoperability. This paper also gives an ontology modeling case based on big data analytical results and achieves accurate inference through SWRL rules and semantic web technologies. This paper focuses on discussions about ontology modeling methodology and semantic deduction. A converter steel making case is given to explain and verify the techniques' validity.

Article available at http://www.matec-conferences.org or http://dx.doi.org/10.1051/matecconf/20164504005

2 BDAKMS

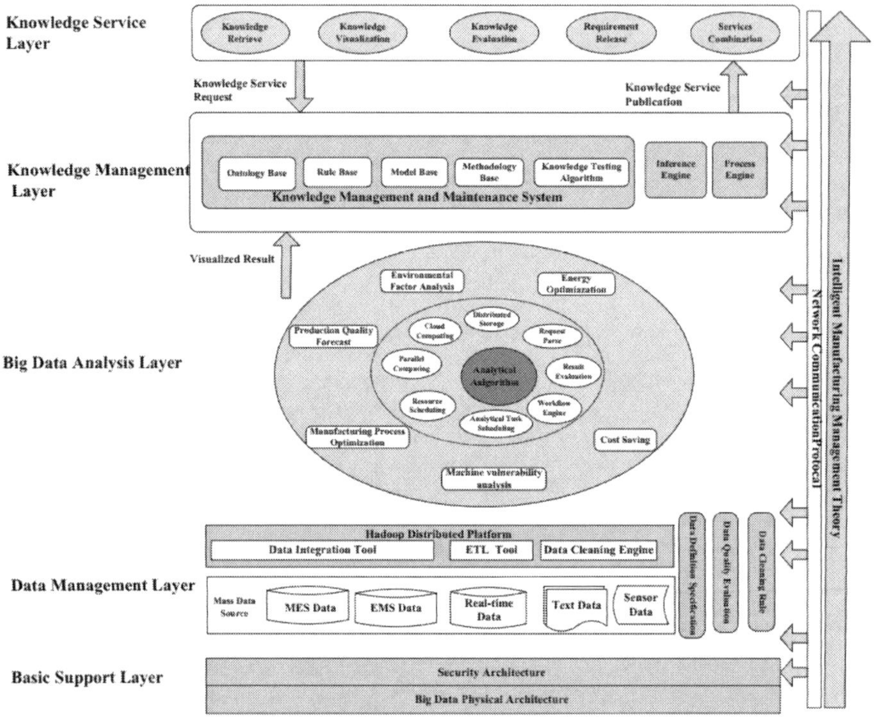

Figure 1. Construction of BDAKMS

2.1 Construction of BDAKMS

BDAKMS adopts data driven, intelligent analysis, service publication, dynamic update pattern to transform data to knowledge: Starting with primary data, knowledge is obtained by intelligent analysis methods. Knowledge is accessed with calling service encapsulation by users, accomplishing the interchange between the knowledge level and data level. The construction of BDAKMS is shown in Fig 1. This structure consists of five layers:

Basic Support Layer. The physical platform adopts Hadoop distributed environment with master slaver structure. NameNode is set to task scheduling while DataNode task execution. Traceability is introduced to security architecture to record the data transmission histories and guarantee the data credibility and privacy.

Data Management Layer. Service data object (SDO) produces the data graph according to the characteristics of source data, which is used in MapReduce for datatransformation and analysis. The ETL tool splits the data and balances the load to integrate preprocessed data into distributed data storage environment. Data access service (DAS) APIs provide data access. DAS is made up of meta data ETL, data registry and data reading. In fact, the data that DAS reads should match the mapping mode which is a set of mapping rules between meta data and data to be loaded. In fact, data information structure, data mapping relations and customized filter are designed in HBase for data unification. For oxygen gas consumption in EMS, the information storage structure is shown in Table1. Data source IP: port is saved in Row key while column description in Column family.

Table 1. HBase storage structure example

Table Name	EMS_Converter_Oxgen	
Row Key	EMS-TP562039:1521	
ColumnFamily	Column Description	
Column Description	Table Name	EMS_
	Data Property	Consumption
	Object Property	_isLinkedFacility
	Mapping Rule	Rule2344
	Unit	IU

2.1.1 Big Data Analysis Layer.

This layer contributes to relationship discovery among features and prediction observation through data mining. The core of big data analysis is to select the appropriate analytical algorithm through programmed Map and Reduce functions. With support tools and middleware like distributed storage, cloud computing, Linux cluster, analytical task scheduling, resource dispatch and data fusion, the goal of this layer is to achieve environmental factor analysis, product quality forecast, manufacturing process optimization, machine vulnerability, cost saving and energy optimization, etc.

2.1.2 Knowledge Management Layer.

Ontology base, rule base, methodology base and model base express formalized knowledge resulted from big data analysis. Knowledge testing algorithm will verify the consistency and integrity of the knowledge system. The knowledge is managed in lifecycle method: The knowledge system only incorporates the valued knowledge with ontology description language. The new knowledge entities combined with the existing have to be inspected by knowledge testing. Knowledge evaluation module assesses the knowledge items periodically to eliminate the outdated or inaccurate knowledge and make a request to big data analysis for update if possible. Moreover, a completed knowledge management system also needs inference engine and workflow engine to circulate the knowledge resources.

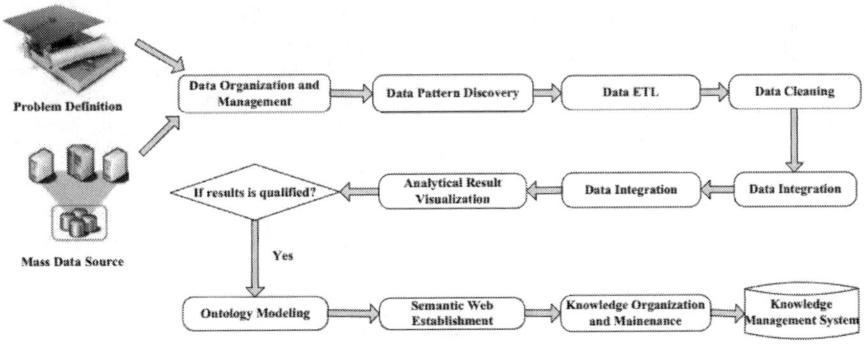

Figure 2. Knowledge acquisition flow based on big data analysis

2.1.3 Knowledge Service Layer.

In addition that this layer has knowledge retrieval and inference modules like traditional domain knowledge base, it enables new knowledge publication interface, service composition according to certain business needs and user analysis requests that act as big data analysis objectives.

2.2 Knowledge Acquisition Based on Big Data Analysis

Based on big data analysis, knowledge acquisition is the prerequisite for BDAKMS. As knowledge is organized in lifecycle management which indicates that the process from data to knowledge is iterative and repetitive. Dynamic external and internal environment of enterprises also contribute to the iteration. In general, big data analysis layer keeps on extracting data from MES, ERP, SCADA, sensors and real-time database to do massive analytical works. The knowledge evaluation module assesses the knowledge quality of both newfound and existing knowledge items, and replaces the outdated information with the most appropriate. The knowledge acquisition flow is shown in Fig 2 and it is summarized as three phases.

2.2.1 Phase1 Data preparation.

Mass data source can be divided into SQL data and NoSQL data(log, web, video and sensor data, etc.) that have been stored in distributed memory system or cloud storage infrastructure. The storage system is equipped with fault tolerance, load balancing and cluster mechanism, and it is designed to provide the multi-source and heterogeneous data with uniform environment. Data ETL (Extraction, Transform, Load) extracts data across diverse platforms in preparation for data integration. For SQL data, ETL tool accesses data easily in SQL query. For NoSQL, ETL is implemented in Hadoop for data search. Furthermore, data cleaning improves the data consistency, integrity, usability and accuracy. For instance, tuple neglect or default fill deals with data missing. Clustering or regression method handles noisy data[10]. To provide the heterogeneous data access interfaces, data integration unifies the specification of data definition and constraint.

2.2.2 Phase2 Data analysis and result evaluation.

Data analysis methods are not confined to traditional strategies like reports, statistics, and combination models. Instead, advanced analysis methods like prediction models, data mining, text analysis, page rank, collaborative filtering have been a research hotspot and widely used. Data mining is used to find out predicative, practical and novel results[11], while relation schema and cloud tags visualize the results. However, the analytical results should also match the business process and get merged in knowledge base if they pass the assessment.

2.2.3 Phase3 Knowledge Management.

Qualified knowledge is expressed by concepts, relations, properties, axioms, rules and instance[12] of ontology. On the basis of ontology, semantic web not only makes knowledge service-oriented but also offers an excellent management system which improves knowledge inference efficiency and eliminates Once updated. Above all, the goal of knowledge management is to form the KMS and publish knowledge service.

3 Ontology modeling and application

Results from big data analysis in BDA KMS should be formalized by ontology and enriched with semantic web. A knowledge management system is finally developed to render knowledge services.

3.1 Ontology Modeling Example

A case of converter steel making process is presented in this paper. Ontology classes are classified into seven groups: environmental factors, product, power, facility, ingredient, energy and process schedule as Fig 3 shows. isA represents a relationship between class and instance. isPartOf represents a parent-child relationship among classes. For example, limestone is one instance of slag making material, while slag-making material is a subclass of non-metal materials. Lime quality is related to humidity, which is an instance of environmental factor.
The ingredients of limestone are CaO, MgO and SiO2, etc. K-means cluster analysis was applied in 500 crude ore blending samples to distinguish the better plans form the invalid. One of the better plans is expressed in

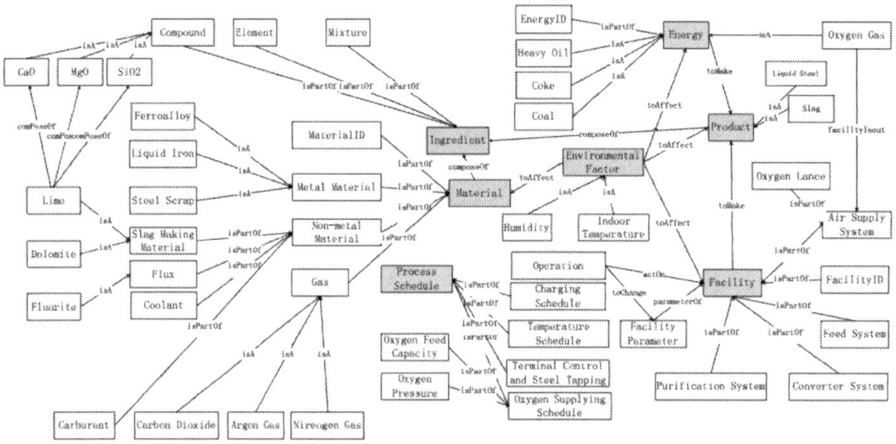

Figure 3. Ontology structure of converter steel making process

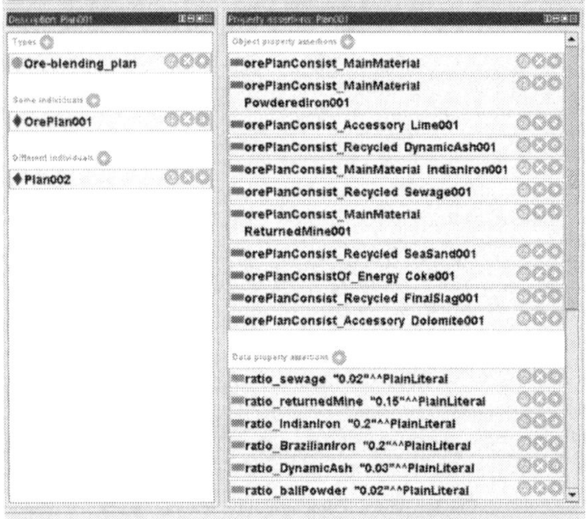

Figure 4. Analysis result formalization in Protégé

Protégé as Fig 4 shows.

3.2 Knowledge Inference

SWRL describes the correlations of distributed and heterogeneous information so that the interoperability is facilitated[13]. Principle of SWRL inference is shown in Fig 5 SWRL should meet various levels of ontologygranularity (top-level ontology, domain ontology, task ontology and application ontology), and it describes rules matching these ontologies.

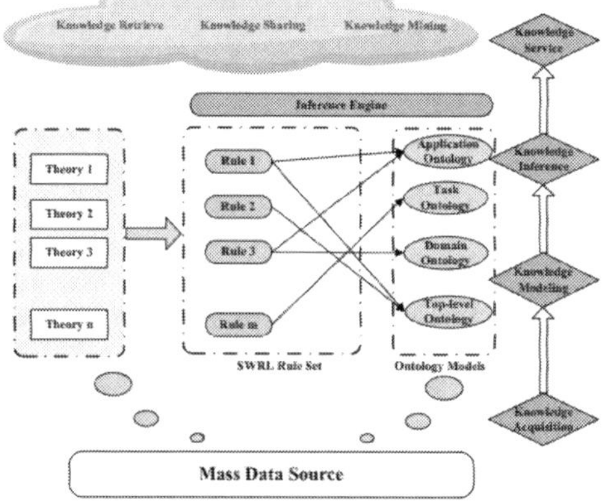

Figure 5. SWRL rule inference frame

Table 2. Parameter and its description

Parameter	Paramete description	Value
IW_{Si}	Si content of iron water	0.85%
IW_{p}	Pi content of iron water	0.2%
LM_{CaO}	CaO content of lime	89%

LM_{SiO2}	SiO2 content of lime	1.2%
LM_{MgO}	MgO content of lime	3.0%
DM_{CaO}	CaO content of dolomite	32%
DM_{MgO}	MgO content of dolomite	21%
DM_{SiO2}	SiO2 content of dolomite	1.3%
SG_{MgO}	MgO content of final slag	6%
μ	Basicity of final slag	3.5
LMneed	Demand of lime	To solve

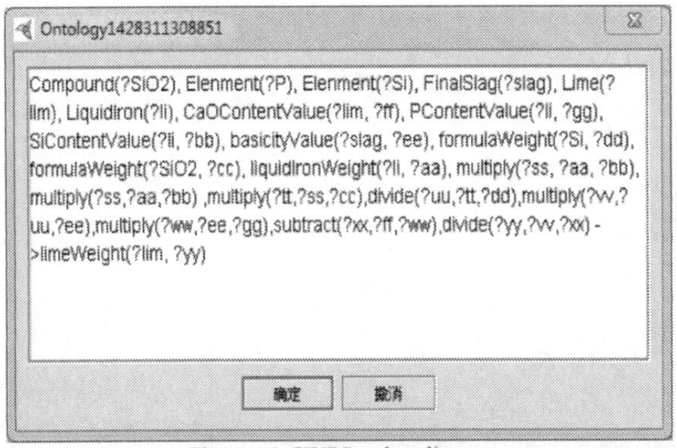

Figure 6. SWRL rule editor

Then, the inference engine(Jess, Pellet, Racer or Jena, etc.) combines
ontologies with rules, reasoning out related knowledge. This paper takes a converter steel-making as an example to illustrate the SWRL rules establishment. The formulation is:

$$LMneed = \frac{IWweight \times IW_{Si} \times M_{SiO_2} / M_{Si} \times \mu}{LM_{CaO} - \mu \times IW_P} \quad (1)$$

Parameters specification in this formula is explained in Table 2. Fig 6 shows the edited rule and the atoms in the edited SWRL atoms are explained as Table 3 shows. Similarly, lime addition, dolomite demand and dolomite addition can be calculated in other formulas. Pellet, as inference, is adopted and initializes the following instance shown in Table 3. This paper also develops a prototype knowledge base system in B/S pattern to the inference interface and the results of this case are shown in Fig 7.

Table 3. SWRL atom description

Atom	Description
Compound(?Si)	?Si is one of compound instance
Element(?P)	?P is one of element instance
LiquidIron(?li)	?li is one of liquid iron instance
Lime(?lim)	?lim is one of lime instance
FinalSlag(?slag)	?slag is one of slag instance
Metalnonmetal(?li,?lim)	?lim mixes with ?lim
liquidIronWeight(?li,?aa)	The weight of ?li is ?aa
SiContentValue(?li,?bb)	The Si content of ?li is ?bb
formulaWeight(?Si,?dd)	The formula weight of ?Si? is ?dd
limeWeight(?lim,?yy)	The quantity demand of ?lim is ?yy

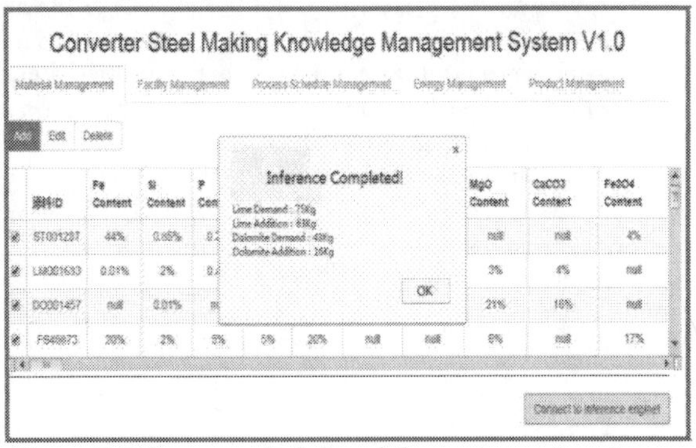

Figure 7. Inference results in knowledge management system

4 Summary

This paper proposes a big data analysis adopting the data driven, intelligent analysis, service publication, dynamic update patterns to transform data to knowledge: integrative service. This paper focuses on ontology modeling in BDAKMS and SWRL rule establishment. Finally, this paper gives a converter steel making case for validity verification. However, before this structure is utilized, three aspects are in need of improvements: 1. More accurate and effective analytical algorithms should be developed catering for steel industries' changing needs. 2. Big data analysis and should be integrated in the whole BDAKMS, but how to translate the user analysis to big data analysis objective is an issue to be considered.

References

1. Anhui Tan, Jinjin Li, Guoping Lin, et al. Fast approach to knowledge acquisition in covering information systems using matrix operations, Knowledge-Based Systems. **79**, 90-98 (2015)
2. Shah Jahan Miah, Don Kerr, John Gammack, et al. A generic design environment for the rural industry knowledge acquisition. Knowledge-Based Systems. **21**, 892-899 (2008)
3. Mara Abel, Luis A. Lima Silva, John A. Campbell, et al. Knowledge acquisition and interpretation problem-solving methods for visual expertise: S study of petroleum-reservoir evaluation, Journal of Petroleum Science and Engineering. **47**, 51-69 (2015)
4. A. Giovannini, A. Aubry, H. Panetto, et al. Ontology-based system for supporting manufacturing sustainability. Annual Reviews in Control. **36**, 309-317 (2012)

5. Qiu Ming, Chen Gang, Dong Jinxiang. Ontology based adaptive product lifecycle management system.Computer Intergrated Manufacturing Systems. **11**, 826- (2005)
6. S.Lemaignan, A.Siadat, J. Y Dantan, et al. MASON: a proposal for an ontology of manufacturing domain. IEEE Workshop on Distributed Intelligent Systems: Collective Intelligence and Its Applications. 195-200 (2006)
7. H.Panetto, M.Dassisti, A.Tursi. ONTO-PDM: product-driven ontology for product data management interoperability within manufacturing process environment. Advanced Engineering Informatics. **2**, 334-348 (2012)
8. Grit Denker, Lalana Kagal,Tim Finin. Security in the semantic web using OWL.Information Security Technical Report. **1**, 51-54(2005)
9. Feng Weizhi,Cen Honghong.The realization of grade II computer controlling software in making steel, Metallurgical Collections. **4**, 35-36 (2004)
10. J.Fan, H.Liu, Statistical analysis of big data and opportunities with big data. Proceeding of the VLDB Endowment. **12**, 2032-2033 (2012)
11. Muhammad Usman.Rusesel Pears, A.C.M.Fong. A data mining approach to knowledge discovery from multidimensional cube structures. Knowledge Based Systems. **40** ,39-41 (2013)
12. Berners L T,Hendler J,Lassila O. The Semantic Web. Scientific American. **284**, 34-43. (2001)
13. Lezcano, Leonardo, Sicilia, Miguel Angel, et al. Integrating reasoning and clinical archetypes using OWL ontologies and SWRL rules. Journal of Biomedical Informatics. **2**, 343-350 (2010)

MATEC Web of Conferences **45**, 04006 (2016)
DOI: 10.1051/matecconf/20164504006

Preliminary Study on SED Distribution of Tactile Sensation in Fingertip

Huiling Chen[1], Liguo Shuai[1], Weihang Zhu[2]

[1]School of Mechanical Engineering, Southeast University, Nanjing 211189, China
[2]Department of Industrial Engineering, Lamar University, Beaumont, TX 77710, USA

Abstract. For investigating the effects of stimuli on the deformations within the soft tissues of fingertips and the dependence of the tactile sensation on the deformations under pressure, in this paper, a finite element model is developed, to simulate the procedure of a fingertip touching a sharp wedge. Characteristics of the strain energy density (SED) distribution within the soft issues are analyzed. Simulation results show that the soft tissues of fingertips are very sensitive to stimuli, and the spatial distribution characteristics of strain energy density within soft tissues can best explain the evoked charging rate of mechanoreceptors.

1 Introduction

Human being skin has the ability of sensing physical stimulation such as pressure, vibration, temperature etc. When the fingertip touches an object, the tactile receptors under the skin will respond to the deformation of the skin and transmit electrical impulses to the brain through the peripheral nerve system. The main features of the impulses, and the frequencies of the nerve impulses, are associated with the mechanical environment of the tactile receptors, e.g. stress and strain within the skin, the strain energy density around the receptors. The study on the biomechanics of the fingertip perception is significant to understand the mechanism of tactile sensation and sensory threshold.

As to the discrimination of one or two-point tactile stimuli and pathology of function disorders within soft tissue in the fingertip, there are two classical models, the physical model [1-2] and the structural model. The structural model refers to a linear elastic half-space models [3-4] and multilayer finite element (FE) models. The physical model is based on the quasi-viscoelastic constitutive equation. Similar with the physical model, linear elastic half- space models also simulate the total deformation characteristics of the skin in fingertip under load. However, it always neglects the multilayer performance of the fingertips, in this case, the model predictions may not agree well with the experiments, so linear elastic half-space model is unsuitable to simulate the local deformation of the soft tissues around to mechanoreceptors. The finite element model based on anatomical structure of fingertip can overcome the main shortcomings in other models. With the finite element model, Wu [5] made a research on soft tissue and bone injury in the fingertip. Maeno [6] also analyzed the complexity and the fine structure of finger skin (such as epidermal ridges, papillate) under deformation within the soft tissue. Simulation results by FE models show the feasibility of FE models to simulate the response of soft tissues to loading. In most FE models, the skin and subcutaneous tissue layer is defined as the linearly elastic. Although the linearly elastic model can get similar results with experimental data and reach a high efficiency in the case of small deformation mechanical stimulation, the skin layers and the subcutaneous tissues are factually nonlinear and time-dependent [7-9] and the response to the mechanical load simulation of the fingertip should also be nonlinear and time-dependent, respectively. Thus, the simple linear model has some limitations.

In this paper, a nonlinear finite element model of a multi-layer fingertip structure is proposed. With this model, the SED distribution is analyzed under a linear load. This paper analyzed the interaction model of the finger-tip with a sharp wedge (linear load), and the sensitivity of the model to weak stimuli is also discussed.

2 Fingertip Model

Anatomical structure of the fingertip is a multilayer structure including the epidermis, dermis, subcutaneous tissue, bone and nail, with the distribution of different types of mechanoreceptors inside it. Markel's discs are located in the epidermis, Meissner corpuscles and Ruffini endings are in the dermis. Pacini corpuscles are in the dermis and subcutaneous tissue. In previous studies [1,2,10], results showed that the epidermis presents hyper-elastic and dermal tissue is linearly viscoelastic when skin is of large indentation. Zhang J et al. [11] pointed out that the mechanical behavior of the skin and the subcutaneous tissue can be simulated using the nonlinear biphasic constitutive model [12]. On the basis of this theory, Wu [13] built a two-dimensional FEM model for a finger. In the model, the skin tissue is assumed to be hyperelastic and viscoelastic, and the subcutaneous tissue is considered to be a nonlinear, biphasic material composed of a hyperelastic solid and an inviscid fluid phase, and the nail and bone are considered to be linearly elastic. The proposed model by Wu[13] has the advantages in the ability to predict the deflection profile of the fingertip surface, the stress and strain distributions within the soft tissue, and the dynamic response of the fingertip to mechanical stimuli. However, when solving complex problems, the model is prone to be non-convergence and low efficiency, which may be due to its complexity. As we know, when the skin suffers a small deformation, the epidermis can be considered linearly elastic because of its small thickness.

In reference to physiological anatomy of the fingertip, we established a two-dimensional finite element model of a cross-section of the index fingertip. The outline of the model is regarded as a plaster contour based on [6]. The size of the index finger is measured by a reading microscope and the geometry can be mapped to an ellipse with a long axis of 17.44mm and a short axis of 13.06mm, as shown in Figure.1. The bone structure is also simplified as an ellipse, with the center coinciding with the outer contour in X-axis direction and 2mm far in Y-axis direction. Each layer of the tissues is symmetrical with the bone in the width direction, and the tissue thickness between the bone and the pulp skin surface is assumed to be bigger than that between the bone and the nail. From outside to inside , the skin layers are as follows, epidermis with the thickness of 0.2mm; dermis with the thickness 1.0mm, subcutaneous tissue between the dermis and the bone, the nail with the thickness of 1.2mm. The whole model is symmetrical on the Y axis, and in order to improve computational efficiency, we adopt the half model for simulation. As the touch deformation is assumed not over 2.0mm in the later analysis, the epidermis is treated as linearly elastic. In order to reach a result close to the practical mechanical property of the skin in fingertip, the dermis and the subcutaneous tissue are regarded as different hyper-elastic and linearly viscoelastic. The properties of the material constitutive equations are quoted from literature [14], whose feasibility has been verified. The bone and the nail are considered to be linearly elastic with a large Young's modulus. The linearly elastic parameters of the epidermis, the bone and nail are based on literature [5-6], as in Table 1. The sharp wedge for the stimulation is simplified as an analytical rigid, with the widths of 0.05mm and 1.0mm on the tip and base, respectively, and a height of 2.0mm.

Table 1. Constants of linearly elastic tissues.

	Epidermis	**Bone**	**Nail**
Young's Modulus (MPa)	0.136	17 000	170
Passion Rate	0.48	0.3	0.3

Table 2(a). Material parameters characterizing hyperelasticity and viscoelasticity for the skin (v=0.40).

	i =1	i =2	i =3
α_i	1.3416	25.000	-7.8571
μ_i (MPa)	6.3754×10^{-3}	1.8874×10^{-4}	-3.7709×10^{-4}
g_i	0.0864	0.2136	-
τ_i(s)	0.2136	8.854	-

Table 2(a). Material parameters characterizing hyperelasticity and viscoelasticity for the skin (v=0.40).

	i =1	i =2	i =3
α_i	-2.0623	25.000	-3.4784
μ_i (MPa)	2.3583×10^{-2}	4.5408×10^{-5}	4.0878×10^{-2}
g_i	0.2566	0.2225	-
τ_i(s)	0.3834	4.6731	-

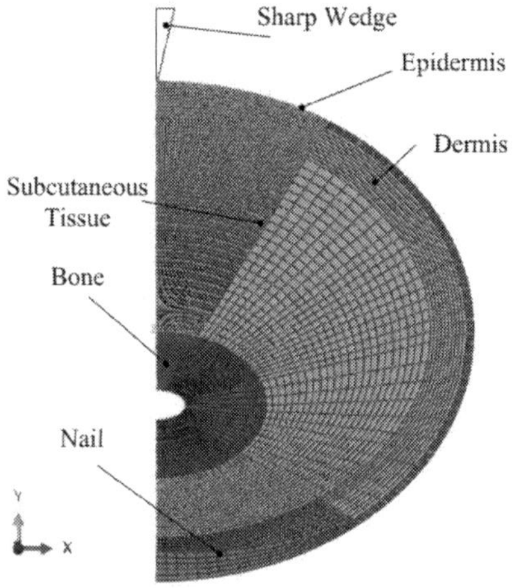

Figure 1. The contact model between the finger and the sharp wedge.

The fingertip was considered to be in contact with the sharp wedge. The biquadral, plain-strain elements were used in the FE modeling and the commercial finite element software package, ABAQUS/Standard (Version 6.11), was utilized for the analysis. The finite element model was shown in Figure. 1. The fingertip was fixed at the center on the nail surface, while the wedge was moved towards the skin surface in a ramp manner, such that a skin deformation of 1 mm was realized within a ramping period of 1s. The displacement of the wedge then remains constant. With this model, the nodes displacement, stress and strain distribution, strain energy density distribution and other field variables within the fingertip were calculated with the finite element method.

3 Simulation Results and Discussion

The electrophysiological experiments suggest that the evoked discharge rate can best be explained by using of SED [16-19]. The spatial distribution characteristic of the SED within the tissues of fingertip is shown in Figure. 2(a) and (b). In the figures, A, B, C and D are the representatives of four different locations adjacent to the interfaces between epidermis and dermis, and that between dermis and subcutaneous tissue, respectively. Figure. 2(a) shows the distribution characteristics of SED along y-direction, from out-edge of the bone to the contact interface. It is seen that the SED has a sudden change from a layer to the next one, e.g, from A to B or from C to D, and a gradual change within the same tissue layer, e.g, from B to C, and then reaches a maximum at point B that locates at the interface between epidermis and dermis. From point C to the out-edge of the bone, a maxima occurs at point D that locates at the interface between dermis and subcutaneous tissue. Because the load is very small, the SED mainly appears in the segment of AD line, while the mechanoreceptors are mainly concentrated in the AD segment and its vicinity as well. Figure. 2(b) shows the range of spatial response of the mechanoreceptors produced by the load. As it can be seen, the individual mechanoreceptor located in different positions has distinctive level of response, and the range of response is consistent to that of stimuli, i.e. contact outline. That's to say, the SED spatial distribution characteristics of the SED can help mechanoreceptors feel the stimuli outline feature and is related closely to the soft tissues' structural properties and stimuli features.

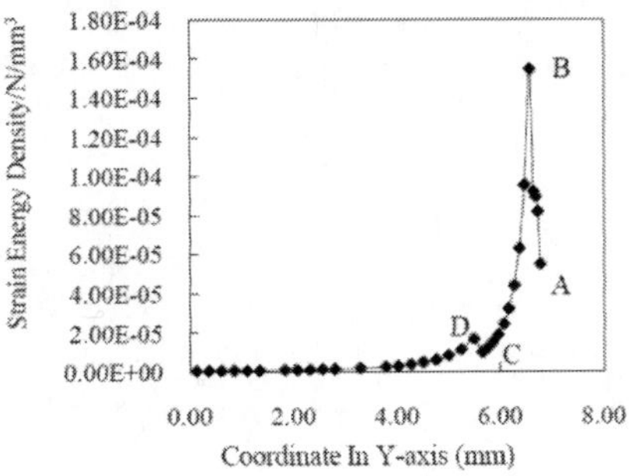

(a)

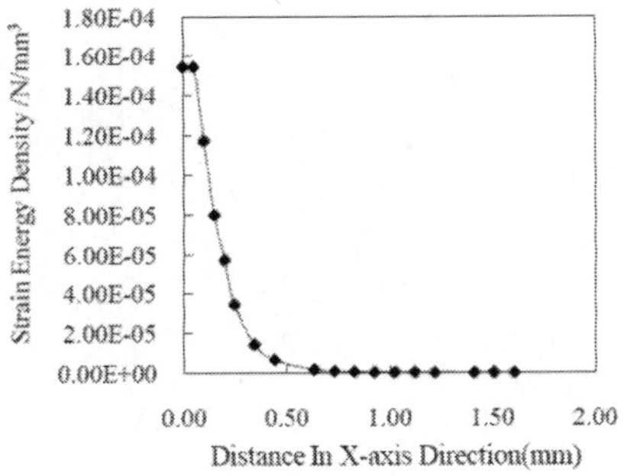

(b)

Figure 2. The distribution of SED within the tissues of fingertip.

The distributions of SED at depths of 0.20, 0.70, and 1.20 mm from the skin surface are shown in Figure. 3.

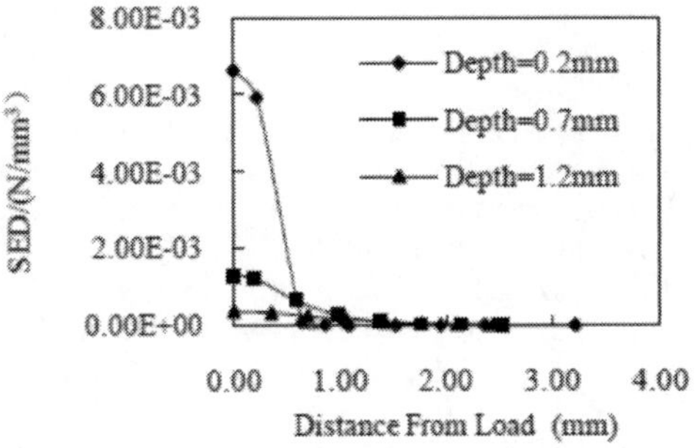

Figure 3. Predicted distributions of the SED in the soft tissue of the fingertip indented using a sharp wedge (at depths of 0.20, 0.7, and 1.20 mm within the tissue).

The value of SED is represented by the logarithmic strain. From Fig. 3 we can see that the larger the depth, the smaller the SED in value, within the 1mm. The SED are all approaching to 0 at the distance of over 1mm from the stimulation location in the X direction.

4 Conclusions

The tactile discrimination of fingertip and distribution characteristics of contact strain is closely related to the performance of tactile sensation. However, because of the difficulty in implementing real-time measurement of the stress and strain within the fingertip, model simulation is a frequently used method to study the response to stimuli, based on the data from in-vitro experiments. This paper simulates the finger load response of the skin tissue in fingertip under the lineal pressure, the results show the tactile strain energy density spatial distribution can be well used to explain mechanoreceptors stimulated discharge characteristics; In addition, the result of the proposed finite element model under lineal load is consistent with the vitro experiments [6] , and the strain energy density distribution of the epidermis and dermis around the mechanical receptors are mapped external shape of the wedge .this model can be used to explore the mechanics of receptors under different types of mechanical stimuli.

Acknowledgements

This work is supported by the National Natural Science Foundation of China (61175069), the National Science Foundation's Course, Curriculum, and Laboratory Improvement (CCLI) program (0737173), and the Natural Science Foundation of Jiangsu Province of China (BK20131205).

References

1. D. Jindrich, Z. Yan, B. Theodore, et al. Non-linear viscoelastic models predict fingertip pulp force

displacement characteristics during voluntary tapping. J Biomech, 2003, 36(4): 497-503.
2. J. Wu, R. Dong, W. Smutz, et al. Dynamic interaction between a fingerpad and a flat surface: experiments and analysis. Med Eng Phys, 2003, 25(3): 397-406.
3. J. Phillips and K. Johnson. Tactile spatial resolution III. a continuum mechanics model of skin predicting mechano receptor responses to bars, edges, gratings. J Neurophysiol, 1981, 46(6): 1204-1225.
4. M. Srinvasan, D. Dandekar. An investigation of the mechanics of tactile sense using two-dimensional models of the primate fingertip. J Biomech Eng, 1996, 118(1): 48-55.
5. J. Wu, R. Dong. Analysis of the contact interactions between fingertips and objects with different surface curvatures. Proc IMech E, 2005, 219(9): 89-103.
6. T. Maeno, K. Kobayashi and N. Yamazaki. Relationship between the structure of human finger tissue and the location of tactile receptors. JSME International Journal, 1998, 41(1): 94-100.
7. Y. Zheng and A. Mak. An ultrasound indentation system for biomechanical properties assessment of soft tissues in vivo. IEEE Biomed Eng, 1996;43:912-918.
8. M. Rubin, S. Bodner and N. Binur. An elastic viscoplastic model for excised facial tissues. ASME J Biomech Eng , 1998;120:686-9.
9. A.W. Wan . Biaxial tension test of human skin in vivo. Biomed Mater Eng ,1994;4:473-86.
10. Y. C. Fung. Biomechanics: Mechanical Properties of Living Tissues. Springer, New York, 1984.
11. J. Zhang, A. Mak, Huang L. A large deformation biomechanical model for pressure ulcers. ASME J Biomech Eng, 1997,119: 406-8.
12. V.C. Mow, S.C. Kuei and W.M. Lai, Armstrong CG. Biphasic creep and stress relaxation of articular cartilage: theory and experiment.ASME J Biomech Eng, 1980, 102:73-84.
13. J.Z. Wu, R.G. Dong, S. Rakheja, et al. A structural fingertip model for simulating of the biomechanics of tactile sensation. Medical Engineering & Physics 26 (2004): 165-175.
14. J.Z. Wu and R.G. Dong, D.E. Welcome. Analysis of the point mechanical impedance of fingerpad in vibration. Medical Engineering & Physics 28 (2006): 816-826.
15. M. Srinivasan. Surface deflection of primate fingertip under line load. J Biomech, 1989, 22:343-9.
16. K. Dandekar, B. Raju and M. Srinivasan. 3-d finite-element models of human and monkey fingertips to investigate the mechanics of tactile sense. J Biomech Eng, 2003, 125(5): 682-691.
17. K. O. Johnson. The role and functions of cutaneous mechanoreceptors. Curr Opin Neurobiol, 2001, 11(4): 455-461.
18. K. O. Johnson, T. Yoshioka and F. V. Bermudez. Tactile functions of mechano receptive afferents innervating the hand.J Clin Neurophysiol, 2000, 17(6): 539-558.
19. A. B. Vallbo and R. S. Johansson. Properties of cutaneous mechanoreceptors in the human hand related to touch sensation. Human Neurobiology, 1984, 3(1): 3-14.

MATEC Web of Conferences **45**, 04007 (2016)
DOI: 10.1051/matecconf/20164504007

Fluid-Structure Interaction Analysis of Hydrofoils in a Pulsating Flow

Jiasheng Li [1], Paul Croaker [2], Jin Tian [1], Mahmoud Karimi [2] and Hongxing Hua [1]

[1] *State Key Laboratory of Mechanical System and Vibration, Shanghai Jiao tong University, Shanghai 200240, P R China*
[2] *School of Mechanical and Manufacturing Engineering, UNSW Australia, Sydney, Australia*

Abstract. The reduction of noise and vibration are very important in the design of hydrofoils. The current study focuses on establishing a theoretical and numerical model to investigate fluid-structure interaction caused by elastic hydrofoils in a pulsating flow. A fully coupled three dimensional boundary element method (BEM) and finite element method (FEM) code is applied to analyze the hydrodynamic performance. The numerical results show that the peak frequencies of the support reactions are related to the natural frequency of the hydrofoil. The natural frequencies and support reaction amplitudes are reduced significantly by including the fluid-structure coupling.

1 Introduction

1.1 Previous work

Fluid-structure interaction has become an important research area in the field of underwater structures and vehicles. Although determining the hydrodynamic loads acting on an elastic hydrofoil is still quite challenging, a number of experimental and numerical studies have been devoted to this topic. Jiho You et al. [1] and Suzanne [2] investigated the performance of elastic hydrofoil by CFD. Young [3, 4] and Lin et al. [5] used potential flow theory to predict the coupled fluid-structural response of elastic propellers under cavitating and non-cavitating conditions. Liu et al. [6] and Torre et al. [7] conducted an experimental study into the hydrodynamic characteristics of an oscillating hydrofoil.

1.2 Objective of the present work

The present work aims to investigate the fluid-structure interaction of a hydrofoil in a pulsating flow. The methodology is presented in Section 2, hydrodynamic performances are given in Section 3 and the main findings are summarised in Section 4.

2 Mathematical model

2.1 Structural model

A hydrofoil with one end free and the other end fixed is considered in this model. By using classical finite element theory, the discrete equation of motion for the hydrofoil can be written as:

$$[M]\left|\ddot{\Delta}_i\right|+[C]\left|\dot{\Delta}_i\right|+[K]\left|\Delta_i\right|=\left|F_w\right| \tag{1}$$

where $\left|\ddot{\Delta}_i\right|$, $\left|\dot{\Delta}_i\right|$ and $\left|\Delta_i\right|$ are the nodal acceleration, velocity, and displacement vectors, respectively. $[M]$, $[C]$ and $[K]$ are the global structure mass, damping, and stiffness matrices. $\left|F_w\right|$ is the hydrodynamic force vector generated by the pulsating flow.

2.2 Pulsating flow

The pulsating flow is generated by assigning the following time-dependent velocity

$$\boldsymbol{V}(x,y,z,t)=\boldsymbol{V}_0+\boldsymbol{A}_p(x,y,z)\cos(\omega t) \tag{2}$$

where, V_0 represents the uniform component of the inflow. $\boldsymbol{A}_p(x,y,z)$ and ω are respectively the amplitude and frequency of the pulsating flow. x, y, z are spatial coordinates, and t denotes time.

2.3 Hydrodynamic model

In order to calculate the hydrodynamic force $\left|F_w\right|$, a BEM method is used to solve the boundary-value problems given by:

$$\nabla^2\phi=0(in\ \Omega) \tag{3}$$

$$-\frac{\partial\phi}{\partial \boldsymbol{n}}=\boldsymbol{V}\cdot\boldsymbol{n}-\frac{\partial\boldsymbol{\delta}}{\partial t}\cdot\boldsymbol{n}(on\ \Gamma) \tag{4}$$

$$\Delta\phi_w\left(\boldsymbol{R}_{wake},t\right)=\Delta\phi\left(\boldsymbol{R}_{te},t-t'\right)(Kutta\ condition) \tag{5}$$

Here, ϕ is the hydrofoil-induced perturbation potential, $\boldsymbol{n}$ is the outward unit normal vector, $\boldsymbol{\delta}$ is the displacement vector of nodes on the hydrofoil surface. $\Delta\phi_w(\boldsymbol{R}_{wake},t)$ represents the potential jump across the wake sheets, $\Delta\phi(\boldsymbol{R}_{re},t-t')$ is the potential jump across the hydrofoil surface at trailing edge. t' denotes the time required for the fluid to travel along the wake surface from the hydrofoil trailing edge $\boldsymbol{R}_{re}$ to the wake point $\boldsymbol{R}_{wake}$.

The Laplace equation is combined with the Morina-Kutta condition to determine the hydrofoil-induced perturbation potential, and kinematic boundary conditions under the small deformation hypothesis (see more details in [3, 4, 10]).

2.4 Hydrodynamic force

The perturbation potential, ϕ, obtained in the preceding section is applied to the linearized Bernoulli equation :

$$p=-\rho\left(\frac{\partial\phi}{\partial t}+\boldsymbol{V}_0\cdot\nabla\phi\right) \qquad (6)$$

to predict the pressure fluctuation on the hydrofoil surface. These hydrodynamic forces are then decomposed into two parts :

$$|F_w|=|F_w^1|+|F_w^2| \qquad (7)$$

Here, $|F_w^1|$ is a fuction of $|\ddot{\Delta}_i|$, $|\dot{\Delta}_i|$ and $|\Delta_i|$, and can be expanded as $-([M_w]|\ddot{\Delta}_i|+[C_w]|\dot{\Delta}_i|+[K_w]|\Delta_i|)$ which describes the interaction between fluid and structure. $[M_w]$, $[C_w]$ and $[K_w]$ are the added-mass, added-damping and added-stiffness matrices. Meanwhile, $|F_w^2|$ has no relationship with the structure and provides excitation for the fluid-structure coupling system.

2.5 Governing equation

By moving $|F_w^1|$ to the left side of equation (1), the governing equation becomes

$$([M]+[M_w])|\ddot{\Delta}_i|+([C]+[C_w])|\dot{\Delta}_i|+([K]+[K_w])|\Delta_i|=|F_w^2| \qquad (8)$$

By Appling a FEM code combined with a BEM code, $[M],[M_w],[C],[C_w],[K],[K_w],|F_w^2|$ can be calculated, with the dynamic response predicted using Wilson-theta method in classical vibration theory.

3 Results and discussions

3.1 Symmetric hydrofoil

The combined BEM [8, 9, 11-13] and FEM method has been used to calculate a NACA 0015 hydrofoil (Fig.1) moving in a pulsating flow with various frequencies given by $\omega=n\pi$, $n=2,4,\ldots,28,30$. The hydrofoil has one end fixed (shown in blue in Fig. 1) and the other end free. The chord length is 1 m and its span length is 4 m. The density, Poisson's ratio and elastic modulus of the hydrofoil are 7800 kg/m^3, 0.3 and 2.1x10^{11} Pa, respectively. In this work, the damping matrix of the hydrofoil is neglected. The incoming uniform velocity is 10m/s, angle of attack is $\alpha=10^{\circ}$, and amplitude of the pulsating flow is 0.1 m/s in the z direction. The first natural frequency of the hydrofoil is 7.71Hz without considering the fluid-structure coupling matrices and 5.59Hz when those matrices are considered. As the support reactions in the x and y directions are very small, only the support reactions in the z direction are presented. Figure 2 shows the impact of the fluid-structure coupling matrices on the reaction force amplitudes as a function of pulsating flow frequency.

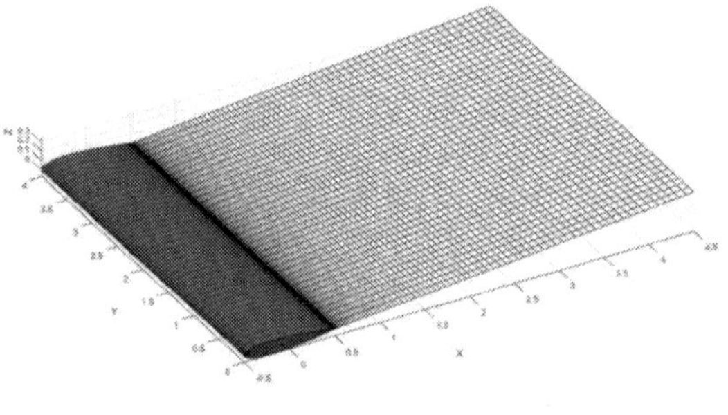

Figure 1. Discretization of symmetric NACA 0015 hydrofoil

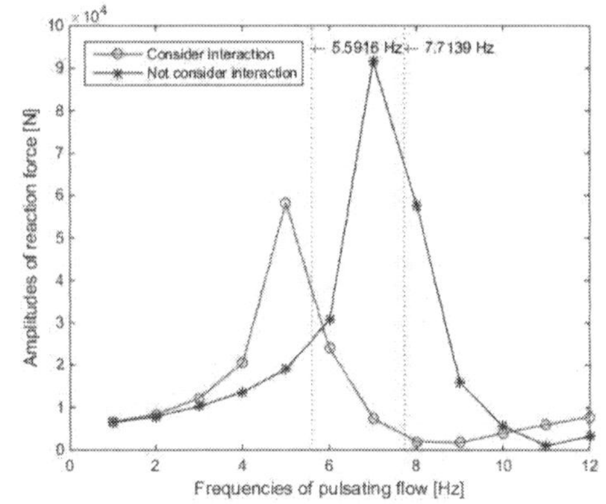

Figure 2. Reaction amplitudes for varying pulsating flow frequency for the symmetric NACA0015 hydrofoil.

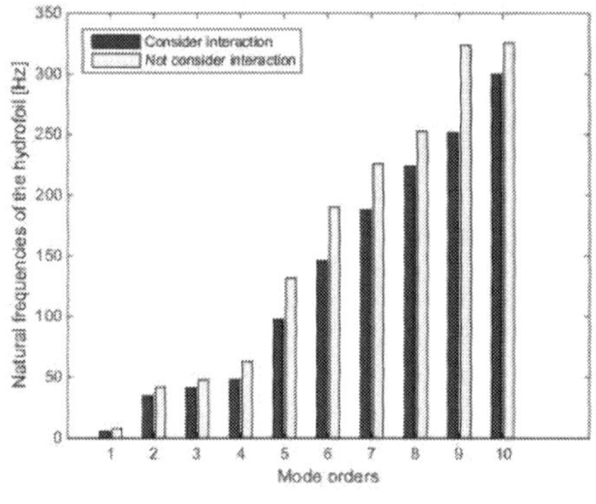

Figure 3. Natural frequency comparison for the symmetric NACA 0015 hydrofoil.

Figure 2 shows that:

1. The maximum reaction force occurs when the pulsating flow frequency matches the first natural frequency of the hydrofoil. This is true regardless of whether the fluid-structure coupling matrices are included. There is no contradiction with classical vibration theory.
2. Including the fluid-structure coupling matrices reduces the first natural frequency by approximately 27.5%. As Figure 3 shows, the other natural frequencies are also considerably reduced, by including the fluid-structure coupling matrices.
3. The amplitude of the support reaction at the first natural frequency is 36% less when the fluid-structure coupling matrices are considered.

3.2 Asymmetric hydrofoil

For comparison, the combined codes are applied to an asymmetric NACA 0015 hydrofoil (Fig.4), with a mean camber line given by $y_c = 4x - 4x^2$. The ratio of sagitta to chord length is 0.1. Figures 5 and 6 show the results of this case. The obtained results show that:

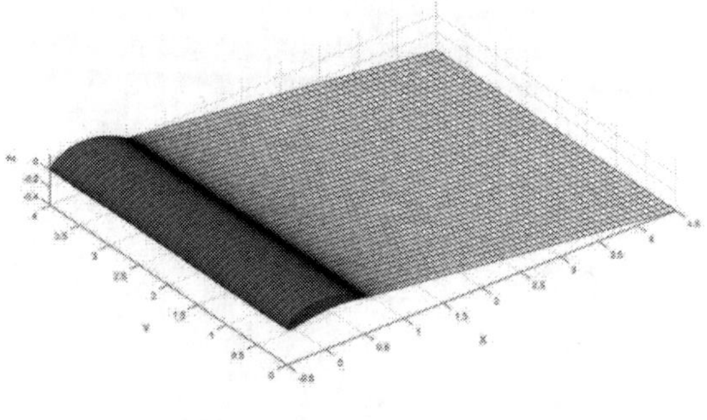

Figure 4. Discretization of asymmetric NACA-0015 hydrofoil

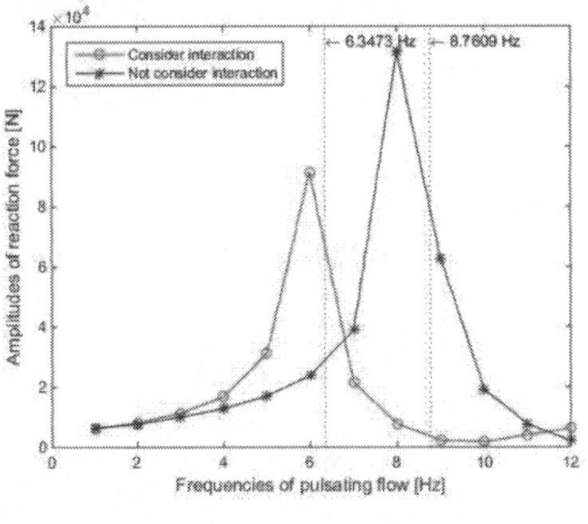

Figure 5. Reaction amplitudes for varying pulsating flow frequency for the asymmetric NACA0015 hydrofoil.

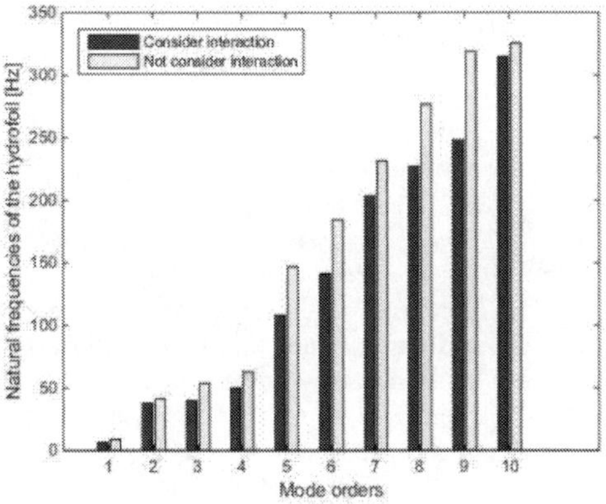

Figure 6. Natural frequency comparison for the asymmetric hydrofoil.

1. No contradiction has been found with the classical vibration theory.
2. The natural frequencies decrease when the fluid-structure coupling matrices are considered.
3. The amplitude of the support reaction at the first natural frequency is less when the fluid-structure coupling matrices are considered.

4 Conclusions

This paper has presented a theoretical and numerical method to study fluid-structure interaction of hydrofoils in a pulsating flow. Both symmetric and asymmetric hydrofoils have been investigated by the combined codes. Results reveal that: 1) natural frequencies of hydrofoils have a significant reduction when the effects of fluid-structure coupling are considered. 2) the amplitude of the support reaction at the 1st natural frequency is approximately 33% less when the fluid-structure coupling matrices are considered. 3) fluid-structure interaction plays an important role in the performance of elastic hydrofoils, even though the elastic deformations are small.

References

1. Jiho You, Jinmo Lee, Donghyun You, Optimization of propulsion kinematics of a flexible foil using integrated CFD-CSD simulations, International Mechanical Engineering Congress & Exposition, (2012)
2. Suzanne Ruth Hutchison, Numerical Modelling of Hydrofoil Fluid-Structure Interaction. Doctoral dissertation (2012)
3. Y.L. Young. Time-dependent hydro-elastic analysis of cavitating propulsors. Journal of Fluids and Structures, 23 :269-295 (2007)
4. Y.L.Young. Fluid – structure interaction analysis of flexible composite marine. Journal of Fluids and Structures, 24 :799 – 818 (2008)

5. Huei-Jeng Lin, Jiun-Jyi Lin, Nonlinear hydroelastic behavior of propellers using a finite element method and lifting surface theory. Mar Sci Technol, 1 :114-124 (1996)
6. Liu Zhen, Hyun Beom-soo, Kim Moo-rong, Jin Ji-yuan. Experimental and numerical study for hydrodynamic characteristics of an oscillating hydrofoil. Journal of Hydrodynamics, Ser. B, Volume 20, Issue 3, Pages 280-287 (2008)
7. O. De La Torre,X. Escaler ,E. Egusquiza,M. Farhat. Numerical and experimental study of a nearby solid boundary and partial submergence effects on hydrofoil added mass. Computers & Fluids 91 (2014)
8. Hess, J. L. and Smith, A. M. O. Calculation of non-lifting potential flow about arbitrary three-dimensional bodies, The Journal of Ship Research,No. 8, pp. 22-44 (1964)
9. Hess, J. L. and Smith, A. M. O. Calculation of potential flow about arbitrary bodies, Progress in Aeronautical Sciences, Vol. 8, pp. 1-138 (1967)
10. Kuo, J. Analysis of propeller blade dynamic stresses. Doctoral dissertation, Department of Naval Architecture and Marine Engineering, The University of Michigan, Ann Arbor, MI. (1984)
11. Mohammad Shahjada Tarafder*, Goutam Kumar Saha*, and Sayeed Tanvir Mehedi. Analysis of Potential Flow around 3-dimensional Hydrofoils by Combined Source and Dipole Based Panel Method. Journal of Marine Science and Technology, Vol.18, No.3, pp.376-384 (2010)
12. Hassan Ghassemi, Ahmad Reza Kohansal, Numerical evaluation of various levels of singular integrals, arising in BEM and its application in hydrofoil analysis, Applied Mathematics and Computation, 213: 277–289 (2009)
13. Xiao-Wei Gao, Evaluation of regular and singular domain integrals with boundary-only discretization—theory and Fortran code. Journal of Computational and Applied Mathematics, 175: 265-290 (2005)

MATEC Web of Conferences **45**, 04008 (2016)
DOI: 10.1051/matecconf/20164504008

The Influence Study of Ultrasonic honing parameters to workpiece surface temperature

Zhang Xiaoqiang[1], Zhu Xijing[1] and Cheng Linglong[2]

[1]*School of Mechanical and Power Engineering, North University of China, Taiyuan, China*

[2]*Shanxi Jincheng Anthracite Mining King Ding Mining and Machinery CO., LTD., Jincheng,China*

Abstract. Ultrasonic vibration honing (UVH), a machine technology, has a lot of advantages. Lower grinding temperature is a significant character and is beneficial for both processing and workpiece surface. But the high temperature caused by big honing pressure becomes the main factor to produce workpiece heat damage in grinding zone. In various honing parameter combinations, the showing effect is different. Based on the thermodynamics classical theory, established the heat transfer equation for grinding zone, simplified the model and obtained the two-dimenssion temperature field expression for workpiece, then simulated the temperature changing trend in a variety of conditions. It is shown that the main temp is in a range of 700K to 1200K. In addition, the variation is huge for every parameter. The study provides a theoretical basis for deeply seeking reasonable machining parameter and obtaining better workpiece quality.

1 Introduction

Ultrasonic vibration honing (UVH) is a fresh machining technology, especially in machining newly hard brittle materials, and has more widely and practical industrial market[1]. The process is a course that the ultrasonic wave is transformed to mechanical vibration firstly, and transmitted by designing device to honing oilstone, caused the oilstone vibrating in machining the workpiece. The method can promote the working efficiency and the products quality.

The grinding temperature field is a major studying branch, many scholars have devoted in theory analysis and experimental measurement. Jaeger, the first scholar who propose the moving heat source theory, has obtained temperature rising theory for bulk heat source, and put forward the rectangular heat source model[2]. Later, the shear heat source model is presented by Outwater based on Jaeger's theory, the distribution method of heat is defined that a part of energy is deliveried to workpiece and other to abrasive dust[3]. While, it is different for Hahn that major heat is come from the abrasive grit wear area[4]. N.Alagumurthi has studyed the heat generation in grinding process, and proved that the temperature developed at the surface of the workpiece has an important influence to the metal remover[5]. Ding Han promoted the grinding temperature model taken abrasive shape and distribution into consideration[6]. J.L.González-Santander studys the heat transfer in surface dry grinding, considering the linear heat flux along the contact zone between the wheel and the workpiece, the

Article available at http://www.matec-conferences.org or http://dx.doi.org/10.1051/matecconf/20164504008

result showed that the analytical result is useful for the monitoring of the online grinding process[7]. Until now, the primary research on grinding temperature is to establish the model and analysis the effect, the report for the mechanism that how the grinding heat impact workpiece quality is rare.

In honing processing, the oilstones have the rotational and reciprocating axially motion meanwhile, and suffer the radial pressure by reason of rotation which can induce a lot of energy in honing zone[8]. When the energy rise to too high, the workpiece surface will be affected, such as burn. So it is indispensable to analysis the temperature in workpiece surface, choose a group reasonable parameter set, gain the better machining quality.

The purpose of this paper is to present the expression of the workpiece surface which consider the machining method and its character. Two-dimension thermal models are established which is indicated by a difference equation. To simulate the temperature changing trend, the reasonable workpiece and oilstone material are chosen.

2 Honing zone temperature model

2.1 UVH Model

The UVH gains widely employed and developed due to the advantages of lowly temperature, pulse nature, small cutting force and etc. UVH device mainly includes ultrasonic transducer, booster, bending vibration disk, flexible string and oilstone. Fig.1 shows the fundamental principle movement. It can be seen obviously that the thermal damage is occurred in honing zone between the oilstone and workpiece. For further completing the honing mechanism, it has become a essential portion to research the temperature field and thermal

Based on the moving heat source theory, the three-dimensional temperature distribution in honing zone for the model is obtained, and is given by the following difference equation expressions[9]:

$$\frac{\partial^2 T}{\partial x^2}+\frac{\partial^2 T}{\partial y^2}+\frac{\partial^2 T}{\partial z^2}=\frac{1}{\alpha}\left(\frac{\partial T}{\partial t}-v_{ea}\frac{\partial T}{\partial x}\right) \tag{1}$$

$$v_{ea}=\sqrt{v^2+\left(v_a+v_f\right)^2}$$
$$=\sqrt{(\pi dn)^2+\left(v_a+2\pi Af\cos 2\pi ft\right)^2} \tag{2}$$

Where, d is the workpiece inner diameter, $\alpha=\lambda/c\rho$ is the thermal diffusivity of workpiece, λ is the thermal conductivity, c is the specific heat capacity, ρ is the liquid fluid density.

2.2 Temperature filed model of grinding zone

For honing system, the boundary conditions satisfies the following set of equations

$$\begin{cases} T(x,y,z,t)=T_i & t=0 \\ -\lambda\dfrac{\partial T}{\partial z}=\begin{cases} q_m & -b\le x\le b \\ h_f(T_\infty-T) & x<-b\,or\,x>b \end{cases} & z=0 \\ -\lambda\dfrac{\partial T}{\partial z}+h_f T(x,y,z,t)=0 & t=0 \end{cases} \tag{3}$$

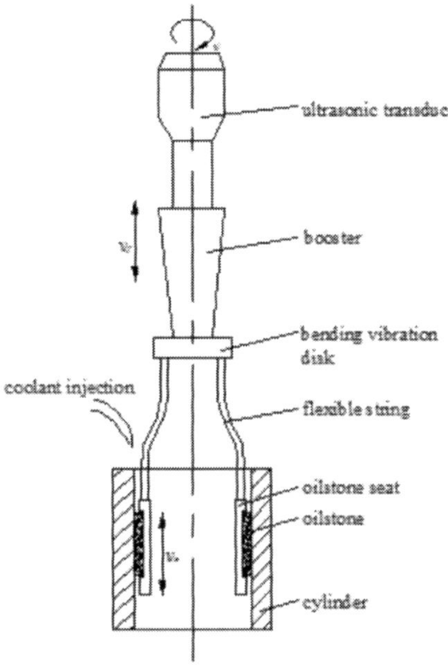

Figure 1. UVH system

Where, b is the oilstone width, h_f is the heat transfer coefficient, T_i and T_∞ are initial temperature and ambient temperature respectively and assumed 20°C.

Then, for any point (x',y',z') in workpiece surface, the temperature field expression is equation (4).

$$T=\frac{\alpha\cdot Q}{\lambda}\left\{\frac{2}{[4\pi\alpha(t-t')]^{3/2}}\cdot\exp\left[-\frac{[x-x'+v_{ea}(t-t')]^2+(y-y')^2+z^2}{4\alpha(t-t')}\right]\right.$$

$$-\frac{h_f}{4\pi\alpha\lambda(t-t')}\cdot erfc\left[\frac{z}{\sqrt{4\alpha(t-t')}}+\frac{h_f}{\lambda}\sqrt{\alpha(t-t')}\right]\cdot erfc\left[-\frac{[x-x'+v_{ea}(t-t')]^2+(y-y')^2+z^2}{4\alpha(t-t')}\right]$$

$$\left.\cdot\exp\left[\frac{zh_f}{\lambda}+\alpha\left(\frac{h_f}{\lambda}\right)^2(t-t')\right]\right\} \tag{4}$$

Table 1. UVH basic parameter

Parameter	Value
Rotation speed of honing head (r/min)	80～240
Honing pressure(MPa)	0.24～1.5
Ultrasonic amplitude(μm)	5～10
Reciprocating speed of honing oilstone (m/s)	0.05～0.4
Ultrasonic frequency(KHz)	17～20
Wave pressure amplitude(MPa)	0.797～2.058

The honing model can be converted from three-dimension to two-dimension, shown in Fig.2.

For simplifying the analysis and calculation, the the 2D steady-state temperature distribution expression should be dimensionless and can be indicated

$$T(X,Z,\tau)=\frac{2Q\alpha}{\lambda v_{ea}}\int_0^\infty d\tau\left[\exp\left(-\frac{Z^2}{4\tau^2}\right)-K_h(Z,\tau)\right]$$

$$\cdot\left[erf\left(\frac{X+L}{2\tau}+\tau\right)-erf\left(\frac{X-L}{2\tau}+\tau\right)\right] \tag{5}$$

Where,

$$K_h(Z,\tau)=\pi H\tau\exp\left(HZ+H^2\tau^2\right)\cdot erfc\left(\frac{Z}{2\tau}+H\tau\right)$$

$$X=\frac{v_{ea}x}{2\alpha};\quad Z=\frac{v_{ea}z}{2\alpha};\quad L=\frac{v_{ea}b}{4\alpha};\quad H=\frac{2\alpha h_f}{\lambda v_{ea}};$$

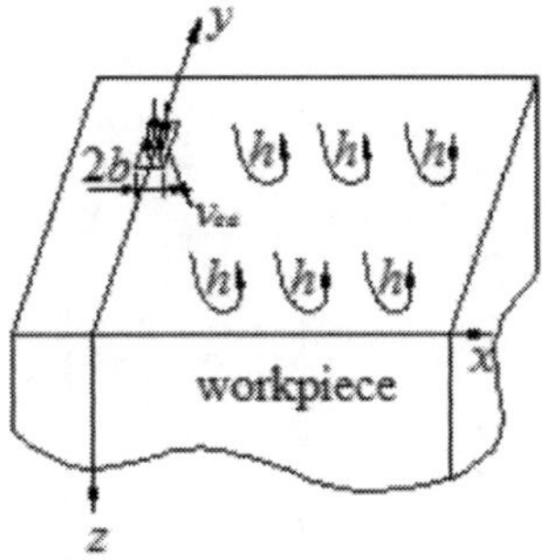

Figure 2. The two-dimension model of honing head

$$K=\sqrt{\frac{v_{ea}^2 t}{4\alpha}}$$

3 Honing parameters

For UVH, there is a lot of relative parameters[10]. For analyzing the grinding temperature, it is computed to choose six important, shown in table 1.

The hard brittle material is the main workpiece quality, so the oilstone needs to choose available material. For the research, the workpiece chooses No.45 steel, and the oilstone chooses PCD. Table 2 shows the detailed material character.

Table 2. The character of materials

	Density (kg/m3)	Young's modulus (GPa)	Possion's ratio	Thermal conductivity (W/(m·K))	Specific heat capacity (J/(kg·°C))
Workpiece	7.85	210	0.31	51.42	0.45
Oiltone	3.52	1050	0.18	136.16	6.17

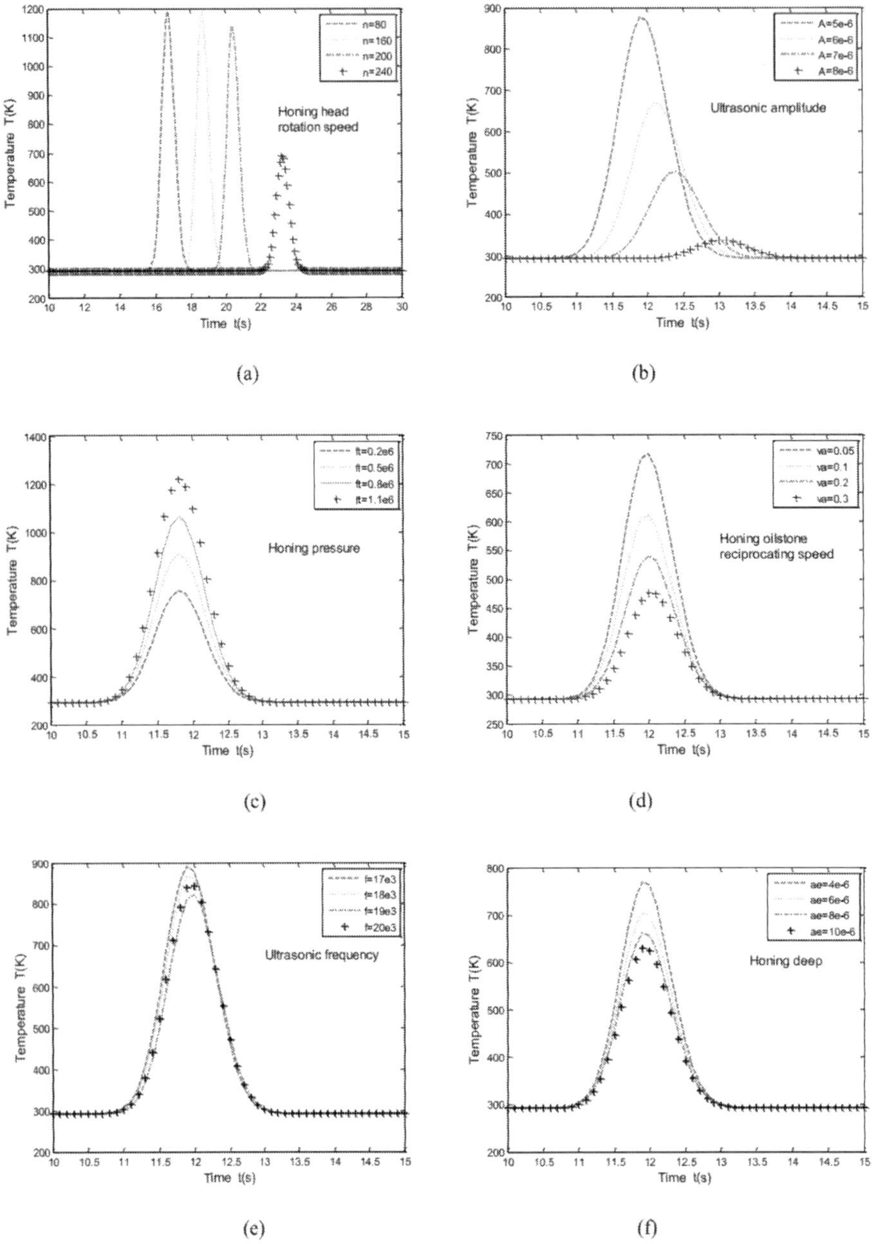

Figure 3. the temperature field of workpiece surface in different parameter

4 Results and discussion

Base on above analysis and choosing parameters, it can be obtained the temperature trend of workpiece surface, shown in Fig. 3.

From (a)~(f), it is obviously shown that the trend is increase firstly, and then decrease. But for different parameters, the trend is various and the maximum temperature is different. For honing head rotational speed and ultrasonic amplitude, the time to reach the maximum is out of step, it is later by the speed enlargement. And the maximum temp will be felled suddenly when the speed reach a value above 200. But for ultrasonic amplitude, it desends uniformly and the rise and fall speed is also very gentle. In addition, it is shown that the peak temp is diverse for every parameter, but mainly located in the range 700K to 1200K. Of course, the temp may be lower for some conditions from Fig.3(b)-(d), and the value is about 400K. The reason is that the heat is various for different parameter combinations, the behavior of the grinding heat is difference in honing zone. All the figure shows that the lowest temp isn't 0K, it can be explained that there is the frictional heating generating when the honing began. And the heat would be transfered quickly to the whole honing head device. Compared (c), (d) and (e), (f), it can be seen that the gradient is apparent, especially for honing pressure and reciprocating speed of honing oilstone.

Otherwise, from Fig.3(c) and (e), it can been seen that the temperature is enlarged by the honing pressure accretion, but reduced for others. And is firstly decreased and then increased for ultrasonic frequency from (e), but the change range is smaller. The time for every factor to reach the extrem value has a big difference. It is abot 17s for rotational speed of honing head to attain, and the time interval for every value of rotational speed is or 2s. While it is about 12s for others, and almost at the same time to reach. From time to explain, it is very short to machine for honing, and the total processing may be within 100s, so the thermal transmission is also extremly short.

In summary, the workpiece surface temperature is contronlled with the limits of 700K to 1200K. The study is basis for futher seeking avialable set to promote the conducts quality. For optimizing the honing process and obtaining better machin quality, the analyzing for temperature field in grinding zone is more significant.

5 Summary

The paper discusses the temperature field of workpiece surface in ultrasonic honing zone, based on the UVH mechanism and the moving heat source theory, established the temperature field expression for workpiece surface firstly, and then simulated the changing trend.

-The rotation speed of honing head and ultrasonic amplitude have more grivous impact for temperature field, not only for the maximum temp but the time to reach.

-In honing process, the temperature is basically increased at the begining and decended later, the gradient is very different for various parameter.

- The maximum temperature can reach about 700K to 1200K for all of parameter combinations, and the various machining parameter set will affect the value.

-The reasonable parameter set could be choosed for futher studying, n=80, A=6 μm, ft=0.5 MPa, va=0.1 m/s, f=19 KHz, ae=6 μm.

Ackowledgements

The authors would like to acknowledge the financial supported by the National Natural Science Foundation of China (No.51275490), the Natural Science Foundation of Shanxi Province (No.2013011024-5) and the 12th Graduate Students of Science and Technology Foundation of North University of China (No.20151214).

References

1. X. J. Zhu, H. J. Xu, A. L. Wang, et al. Research on the cutting principle of new non-conventional technology——Efficiency Ultrasonic Honing[J]. Key E M. **259-260**: 640-643(2004).
2. J.C. Jaeger, Moving sources of heat and the temperature at sliding contacts.[J]. *Proceedings of the Royal Society of New South Wales.* **76**:203(1942).
3. J. O. Outwater, Shaw M C. Surface temperature in grinding Trans ASME. **76**:203(1952).
4. R. S. Hahn On the nature of the grinding process [J]. *Proceedings 3rd Machine Tool Design and Research Conference*. **1**:129-154(1962).
5. N. Alagumurthi, K. Palaniradja, V. Soundararajan. Heat generation and heat transfer in cylindrical grinding process-a numerical study [J]. Int J Adv Manuf Technol. **34**:474-482(2007).
6. Dahu Zhu, Beizhi Li, Han Ding. An improved grinding temperature model considering grain geometry and distribution [J]. Int J Adv Manuf Technol. **67**:1393 -1406 (2013).
7. J.L.González-Santander. Analytic solution for

maximum temperature during cut in and cut out in surface dry grinding [J]. Applied Mathematical Modelling. **10**:1-12. (2015).

8. Zhu Xi Jing, Guo Ce, Wang Jian Qing. The pressure field radiated by cavitation bubble in the grinding area of power ultrasonic honing [J]. Adv Mat Res. **1027**:44-47 (2014).
9. N. R. Des Ruisseaux, R. D. Zerkle. Temperature in semi-infinite and cylindrical bodies subjected to moving heat sources and surface cooling [J]. Journal of Heat Transfer. **92**(3): 456-464(1970).
10. Guodong Liu, Xijing Zhu, Ce Guo. Research on modelling and simulation of cavitation sound field in the grinding zone of the power ultrasonic honing [J]. ACTA ACUSTICA. **38**(6): 663-668(2013).

MATEC Web of Conferences **45**, 04009 (2016)
DOI: 10.1051/matecconf/20164504009

Procedure for the automatic mesh generation of innovative gear teeth

Andrea Chiaramonte Radicella[1], Giulio Di Francesco[2], Vincenzo La Battaglia[2], Stefano Marini[2]

[1] *Toyota Motor Europe, Avenue Du Bourget 60, 1140 Ville de Bruxelles, Belgium*
[2] *Università degli Studi Roma Tre - Dipartimento di Ingegneria, Via della Vasca Navale 79, 00146 Roma, Italy*

Abstract. After having described gear wheels with teeth having the two sides constituted by different involutes and their importance in engineering applications, we stress the need for an efficient procedure for the automatic mesh generation of innovative gear teeth. First, we describe the procedure for the subdivision of the tooth profile in the various possible cases, then we show the method for creating the subdivision mesh, defined by two series of curves called meridians and parallels. Finally, we describe how the above procedure for automatic mesh generation is able to solve specific cases that may arise when dealing with teeth having the two sides constituted by different involutes.

1 Introduction

In recent years the studies and applications of gear wheels with teeth having the two sides constituted by different involutes are increasing [1-5].

These innovative wheels (different pressure angles for the two sides of the tooth: α_{01} and α_{02}) have some considerable advantages [6]; the tooth is more robust: with other factors being equal, they can transmit a higher torque; on equal torque to be transmitted, their overall size is reduced.

Their applications are manifold, particularly in transmissions in which there is only one direction of rotation: speed gear box in automotive transmissions, gear pumps, etc.

By way of example, the following figures 1 and 2 show the traditional profile with $\alpha_{01} = \alpha_{02} = 20°$ compared with the profiles of innovative teeth characterized by α_{01} and α_{02} equal to, respectively: 20° - 14°, 20° - 16°, 20° - 24°, 20° - 26°.

As you can see, the geometrical difference varies as the angles α_{01} and α_{02} of the two sides of the tooth, the left side and the right side, vary.

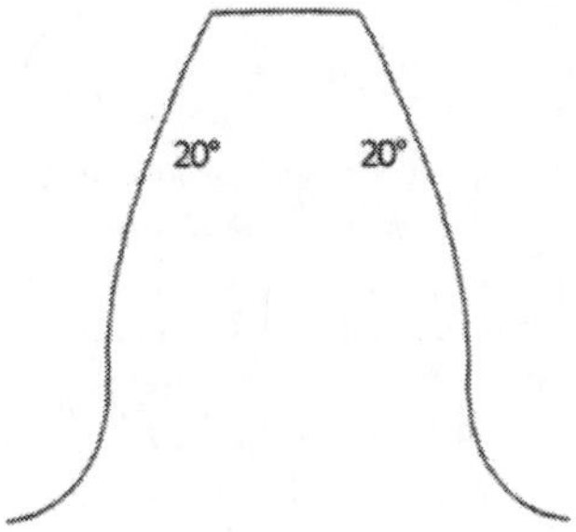

Figure 1. Traditional profile with $\alpha_{01} = \alpha_{02} = 20°$

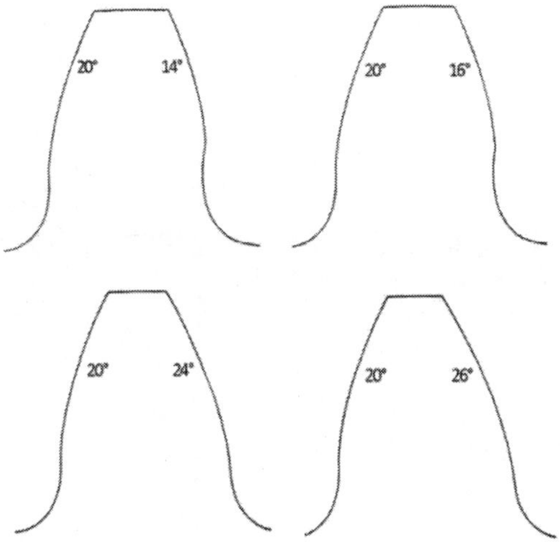

Figure 2. innovative teeth profile with $\alpha_{01} = 20°$ and different value of α_{02}

The variation in the second pressure angle, permits to vary the robustness of the tooth and hence to increase transmission performance [7] [8]. However, the quantification of this increase in performance cannot be estimated other than through a prior FEM analysis of the tooth. This analysis is in any case limited by the absence of an automatic and efficient mesh generator for the tooth structure [9-13].

The meshing system that we intend to propose is based on the subdivision of the tooth through a network of "meridians" and "parallels": the "meridians" are roughly composed of circular arcs, the "parallels", on the other hand, are curves with a trend similar to that of the tooth profile.

2 Subdivision of the tooth profile

The first step in the meshing of the tooth consists in the subdivision of its profile.

The tooth profile presents itself as the succession of various curves with a variable radius of curvature. The subdivision of this profile must follow said variability increasing the number of elements in the areas with a smaller radius of curvature so as to thin it out as the same radius of curvature increases (see figure 3).

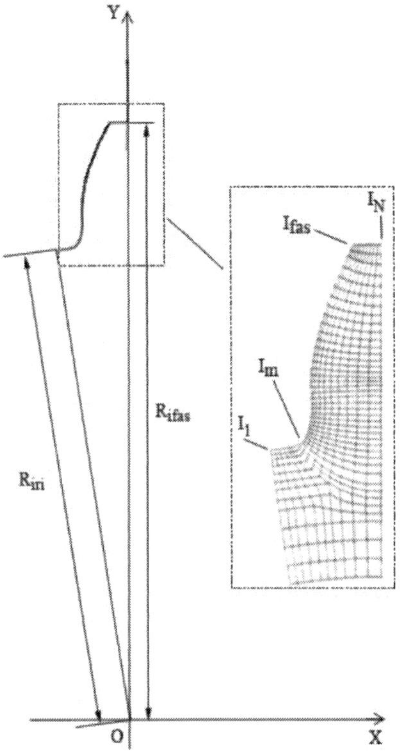

Figure 3. Tooth profile with relative indexes.

2.1 Initial subdivision

Once the characteristic parameters of the tooth[a] are known, the tooth profile is delimited by the outside and root circles. Where N, taken by the designer, is the number of subdivisions along the outer profile of the tooth, we make an initial subdivision with the following reservation:

$20 \cdot N/150$ subdivisions for the internal circular arc
$50 \cdot N/150$ subdivisions for the fillet profile
$75 \cdot N/150$ subdivisions for the active flank
$5 \cdot N/150$ subdivisions for the outside circular arc

The side of the tooth (fillet profile + active flank) is therefore characterised by the following number of subdivisions: $I_{tot} = (50 + 75) \cdot N/150$

The results of this first subdivision will form the basis for the subsequent refinement based on the area, if any, where we will want the highest concentration of elements in order to thicken the mesh in a specific area.

Each subdivision of the profile is identified by an index I_i, with $1 \leq i \leq N$ increasing as the distance from the center of the wheel increases.

After having indicated with I_m the index corresponding to the area in which we want the highest concentration of subdivisions, that is, the area where the mesh must be thickest and with ω the relationship, also defined by the designer, between maximum and minimum concentration, we identify the section in which the intervals are more sparse. If we obtain:

$I_m > I_{tot}/2$ the distance of this section from the center of the wheel is lower than the distance of the interval at maximum concentration.

If, on the other hand, we obtain:

$I_m < I_{tot}/2$ then the distance of this section from the center of the wheel is greater than that of the interval at maximum concentration.

2.1.1 First case: Im > Itot / 2

In case of a section with a minimum concentration of subdivisions external to the section at maximum concentration, the following system of equations is developed:

$$\begin{cases} R_{(Ifas)} - R_{(Ifas-1)} = R_{(Im+1)} - R_{(Im)} + K_y \cdot (I_{fas} - I_m - 1) \\ (R_{(Im+1)} - R_{(Im)}) \cdot \omega = R_{(Ifas)} - R_{(Ifas-1)} \end{cases} \quad (1)$$

$R_{(i)}$ is the value of the radius at the i-th subdivision, "I_{fas}" the index "I" corresponding to the outside radius, "K_y" is the constant radius increase between any two adjacent intervals[b]. The system (1) expresses the consistency of growth between adjacent intervals and the existence of a predetermined ratio, equal to ω between the interval of larger size and the interval of smaller size. From the previous system we obtain, for the increase K_y, the following expression:

$$K_y = \frac{(\omega - 1) \cdot (R_{(Im+1)} - R_{(Im)})}{(I_{fas} - I_m - 1)} \quad (2)$$

The condition of constant increase between two consecutive intervals makes it possible to define the interval to be subdivided as:

$$R_{(Ifas)} - R_{(Iri)} = (I_{fas} - I_{ri}) \cdot (R_{(Im+1)} - R_{(Im)}) + K_y (S_1 + S_2) \quad (3)$$

The index I_{ri} identifies the first value of the index I on the fillet profile; $R_{(Iri)}$ therefore coincides with the value of the root radius and $R_{(Ifas)}$, on the other hand, coincides with the value of the outside radius. The parameters S_1 and S_2, on the other hand, are defined as follows:

$$S_1 = 1 + 2 + 3 + 4 + 5 + \ldots + (I_{fas} - I_m - 1) \quad (4)$$
$$S_2 = 1 + 2 + 3 + 4 + 5 + \ldots + (I_m - I_{ri} - 1)$$

[a] For modular proportion gear teeth: number of teeth, module, pressure angle on the two sides.

[b] With the exception of those which share the point of subdivision with the index I_m: intervals ranging, that is, between $R_{(Im+1)}$ and $R_{(Im)}$ and the interval between $R_{(Im)}$ and $R_{(Im-1)}$, equal intervals.

As proposed, each interval differs from the previous one by a constant quantity K; indicating with η the minimum distance between two successive points of subdivision ($\eta = R_{(Im+1)} - R_{(Im)} = R_{(Im)} - R_{(Im-1)}$), the thickness of each interval can be expressed as the sum of η and of as many increments K_y as the intervals separating it from the thickness η interval.

In other words, for any value $X > I_m$, with $I_{ri} > X > I_{fas}$, we have the following relationship:

$$R_{(X+1)} - R_{(X)} = \eta + K_y (X - I_m - 1) \tag{5}$$

While for $X < I_m$ the corresponding one applies:

$$R_{(X)} - R_{(X-1)} = \eta + K_y (X - I_m - 1) \tag{6}$$

There follow the (4) referred to above.

Relation (3) expresses the fact that the total height of the tooth can be expressed as the sum of ($I_{fas} - I_{ri}$) intervals and of many increments K_y obtained precisely from the sum of S_1 and S_2 as shown above.

From (2) and (3) we obtain the following expression :

$$R_{(Ifas)} - R_{(Iri)} = (I_{fas} - I_{ri})\,\eta + \eta\,(S_1 + S_2) \cdot (\omega - 1) / (I_{fas} - I_m - 1) \tag{7}$$

From this it is possible to obtain the expression for η, that is, the size of the minimum interval:

$$\eta = R_{(Im+1)} - R_{(Im)} = \frac{(R_{(Ifas)} - R_{(Iri)})}{(I_{fas} - I_{ri}) + \frac{(S_1 + S_2)\cdot(\omega - 1)}{(I_{fas} - I_m - 1)}} \tag{8}$$

All the second member sizes are and can be defined quickly based on the characteristics of the tooth and on the data entered by the user; after having obtained η it is very simple to calculate K_y from (2):

$$K_y = \frac{\frac{(\omega - 1)}{(I_{fas} - I_m - 1) \cdot (R_{(Ifas)} - R_{(Iri)})}}{\frac{(I_{fas} - I_{ri}) + (S_1 + S_2) \cdot (\omega - 1)}{(I_{fas} - I_m - 1)}} \tag{9}$$

This expression, duly simplified, takes on the following form:

$$K_y = \frac{(\omega - 1) \cdot (R_{(Ifas)} - R_{(Iri)})}{(I_{fas} - I_{ri}) \cdot (I_{fas} - I_m - 1) + (S_1 + S_2) \cdot (\omega - 1)} \tag{10}$$

Expression (10) defines the increase between two successive intervals.

2.1.2 2nd case: Im > Itot / 2

Similarly, the same sizes are defined, η and K_y, in the case of a section with a minimum concentration of subdivisions internal to the section at maximum concentration; in this case, obviously, the starting equations differ:

$$\begin{cases} R_{(Iri+1)} - R_{(Iri)} = R_{(Im)} - R_{(Im-1)} + K_y(I_m - I_{ri} - 1) \\ (R_{(Im)} - R_{(Im-1)}) \cdot \omega = R_{(Iri+1)} - R_{(Iri)} \end{cases} \tag{11}$$

From which:

$$K_y = \frac{(\omega - 1) \cdot (R_{(Im+1)} - R_{(Im)})}{(I_m - I_{ri} - 1)} \tag{12}$$

The latter replaced in (3), which remains unvaried, provides the expression of η in the second case:

$$\eta = R_{(Im)} - R_{(Im-1)} = \frac{R_{(Ifas)} - R_{(Iri)}}{(I_{fas} - I_{ri}) + \frac{(S_1 + S_2) \cdot (\omega - 1)}{(I_m - I_{ri} - 1)}} \tag{13}$$

and, hence, of K_y:

$$K_y = \frac{(\omega - 1) \cdot (R_{(Ifas)} - R_{(Iri)})}{(I_{fas} - I_{ri}) \cdot (I_m - I_{ri} - 1) + (S_1 + S_2) \cdot (\omega - 1)} \tag{14}$$

3 Definition by points of the tooth profile

The subdivision of the interval identified by outside and root radiuses permits to identify N subdivision points on each of the two tooth profiles; except for symmetrical toothings, these points lie at different levels, that is, the distance of the i-th point of the first profile from the center of the wheel (radius) is generally different from that of the corresponding i-th point of the second profile. For a rational mesh of the tooth, in particular for the tracking of the "parallels", however, the corresponding points on the opposite profiles should lie at the same radius; to that end, an iterative process is developed in order to identify the point of equal radius of the opposite profile, thus making it possible to create, on toothings having the same module and number of teeth, although a different value of the second pressure angle, meshes comparable between them as they have been realized based on the same first profile. The point of equal radius on the second profile is identified by seeking the first point of radius greater than the corresponding point on the first profile. The radius of a quantity Δr is therefore reduced and this operation is repeated until the radius drops below the radius on the first profile; when this occurs Δr is halved and is added to the obtained value, repeating and changing the sign whenever the radius on the first profile is exceeded. The process continues until Δr drops below a threshold equal to the set degree of accuracy.

4 Identification of the circular arcs perpendicular to the profile in the points with equal radius

After having identified the points of equal radius on the opposite profiles, said points with perpendicular circular arcs are joined with both profiles at the aforesaid points of equal radius. In order to interpolate a circular arc between two points of equal radius, it is first necessary to determine the tangents to the two profiles at the aforesaid points (in the case of symmetrical teeth the two tangents have the same angular coefficient, although of the opposite sign). The tangent of the i-th point of the profile is defined as the straight line with angular coefficient given by the average of the angular coefficients of the two straight lines passing, respectively, through points (i, i+1) and (i, i-1). Having thus defined the tangents on the two opposite profiles at the points of equal radius, we obtain the average value of the tangent to the two profiles and, from this, the center of the circular arc sought, which is therefore almost perpendicular to the two profiles.

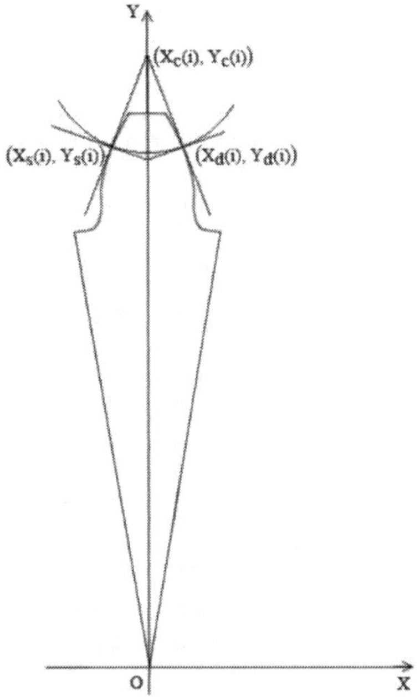

Figure 4. Coordinates of a circular arc center.

We assume an XY coordinate system originating from the center of the wheel and axis Y coinciding with the radial axis of the tooth.

The ordinate $X_c(i)$ of the i-th center of the arc is given by the mean value of the ordinates $X_s(i)$, relating to the point belonging to the left profile and $X_d(i)$, relating to the point belonging to the right profile, while to identify the abscissa $Y_c(i)$, the belonging of the center to the tangent of the profile is set (see figure 4).

After having determined the coordinates of the center, we determine the radius of curvature of the arc, hence the equation of the circle of which the arc is part.

The analytical procedure is developed on the basis of the coordinates $X_s(i)$, $Y_s(i)$ and $X_d(i)$, $Y_d(i)$ of the points of equal radius.

First we calculate the angular coefficients of the tangents to the left and right profiles in proximity of the i-th point:

$$m_{s1i}=\frac{Y_s(i+1)-Y_s(i)}{X_s(i+1)-X_s(i)} \quad (15) \qquad m_{s2i}=\frac{Y_s(i)-Y_s(i-1)}{X_s(i)-X_s(i-1)} \quad (16)$$

$$m_{d1i}=\frac{Y_d(i+1)-Y_d(i)}{X_d(i+1)-X_d(i)} \quad (17) \qquad m_{d2i}=\frac{Y_d(i)-Y_d(i-1)}{X_d(i)-X_d(i-1)} \quad (18)$$

The angular coefficients of the tangents to the right and left profiles are determined as the average of the previous ones:

$$m_{si}=\frac{m_{s1}(i)+m_{s2}(i)}{2} \quad (19) \qquad m_{di}=\frac{m_{d1}(i)+m_{d2}(i)}{2} \quad (20)$$

The angular coefficient of the average tangent is given by:

$$m_i=\frac{|m_s(i)|+|m_d(i)|}{2} \quad (21)$$

which, on the two opposite profiles, will acquire the corresponding sign.

The equations of the tangents to the profiles are therefore the following:

$$Y_i= m_i X + q_s \quad (22) \qquad Y_i = -m_i X + q_d \quad (23)$$

where:

$$q_s = Y_{s(i)} - m_i X_{s(i)} \quad (24)$$

$$q_d = Y_{d(i)} - m_i X_{d(i)} \quad (25)$$

We then obtain the ordinate Xc(i) of the center as:

$$X_c(i)=\frac{X_s(i)+X_d(i)}{2} \quad (26)$$

and the ordinate $Y_c(i)$, imposing the belonging of the center to the tangent to the profile; once the tangent has been determined, by imposing the passage through the tangent point we get:

$$Y_{c(i)} = m_i X_{c(i)} + q_s \quad (27)$$

Having thus defined the coordinates of the center, we then define the radius as the distance of one of the tangent points from the center:

$$R(i)=\sqrt{(X_c(i)-X_s(i))^2+(Y_c(i)-Y_s(i))^2} \quad (28)$$

At this stage, the equation of the circumference can be easily obtained in the form

$$X^2 +Y^2 +aX + bY +c = 0 \quad (29)$$

where:

$$a = - 2X_c(i) \quad (30)$$

$$b = - 2Y_c(i) \quad (31)$$

$$c = [X_c(i)]^2 + [Y_c(i)]^2 - [R(i)]^2 \quad (32)$$

Given the need to identify the arc included between the points of equal distance from the center of the wheel, the equation of the circle must be accompanied by the following restrictions on the Y coordinate:

$$Y_i < Y_s(i) \text{ if } Y_c(i) > Y_s(i) \quad (33)$$

$$Y_i > Y_s(i) \text{ if } Y_c(i) < Y_s(i) \quad (34)$$

By applying the procedure described above to all the points of subdivision of the profile, we get the system of "parallels" of the mesh (see figure 5).

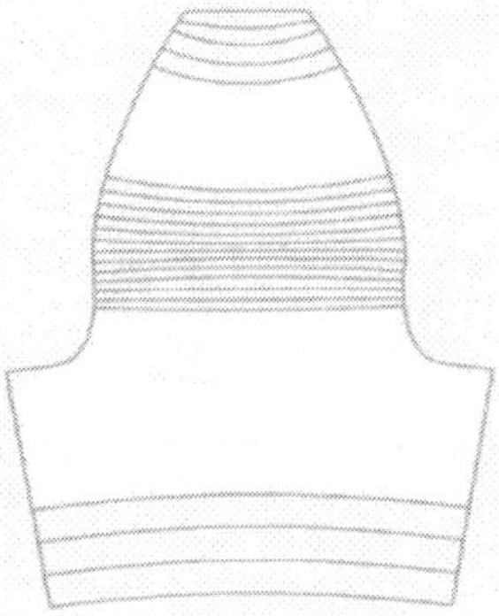

Figure 5. System of "parallels" of the mesh.

This procedure may present problems in the case of asymmetrical teeth: the fact that there may be areas of the profiles with a tangent of the same sign makes it impossible to interpolate circles that satisfy the set requirements; the solution to this problem will be addressed separately.

5 Principle of subdivision of the arcs

The subdivision of arcs makes it possible to create the mesh, generating a series of points on each "parallel", each of which will be crossed by one of the "meridians".

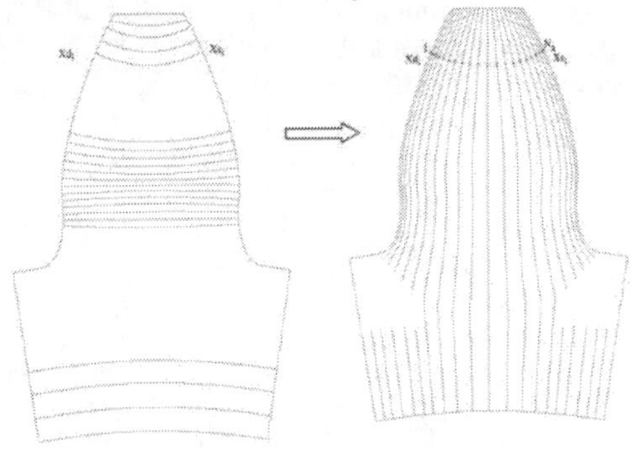

Figure 6. Subdivision of the arcs.

The basic principle of arc subdivision (see fig. 6) is based on the requirement of the most "squared" elements possible in the area with maximum concentration; this requirement will define the increase to be given to the minimum size of the interval on the arc, which size is known and equal to the value already defined for the vertical subdivision of the profile. The first check to be carried out concerns the number of set subdivisions: this will permit a number of intervals sized at least; once this has been achieved, we then look for the increase to be attributed to each interval so as to cover the entire arc with gradually increasing elements of the constant quantity K_x to be determined. After having determined the coordinates of the midpoint of the arc, we then calculate the value of the increase K_x as follows: where N_x is the number of subdivisions for the single arc and X_m the ordinate of the midpoint of the arc, having indicated as X_d and X_s the ordinates of the points on the right and on the left profiles, respectively, we get:

$$X_d - X_s = \eta\, N_x + 2\, K_x\, S \tag{35}$$

From which we obtain:

$$K_x = \frac{X_d - X_s - N_x \eta}{2\cdot s} \tag{36}$$

which provides the expression sought for K_x ; in it S obviously takes on the meaning already seen in the case of the vertical subdivision of the tooth for S_1 and S_2: after the user has set the value J_m where the maximum concentration is desired, S is defined as: $S = S_a + S_b$, where:

$$S_a = 1 + 2 + 3 + \cdots + (j_m - 1) \tag{37}$$

$$S_b = 1 + 2 + 3 + \cdots + (\tfrac{N_x}{2} - j_m - 1) \tag{38}$$

6 Identification of the mesh in areas with profiles having a tangent of equal sign

Especially in the case of asymmetrical teeth, the problem may arise of meshing areas of the tooth whose opposite profiles have a tangent of equal sign; in that case it is not possible to interpolate a circular arc: a different strategy is therefore identified. The identification of the arcs is suspended, the suppressed arcs are counted and equal number of subdivisions in a vertical direction is introduced anew, repeating the procedure for all the vertical intervals identified by the pairs of corresponding points on the arcs facing the area in question. Once the vertical subdivision of the interval has been carried out, an ordinate is attributed to each of these subdivisions, dividing the interval in the X direction in a number of parts equal to the number of suppressed arcs.

7 Discretization of the profile in the vicinity of angular points

The presence of points on the tooth profile in which there is a sudden change in the radius of curvature of the profile leads to abnormal situations, with the adjacent arcs having an opposite radius of curvature; this generates elements of abnormal size, or arcs intersecting one another.

7.1 Solution in the event of intersecting arcs

In this case the calculation program automatically deletes the intersecting arcs. The area of the tooth thus "emptied" will therefore need to undergo a new mesh process. For this purpose, it is first necessary to check whether the cleared area allows the insertion of parallels, depending on the size of the adjacent elements; if this is not possible, the nodes on the surviving arcs will simply be joined with straight segments thus recreating the meridians in the area in question. If, on the other hand, it is possible to insert parallels (or if the emptied area has multiple "radial" sizes compared with the adjacent elements), a number of parallels equal to the identified will be entered.

7.2 Solution in case of areas with thinning of the mesh due to the opposite curvature of the adjacent arcs

This situation occurs frequently in the fillet area of the tooth, even in the case of symmetrical teeth; the solution to this problem is in this case extremely simple: it will suffice to check the size of the elements, inserting a parallel passing through the midpoints of the radial segments linking the nodes of the adjacent arcs whenever the size of the elements becomes such as to be at least double the previous element. In the case where the ratio between the size of the elements is even greater, similarly to what is set out in the previous paragraph, the programme will enter more parallels.

8 Verification of the size of the elements.

In order to obtain a high level of reliability of the FEM calculation, the discretized elements must have the most regular shape possible; in order to ensure this property, an additional check is carried out by verifying that on all the elements the relationship between the sides of the single element does not exceed the value 1/3, in which case an additional subdivision parallel is introduced, thus avoiding the risk of having disproportioned elements.

9 Conclusions

The algorithms and strategies illustrated above permit to carry out a tooth mesh simply by starting from the parameters of the same tooth. The translation of these algorithms into a calculation software permits to automate the process and to quickly analyse under the FEM any asymmetric gear so as to compare its characteristics with the equivalent traditional tooth, making it possible to predict the performance variation corresponding to the variation of the second pressure angle and, hence, to size the asymmetrical tooth.

References

1. A. Kapelevich, "Direct gear design for asymmetric tooth gears" Mechanisms and Machine Science, **34**, pp. 117-143, (2016)
2. S. Ekwaro-Osire, I. Durukan, F.M. Alemayehu, "Experimental and probabilistic analysis of asymmetric gear tooth. Conference Proceedings of the Society for Experimental Mechanics Series, **5**, pp. 207-212, (2011)
3. A. Sanders, D. R. Houser, A. Kahraman, J. Harianto, S. Shon, "An experimental investigation of the effect of tooth asymmetry and tooth root shape on root stresses and single tooth bending fatigue life of gear teeth." Proceedings of the ASME Design Engineering Technical Conference, **8**, pp. 297-305. (2011).
4. G. Di Francesco, S. Marini, "Asymmetric teeth: bending stress calculation", Gear Technology (Mar-Apr 2007), Randall Publishing inc., Elk Grove Village, IL USA (2007)
5. R. Thirumurugan, C.C.C. Deepak, "Design and analysis of asymmetrical spur gear drive for automobile gear box application", Applied Mechanics and Materials, 592-594, pp. 2277-2281, (2014).
6. G. Di Francesco, S. Marini, "Tensioni di flessione di dentature asimmetriche e dimensionamento", Organi di Trasmissione (April 2005), Tecniche Nuove, Milan (2005)
7. G. Di Francesco, S. Marini, "Structural Analysis of Asymmetrical Teeth: Reduction of Size and Weight" *Gear Technology*, Randall Publishing inc. Sept- Oct. 1997, Elk Grove Village IL. U.S.A.(1997)
8. Di Francesco, S. Marini, "Asymmetrical gear wheel design: determining the bending stress in relation to the degree of asymmetry", TMT 2004, Trends in the development of machinery and associated technology, Neum (2004)
9. G. Di Francesco, S. Marini, "Dentature asimmetriche: evoluzione e procedura analitica per il dimensionamento" *Organi di Trasmissione*, Tecniche Nuove, Milano, (1998)
10. G. Di Francesco, S. Marini, G. Grasso, C. Clienti, "Asymmetric gear wheel thermographic analysis", Colloque Photomecanique, Étude du comportement des matèriaux et des structures, Albi (2004)
11. G. Di Francesco, S. Marini, "Asymmetrical gear wheels: automatized procedure for the design", International Conference on Gears, Munich 2005
12. G. Di Francesco, S. Marini, "Adattamento della procedura ISO C alle dentature asimmetriche", XXXIV Convegno AIAS, Milan (2005)
13. G. Di Francesco, S. Marini, "Ruote a denti asimmetrici: procedura automatizzata per il dimensionamento", Organi di Trasmissione (February 2008), Tecniche Nuove, Milan (2008)

MATEC Web of Conferences **45**, 04010 (2016)
DOI: 10.1051/matecconf/20164504010

Fuzzy model for Laser Assisted Bending Process

Oliviero Giannini , Stefano Guarino

Università degli Studi Niccolò Cusano – via Don Gnocchi 3, 00166 Rome, Italy

Abstract. In the present study, a fuzzy model was developed to predict the residual bending in a conventional metal bending process assisted by a high power diode laser. The study was focused on AA6082T6 aluminium thin sheets. In most dynamic sheet metal forming operations, the highly nonlinear deformation processes cause large amounts of elastic strain energy stored in the formed material. The novel hybrid forming process was thus aimed at inducing the local heating of the mechanically bent workpiece in order to decrease or eliminate the related springback phenomena. In particular, the influence on the extent of springback phenomena of laser process parameters such as source power, scan speed and starting elastic deformation of mechanically bent sheets, was experimentally assessed. Consistent trends in experimental response according to operational parameters were found. Accordingly, 3D process maps of the extent of the springback phenomena according to operational parameters were constructed. The effect of the inherent uncertainties on the predicted residual bending caused by the approximation in the model parameters was evaluated. In particular, a fuzzy-logic based approach was used to describe the model uncertainties and the transformation method was applied to propagate their effect on the residual bending.

1 Introduction

Bending processes are widely used in various manufacturing domains to produce a large variety of geometrical shape starting from raw shape and components. [1-4]. The main drawback of the bending process is the springback phenomenon. The springback is the result of the energy stored inside the processed workpiece when plastically deformed during the bending process. This energy is released when the constraints are removed at the end of the bending process, leading to an unexpected change in the final shape of the workpiece due to residual elasticity kept inside the bent material [1]. This aspect can cause serious error in the final shape of the workpiece compromising the process performance in terms of lack of the closest shape and dimensional tolerances. Several studies in the literature deal with the analysis of the springback in metal sheet forming and its compensation [5-7].

In recent years, the laser forming process has reached a considerable importance as promising technology in many manufacturing domains, such as automobile, microelectronics and aerospace [1-4]. The laser forming process has many advantages. First, it requires no external forces, so the springback may be negligible. Second, it can shape complex surface producing even very small parts. Moreover, many hard and brittle materials, such as titanium alloy, nickel alloy, ceramics, etc., can be processed. Unfortunately laser forming is a time-consuming process, when compared to traditional technology.

In this paper, a diode laser was used to assist the traditional forming process. The laser focused on the critical areas during the sheet forming induces a localized heating of the component that permits the release of the energy stored during the forming process reducing the springback phenomena. This application of laser in forming process is relatively new and the few studies are available in the literature [6-8]. The problems related to this approach are due to the lack of understanding of leading phenomena involved as well as process simulation and modelling.

The present work proposes an experimental investigation and a fuzzy modeling on laser-assisted bending process of aluminium sheet. The study is focused on the evaluation and control of the springback extent by assisting the forming operations with a selective heating of the bent area by irradiation with a high power diode laser. All the experimental tests were aimed at understanding the basic mechanisms involved in the laser-assisted forming process and to study the operational parameters of the process (laser power, scan speed and imposed inflection). Then a fuzzy methodology is applied to a regression model used to predict the residual bending. This third application of the method allows the complete formalization of the procedure based on fuzzy approach to the process. In particular, this approach allowed:

(i) to evaluate how much a given sample, characterized by a combination of process parameters and the corresponding residual bending level, belonged to the fuzzy set defined by the model;

Article available at http://www.matec-conferences.org or http://dx.doi.org/10.1051/matecconf/20164504010

(ii) to evaluate the percentage of samples which were described by the model for a given membership level, and thus ;

(iii) to select operational parameters and predict the fidelity of the model at the operational point chosen.

2 Methodology

2.1 The machining process

Laser-assisted bending process was carried out on specimen of AA6082T6 aluminium alloy with the dimensions of 150x30x1.2 mm. The geometrical tolerances (< 0.05 mm) and uniform starting surface conditions were assured by controls performed with coordinate-measuring machine (Mitutoyo CRISTA APEX C544). An ultrasound cleaning system (SOLTEC model Sonica 2200) was utilized with the aim of having the same cleanliness for all the specimens.

Laser processing was performed using a diode laser source (ROFIN-SINAR, model DL 015) with a wavelength of 940±10 nm and an elliptical spot, with dimensions of 1.9mm along its major axis and 0.6 mm along other axis..For protection and isolation purposes an argon flux was flushed on the substrate surface during the laser treatment. The focal distance of the lens was 63 mm, while the working distance was 32 mm.

The aim of this experimentation was to understand the effects of the laser on the elastic properties of the material. All the tests were performed by bending the specimens in the elastic range and treating them with a diode laser. The heating of the material induces a reduction of the elastic energy that the material is able to store and, as a consequence, the reduction of the springback phenomenon [7,8].

The steps of experimental procedure were as follows:

(i) the specimens were cleaned by using ultrasound cleaning;

(ii) the aluminium sheets were placed in a three points bending load system with which it was possible to induce a fixed starting inflexion in the elastic range. Figure 1 depicts a diagram of the load system used in the experimentation;

(iii) the specimens bent in the elastic range were treated with the diode laser along the medium axis (highlighted with the blue arrow in figure 1)

(iv) After the removal of the applied load, the residual inflection of the aluminium sheets was monitored with a Taylor-Hobson profilometer. Ten profiles along the axial direction and at a distance of 0.5 mm from each other were acquired. The maximum inflection value was measured for all the profiles. The mean of this inflexion values represents the residual inflexion considered. The measure of the straightness of the specimens using the profilometer was taken to assure the same starting conditions before the laser treatment

Table 1 summarizes the experimental plan used. Three operational parameters were considered: laser power (range 500 to 800 W), laser scan speed (range 4 to 8 mm/s), and the starting inflexion (range 0.5 to 2 mm). All the tests reported in Table 1 were repeated at least three times to assure process repeatability and reproducibility.

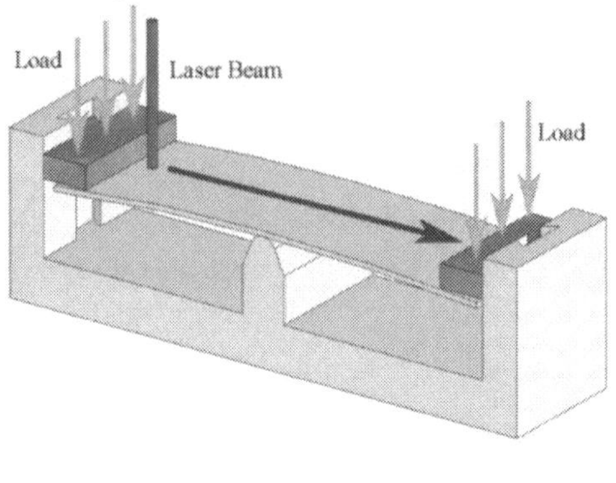

Figure 1. Diagram of the inflexion system used.

A full factorial design of experiment was employed to study the effects of laser treatment on the material behavior during the forming process. Table 1 summarizes the experimental plan developed. Three operational parameters were considered: laser power (range 500 to 800 W), laser scan speed (range 4 to 7 mm/s), and the starting inflexion (range 0.5 to 2 mm). All the tests reported in Table 1 were repeated at least three times to assure process repeatability and reproducibility. The outcome of the experimental plan was the residual inflection.

Table 1. Full factorial experimental plan.

Experimental Factors			
Experimental Levels	**Laser Power, W**	**Laser speed, mm/s**	**Starting Inflexion, mm**
I	500	4	0,5
II	600	5	1
III	700	6	1,5
IV	800	7	2

2.2 Uncertain Model of the process

A standard linear regression model was developed, to link the process parameters (velocity v, laser Power P, and initial bending I) to the achieved residual bending B after the process. In addition, the total energy *E* provided by the laser is considered also. Total energy is proportional to a term *P/v*. As a test case, the model described by a second order model following the equation:

$$B = a_1 + a_2 v + a_3 P + a_4 I + a_5 E + a_6 vP + a_7 vI + a_8 vE + a_9 PI + a_{10} PE + a_{11} IE + a_{12} v^2 + a_{13} P^2 + a_{14} I^2 + a_{15} E^2$$

A practical way to model the uncertainties inherent to the parameters of a model is to represent the model parameter by fuzzy numbers. Forming a special class of fuzzy sets, fuzzy numbers are characterized by a membership function μ (x), 0<μ (x)<1, that defines the degree to which the parameter can take a certain value.

For the evaluation of systems with uncertain the Transformation Method, that is a practical implementation of the fuzzy arithmetic, can be used [9, 10]. The Transformation Method considers the model as a black box providing only an input-output relation that is a mapping between points in the input parameter hyperspace and points in the output space.

Among the advantages of the Transformation Method, its characteristic property of reducing the fuzzy arithmetic to multiple crisp-number operations entails that the Transformation Method can be implemented without major problems into an existing software environment for system simulation, avoiding expensive rewriting of the software. This approach was already used to describe the inherent uncertainties related to a model describing the EFB coating process [10] and FB finishing process [11]. The aim of this paper is to demonstrate the applicability of the method to the process of laser assisted bending, as well as its utility in the prediction of the residual bending variability and in operational parameters setting.

3 Results and discussion

The unknown coefficients $a_1, a_2, ..., a_{15}$ can be obtained from a linear regression algorithm using the experimental data. The dataset dispersion around this model is shown in Figure 2. It reports, for each parameter combination, the measured residual bending (x axis) and the corresponding model results (y axes).

The model is very rough in describing the physics of the process, thus it is essential to assess the inherent model uncertainty through the use of a fuzzy model that can be expressed as:

$$\underline{B} = \underline{a}_1 + \underline{a}_2 v + \underline{a}_3 P + \underline{a}_4 I + \underline{a}_5 E + \underline{a}_6 vP + \underline{a}_7 vI + \underline{a}_8 vE + \underline{a}_9 PI + \underline{a}_{10} PE + \underline{a}_{11} IE + \underline{a}_{12} v^2 + \underline{a}_{13} P^2 + \underline{a}_{14} I^2 + \underline{a}_{15} E^2$$

where the regression coefficients $\underline{a}_1, \underline{a}_2, ..., \underline{a}_{15}$ are triangular fuzzy numbers. Each of them has the modal value coinciding with the results provided by the linear regression, while the support is appropriately chosen, based on the standard deviation provided by the linear regression analysis, so that all the experimental data are included at $\mu = 0$.

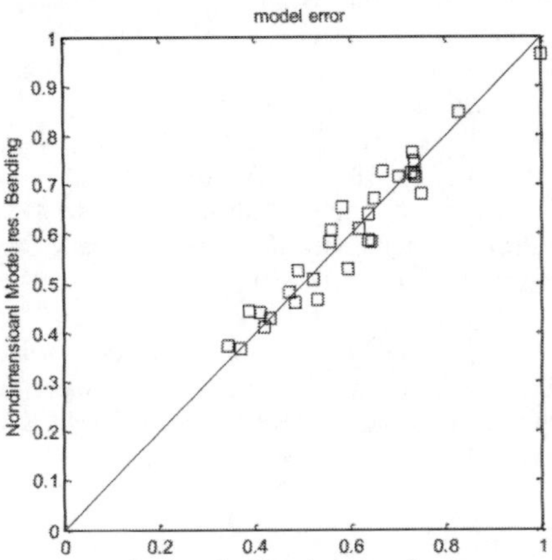

Figure 2. Dataset dispersion around the model. Blue line: model; red circles: experimental results

The fuzzy model can be used to evaluate how much a given experimental sample, characterized by v_i, P_i, I_i, E_i and the corresponding bending B_i, belongs to the fuzzy set defined by the model. In particular, the crisp model does not represent any experimental data. As the level of uncertainty is increased, measured by a decrease in the membership function, the model accommodates a larger number of samples with lower membership level. The Transformation Method is used, given the input fuzzy numbers, to propagate to the residual bending the values of the input parameters: velocity v, power P, initial bending I and energy E. The results are presented in Figure 3.

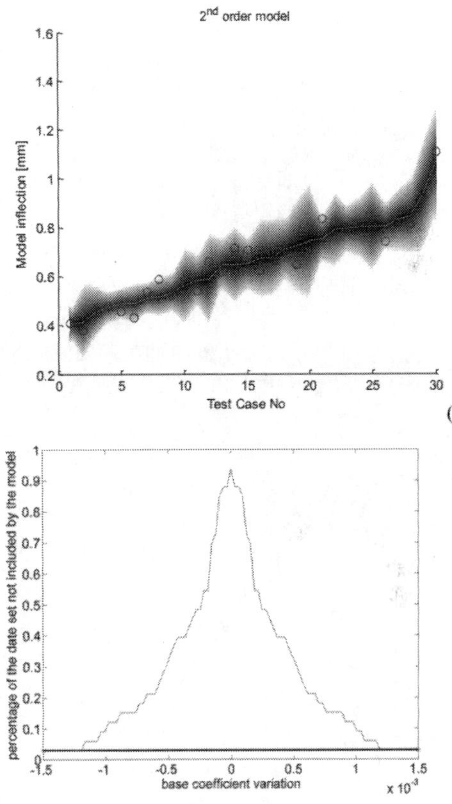

(a) (b)

Figure 3. a) Results of the Transformation Method (grey shaded area), experimental results (blue); b) Percentage of experimental sample described by the model as a function of the membership level.

In Figure 3a the experimental data are represented as blue circles and the fuzzy model results are represented by the shaded area, where lighter zones refer to lower membership level. The graph in Figure 3b suggests how much a given sample, characterized by a combination of process parameters and the corresponding residual bending, belongs to the fuzzy set. The fuzzy model can be used to predict the final bending of the sample, but it is also a useful tool to select operational parameters, as already shown in [8]. By assuming that a fixed bending must be obtained and after fixing a parameter *e.g.* the velocity. Therefore, for a desired value of the residual bending *B*, the inverse model provides the combinations of *P* and *I* leading to that result (intersection between the surface and the plane in Fig. 3a, red line on Fig. 3b). Moreover, it is possible to built a fuzzy process map where, as shaded areas, the uncertainty related to each operational point is reported. The fuzzy model produces a results that cannot be obtained from the crisp regression model that is: among all the points that produce the same desired residual bending, some of them (e.g. I=1.5mm and P = 510W) fall in areas where the model has a large uncertainty, while some of them (e.g. I = 0.5mm P=625W) fall in areas where the model is very precise.

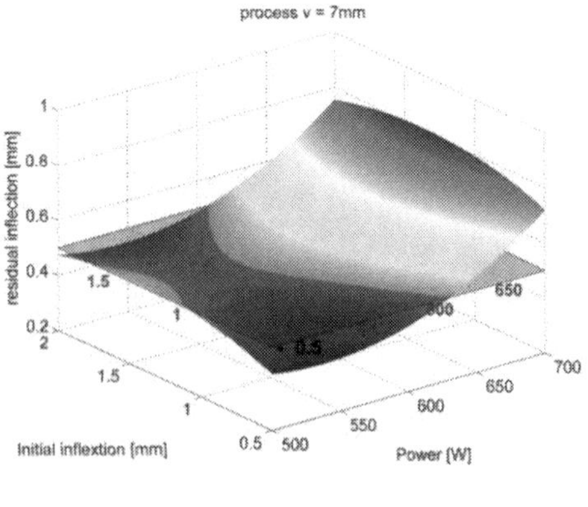

(a)

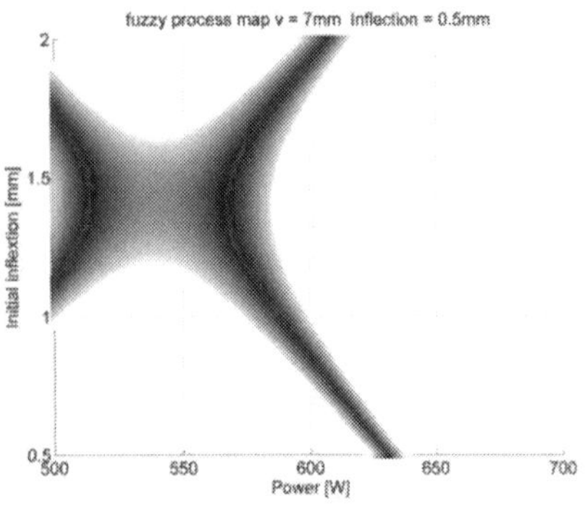

(b)

Figure 4. a) a) possible operational point from the crisp model b) Results of the Transformation Method (grey shaded area).

4 Conclusions

The fuzzy-logic based approach was used to describe the uncertainties of a model generated by an experimental investigation in the field of assisted laser bending. The fuzzy approach is used to evaluate how much a given sample, characterized by a combination of process parameters and the corresponding residual bending, belongs to the fuzzy set defined by the model. At the same time, the fuzzy model can be used to select process parameters and predict the inherent uncertainty related to the adopted model. This approach provides a tool to select, the model with the best fidelity, not just by looking a the "mean model" but also to the level of uncertainty that the model produces. Moreover, given a model and a set of parameters, the fuzzy approach allows the user to select the best operational point i.e. the one that is characterized by the lowest level of uncertainty.

References

1. Kalpakjian S, Schmid S. Manufacturing. Engineering and technology. 5th ed. 2005.
2. Tekiner Z. An experimental study on the examination of springback of sheet metals with several thicknesses and properties in bending dies. Journal of Materials Processing Technology, **145**:109–17 (2004).
3. Vukota B. Sheet metal forming processes and die design. (2004)
4. Firat M, Kaftanoglu B, Eser O. Sheet metal forming analyses with an emphasis on the springback deformation. Journal of Materials Processing Technology, **196**:135–48 (2008)
5. Eggertsen P-A, Mattiasson K. On the modelling of the bending–unbending behaviour for accurate springback predictions. International Journal of Mechanical Sciences, **51**:547–63 (2009)
6. Liu YC. The effect of the restraining force on shape deviations in flanged channels. ASME Journal of Engineering Materials and Technology, **110**: 389–94 (1988)
7. A. Gisario, M.Barletta, C.Conti, S.Guarino. Springback control in sheet metal bending by laser-assisted bending: Experimental analysis, empirical and neural network modeling, Optics and Lasers in Engineering **49**, 1372–1383 (2011).
8. M. Barletta, A Gisario, S. Guarino. Hybrid forming process of AA 6108 T4 thin sheets: modelling by neural network solutions. Proceedings of the Institution of Mechanical Engineers, Part B: Journal of Engineering Manufacture **223**, 535–545 (2009).
9. O. Giannini M. Hanss M. "The component mode transformation method: A fast implementation of fuzzy arithmetic for uncertainty management in structural dynamics", Journal of Sound and Vibrations, **311**, 1340–1357 (2011).
10. O. Giannini, M. Hanss, "An interdependency index for the outputs of uncertain systems", Fuzzy Sets and Systems 159, pp. 1292-1308, (2008)
11. F. Trovalusci, M. Barletta, O. Giannini, "Fuzzy model for electrostatic fluidized bed coating", Proceedings of the 12th Biennial Conference on Engineering Systems Design and Analysis ESDA14, June 25-27, 2014, Copenhagen, Denmark.

MATEC Web of Conferences **45**, 04011 (2016)
DOI: 10.1051/matecconf/20164504011

The Study of Cutting Conditions Effects on the Damping Process Using the Experimental Taguchi Method

Haikel Mejri[1], Ziyed Sahraoui[1] and Kamel Mehdi[1,2]

[1]*URMSSDT, Engineering National High School of Tunis (ENSIT), University of Tunis (UT), 5 Avenue Taha Hussein P.B. 56 Bab Mnara 1008 Tunis, Tunisia*
[2]*Preparatory Institute for Engineering Studies El Manar (IPEIMa), University of Tunis EL Manar (UTM), PB. 24, 2092 Tunis, Tunisia*

Abstract. This article focuses on determining the effects of cutting conditions and their interactions on the cutting process damping in the case of curvilinear milling. The tests were performed using a numerical model simulation that allows the prediction of cutting forces and damping. The effects and interactions are determined using the Taguchi experimental method. Analysis of variance (ANOVA) was performed to know the level of importance of the machining parameters on the cutting damping process. The results revealed that the Depth of cut A_p **"C"** and cutting speed V_c **"B"** have the most significant influence on the **Cxx** and **Cxy** process damping. The variations of tool diameter D **"A"** and clearance angle **α** have remarkable effects on the process damping **Cxx**. The **"BC"** interaction has the greatest effect on the process damping **Cxx** while the **"AC"** interaction has the greatest effect on the process damping **Cxy**.

1 Introduction

The comprehension of the various phenomena related to machining is necessary in order to propose models and methods which can predict the stability area. In most carried out researches, (which aims at the determination of the influence of the various cutting parameters), the researchers simplified or completely excluded the cutting process damping [1]. In fact, the experiments showed that for low cutting speeds, the cutting damping process is stable [2].

An important contribution on knowledge of the physical modeling of the cutting process was provided by Tlusty [3]. He showed with Ismail [4] that the cutting process damping affects the stability during the machining and it increases with the reduction of the cutting speed.

In 2012, Mehdi and Zghal [5] presented a numerical model allowing the prediction of cutting forces in peripheral milling process. They studied the effects of tool parameters (diameter, helix angle, number of teeth) on cutting process damping and cutting force distributions. In this model the cutting damping process was presented by the following matrix:

$$[C_c] = \begin{bmatrix} C_{xx}(t) & C_{xy}(t) \\ C_{yx}(t) & C_{yy}(t) \end{bmatrix}$$

Where:

$$C_{xx}(t) = \sum_{j=1}^{N_t}\sum_{k=1}^{N_d} \delta_{k,j}[c_{t_j}(t) + c_{r_j}(t)]\cos\varphi_j(t)\sin\varphi_j(t) \quad (1)$$

$$C_{xy}(t) = \sum_{j=1}^{N_t}\sum_{k=1}^{N_d} \delta_{k,j}[c_{t_j}(t)\cos^2\varphi_j(t) - c_{r_j}(t)\sin^2\varphi_j(t)] \quad (2)$$

$$C_{yx}(t) = \sum_{j=1}^{N_t}\sum_{k=1}^{N_d} \delta_{k,j}[c_{r_j}(t)\cos^2\varphi_j(t) - c_{t_j}(t)\sin^2\varphi_j(t)] \quad (3)$$

$$C_{yy}(t) = -C_{xx}(t) \quad (4)$$

Where:

$c_{r_j}(t)$ and $c_{t_j}(t)$ represent the cutting damping factors in the thrust and tangential directions,

$\varphi_j(t)$ is the position angle of a point on the cutting edge of the jth helical flute,

N_t is the number of the tool's teeth and N_d is the number of the elementary cutting disks.

$\delta_{k,j}$ is equal to 1, if the jth tooth of the elementary k^{th} disk of the tool is in cut, and 0 if it's out of cut.

It is difficult to determine the cutting process damping in experiments, hence the importance of the numerical model simulation which was presented and validated experimentally. This model allowed us, initially, to determine the cutting process damping while using the experimental cutting values. Secondly, it permitted to study the variation of the damping process in function of cutting parameters thus enhancing the results already presented by researchers over the years by using the experimental plans method. In 2015, Li Xin and all [6] confirmed the importance of taking into account the

Article available at http://www.matec-conferences.org or http://dx.doi.org/10.1051/matecconf/20164504011

cutting dumping process in their research on the machining of thin-walled workpiece.

The experimental plans method, is applicable to all kinds of engineering problems [7-10], and it brings a rigorous approach in the governance and the establishment of the tests, which enables us not to reason more by groping.

This paper comes to complete the previous work of Mehdi and Zghal [5]. It focuses on determining the effects of cutting conditions and their interactions on the cutting process damping in the case of curvilinear milling. The tests were performed using a numerical model simulation that allows the prediction of cutting forces and damping. The effects and interactions are determined using the Taguchi experimental method. Analysis of variance (ANOVA) was performed to know the level of importance of the machining parameters on the cutting damping process.

2 Cutting process damping Cxx

In this part, we will use the parameters of cut for curvilinear machining allowed in table 1and the material properties in table 2.

Table 1. L16 standard table of settings applied to the simulation

N° test	D (A) (mm)	Vc (B) (m/min)	Ap (C) (mm)	α (D) (degrees)
1	6	25	0,25	4
2	6	25	0,25	15
3	6	25	3	4
4	6	25	3	15
5	6	300	0,25	4
6	6	300	0,25	15
7	6	300	3	4
8	6	300	3	15
9	40	25	0,25	4
10	40	25	0,25	15
11	40	25	3	4
12	40	25	3	15
13	40	300	0,25	4
14	40	300	0,25	15
15	40	300	3	4
16	40	300	3	15

Figure 1 presents the results obtained by simulation of cutting forces. During each test, the measurement of cutting damping process Cxx was taken into account and we made the average.

Table 2. Material properties

Steel	A60
Rr (MPa)	570
Re (MPa)	335
E (MPa)	200000
Density ρ (Kg/m3)	7850
Poisson's ratio	0.29

2.1 Calculation of average effects of factors and the interactions effect

The average value response of all factors of a jth level is given equation (5).

$$\bar{A}_j = \frac{1}{n}\sum_{i=1}^{n} R_{ij} \tag{5}$$

Here, we have considered two levels (j = 1, 2). Table 3 presents the average response to each factor level.

The overall average $\bar{T}$ of all the tests corresponds to the central point of average responses for the levels of each factor:

$$\bar{T} = \frac{\bar{A}_1+\bar{A}_2}{2} = \frac{210.961+159}{2} = 184.98\ Ns/m \tag{6}$$

The average effect of each factor level, appreciates relatively to the overall average. For example, the average effect of the factor A at the level 1 is calculated by the following equation:

$$E_{A_1} = \bar{A}_1 - \bar{T} = 210.961 - 184.98 = 25.981 \mu m \tag{7}$$

Similarly, we calculate the average effect of the A at the level 2:

$$E_{A_2} = \bar{A}_2 - \bar{T} = 159 - 184.98 = -25.98 \mu m \tag{8}$$

It is clear that: $EA_1 = - EA_2$

Table 4 gives the average effects of all factors.

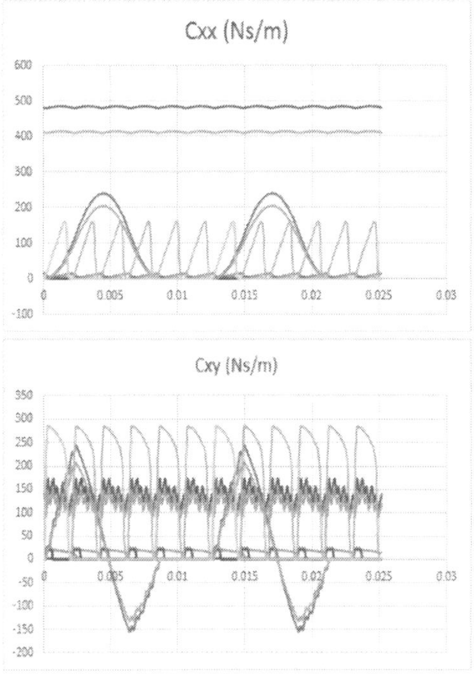

Figure 1. Effect of cutting parameters on cutting process damping (simulation)

Table 3. Average values of factors responses for each level

Fact	Average response for each factor level	
	Lev. 1	Lev. 2
A	210.961	159.000
B	266.469	103.493
C	50.438	319.524
D	151.689	218.273

Table 4. Effect factors on the measured values.

Fact	Effect of factors on the measured values	
	Lev. 1	Lev. 2
A	25.981	-25.981
B	81.488	-81.488
C	-134.543	134.543
D	-33.292	33.292

2.2 Calculation of the interactions

By applying the general formula of calculation of the interactions, we find for IA_1B_1:

$$I_{A_1B_1} = \overline{A_1B_1} - E_{A_1} - E_{B_1} - \overline{T} \qquad (9)$$

The average response related to each of the combinations AB is calculated using the same principle as for the following factors:

$$\overline{A_1 B_1} = \frac{160.86+160.86+204.76+414.11}{4} = 235.148 \; Ns/m \qquad (10)$$

Table 5 gives the average values of the interactions

Table 5. Average values of interactions.

Average values of interactions			
A_1B_1	A_1B_2	A_2B_1	A_2B_2
235.148	186.775	297.790	20.210
A_1C_1	A_1C_2	A_2C_1	A_2C_2
85.905	336.018	14.970	303.030
A_1D_1	A_1D_2	A_2D_1	A_2D_2
153.166	283.071	149.250	168.750
B_1C_1	B_1C_2	B_2C_1	B_2C_2
73.092	493.309	51.642	187.243
B_1D_1	B_1D_2	B_2D_1	B_2D_2
230.550	302.388	72.828	134.158
C_1D_1	C_1D_2	C_2D_1	C_2D_2
50.438	34.663	252.940	386.108

Then, from this table, we get:

$$I_{A_1B_1} = 235.148 - 25.98 - 81.488 - 184.98 = -57.30\,\mu m \qquad (11)$$

And in the same way we calculate the other interactions. The values of these interactions are given in table 6.

Table 6. Effect of interactions on the measured values

Effect of interactions on the measured values			
A_1B_1	A_1B_2	A_2B_1	A_2B_2
-57.302	57.302	57.302	-57.302
A_1C_1	A_1C_2	A_2C_1	A_2C_2
9.487	-9.487	-9.487	9.487
A_1D_1	A_1D_2	A_2D_1	A_2D_2
-24.504	-0.005	23.542	-23.542
B_1C_1	B_1C_2	B_2C_1	B_2C_2
-58.834	92.297	82.692	-50.793
B_1D_1	B_1D_2	B_2D_1	B_2D_2
-2.627	2.627	2.627	-2.627
C_1D_1	C_1D_2	C_2D_1	C_2D_2
33.292	-0.025	-33.292	33.292

2.3 Discussion

The curves illustrated in figures 2 and 3 represent all the effects and interactions on the cutting damping process Cxx. From these figures, we can see that:

-The pass depth (C) and the cutting speed (B) have the most significant influence on the process damping Cxx. Indeed, if we increase the cutting speed, the process damping Cxx decrease.

-By against, if we increase the depth of cutting, the process damping Cxx increase.

-The variation of the tool diameter (D) and the variation of the clearance angle (α) have an effect on cutting damping process Cxx. However, the effect of both factors remains lowers than that the depth of cutting and cutting speed.

-The BC interaction has the greatest effect on the cutting damping process Cxx, AB interaction also exhibits a significant effect. This is a reason for which it must be taken into consideration.

-The CD interaction is intense, although its effect is lower than that of BC and AB.

-Interactions AB, AC, AD and BC are low.

-The BD interaction is null, then it has no effect on the cutting damping process Cxx.

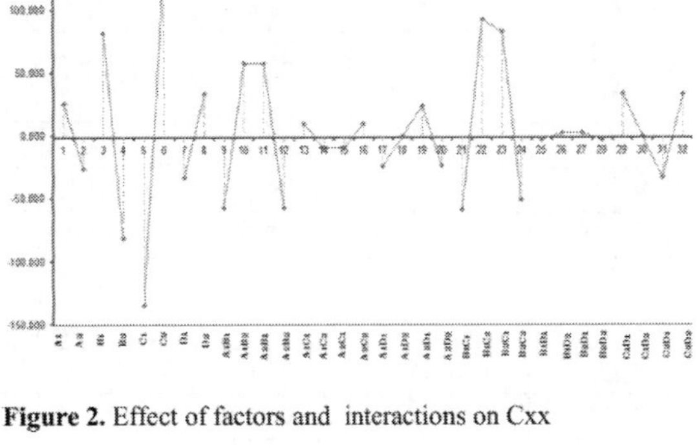

Figure 2. Effect of factors and interactions on Cxx

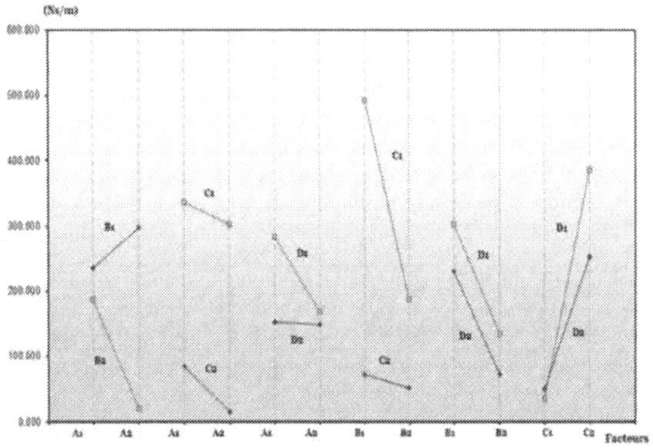

Figure 3. Effect of interactions on Cxx cutting damping

3 Cutting process damping Cxy

Like in the case of cutting process damping Cxx, the same approach has been adopted to determine the tables of the average response, the effect of factors on the measured values, the average values of the interactions and the effect of the interactions on the measured values (tables 7, 8, 9 and 10).

Table 7. Average values of factors responses for each level.

Fact	Average values of factors for each level	
	Lev. 1	Lev. 2
A	168.333	140.233
B	244.599	63.967
C	82.679	225.886
D	152.224	156.342

Table 8. Effect of factors on the measured values.

Fact	Effect of factors on the measured values	
	Lev,1	Lev,2
A	14.050	-14.050
B	90.316	-90.316
C	-71.604	71.604
D	-2.059	2.059

Table 9. Average values of interactions

Average values of interactions			
A_1B_1	A_1B_2	A_2B_1	A_2B_2
231.560	105.106	257.638	22.828
A_1C_1	A_1C_2	A_2C_1	A_2C_2
143.341	193.325	22.018	258.448
A1D1	A_1D_2	A_2D_1	A_2D_2
184.883	151.783	119.565	160.900
B1C1	B_1C_2	B_2C_1	B_2C_2
152.638	336.560	12.721	115.213
B1D1	B_1D_2	B_2D_1	B_2D_2
231.945	257.253	72.503	55.431
C_1D_1	C_1D_2	C_2D_1	C_2D_2
82.678	82.681	221.770	230.003

Table 10. Effect of the interactions on the measured values

Effect of the interactions on the measured value			
A_1B_1	A_1B_2	A_2B_1	A_2B_2
-27.089	27.089	27.089	-27.089
A_1C_1	A_1C_2	A_2C_1	A_2C_2
46.612	-46.612	-46.612	46.612
A1D1	A_1D_2	A_2D_1	A_2D_2
18.609	-18.609	-18.609	18.609
B1C1	B_1C_2	B_2C_1	B_2C_2
-20.358	20.358	20.358	-20.358
B1D1	B_1D_2	B_2D_1	B_2D_2
-10.595	10.595	10.595	-10.595
C_1D_1	C_1D_2	C_2D_1	C_2D_2
2.058	-2.057	-2.058	2.058

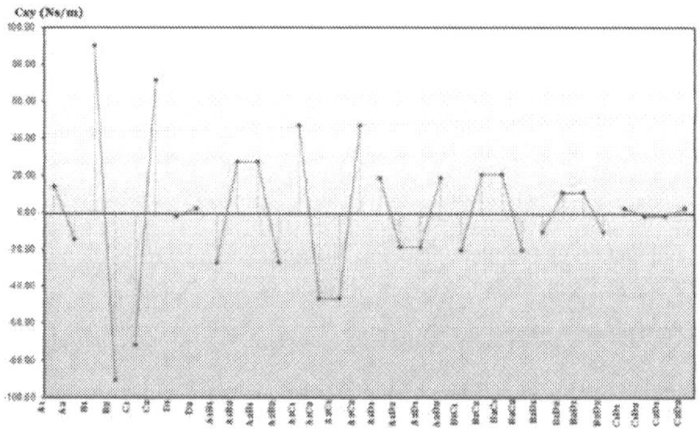

Figure 4. Effect of factors and interactions on Cxy

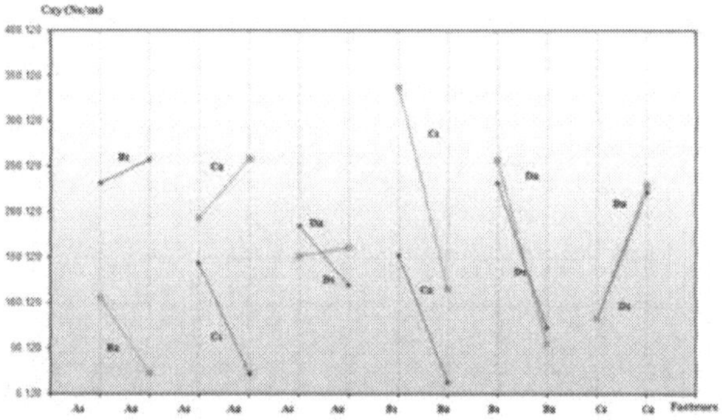

Figure 5. Effect of interactions on Cxy damping

The curves illustrated in figures 4 and 5 represent all the effects and interactions on the Cxy cutting damping process. From these figures we can see that:

-The cutting depth (C) and the cutting speed (B) have the most significant influence on the cutting damping process Cxy. In fact if we increase the cutting speed, the cutting damping process Cxy decrease. On the other hand, if we increase the depth of cut, the cutting damping process Cxy increase. It is noticed that the effect of the cutting speed (B) is more important.

-The cutting damping process Cxy passes by negative values by increasing the cutting depth.

-The interaction AC has the greatest effect on the cutting damping process Cxy, interactions AB, AD and BC also present a significant effect.

-The interactions AD, BD and CD are intense; despite their effects remain lower than that of AC.

-Interactions AB, AC and BC are not very significant.

4 Conclusions

In this paper, a Taguchi experimental plan was used. This method allows the limitation of the number of necessary and sufficient tests to study the cutting damping process in curvilinear milling was used.

We have succeeded in identify the main influential machining parameters on the cutting damping process. Such as the tool diameter (D), the cutting depth (ap), the cutting speed (Vc) and clearance angle (α).

The simulation results have allowed us to conclude that:

-The cutting depth (C) and the cutting speed (B) have the most significant influence on the cutting damping process Cxx and Cxy.

-The variation of the tool diameter (D) and the variation of the clearance angle (α) act in a remarkable manner on the cutting damping process Cxx, but they have a less significant effect on the cutting damping process Cxy.

-The BC interaction has the greatest effect on the cutting damping process Cxx, the AB interaction of same has a significant effect.

-AC interaction has the greatest effect on the cutting damping process Cxy, interactions AB, AD and BC of same have a significant effect. This is one reason why we must take into consideration.

Finally, it is important to insist on the major role of numerical model simulation, with it we have predicted the cutting damping process in curvilinear milling. The future work of this research is the study of the effect of cutting parameters and tool parameters and their interactions on the cutting forces.

References

1. C.Y. Huang , J.J. Wang, "Mechanistic Modeling of Process Damping in Peripheral Milling", *Transactions of the ASME, Journal of Manufacturing Science and Engineering,* Vol. 129, pp. 12-20 (2007).
2. D.Y. Jang , A. Seireg, "Tool Natural Frequency as the Control Parameter for Surface Roughness" *Machine Vibration*, Vol. 1, pp. 147-154). (1992)
3. J. Tlusty, "Dynamics of high speed milling" *Transactions of the ASME, Journal of Engineering for Industry*, Vol. 108, pp. 59-67 (1986).
4. J. Tlusty, F. Ismail, "Special Aspects of Chatter in Milling" *Transactions of the ASME, Journal of Engineering for Industry*, Vol. 105, pp. 24-32 (1983).
5. K. Mehdi, A. Zghal , "Modelling Cutting Force Including Thrust and Tangential Damping in Peripheral Milling Process" *Int. J. Machining and Machinability of Materials,* Vol. 12, No. 3, (2012)
6. Xin Li, Wei Zhao, Liang Li, Ning He, and ShengWei Chi"Modeling and Application of Process Damping in Milling of Thin-Walled Workpiece Made of Titanium Alloy" *Hindawi Publishing Corporation Shock and Vibration* Vol. 2015, Article ID 431476, 12 pages, http://dx.doi.org/10.1155/2015/431476.
7. J. GOUPY. "Plans d'expériences" *Techniques de l'ingénieur. Traité Analyse Chimique et Caractérisation*, P 230, p. 1-20. (1992).
8. J.GOUPY. *"Introduction aux plans d'expériences"* Dunod. Paris. Pages 303. (2001).
9. J.GOUPY. *"Pratiquer les Plans d'Expériences"*. Dunod. Paris. 560 pages. (2005).
10. J. GOUPY, L. CREIGHTON, "*Introduction aux plans d'expériences*", DUNOD, (2006).

MATEC Web of Conferences **45**, 04012 (2016)
DOI: 10.1051/matecconf/20164504012

Optimization and comparison between two 6-DoF parallel kinematic machines for HIL simulations in wind tunnel

Enrico Fiore and Hermes Giberti

Politecnico di Milano, Mechanical Department Via La Masa 1, 20156 Milano, Italy

Abstract. Parallel Robots are for sure the best choice for those applications in which a high accuracy and a high stiffness are required. As a matter of fact the closed-loop kinematic chains that characterize these machines guarantee a very low sensitivity of the end-effector position to the various sources of error. However the working volume of parallel kinematic machines is in general limited with respect to their dimensions. Moreover the reduced mobility range of the passive joints represents a critical aspect that cannot be underestimated. For these reasons PKMs should be optimized for the application they are meant.Given the specifications of the workspace that the machine should be able to cover, it's necessary to identify an architecture that is suitable for this specific task. This paper describes the optimization process and the comparison between two candidates, the Hexaslide and the Gough-Stewart platform, in order to find out which one is the best choice for a task that requires to reproduce the hydrodynamic effect of sea waves at the bottom of scale models of off-shore wind turbines and boats during HIL simulations in the wind tunnel of the Politecnico di Milano.

1 Introduction

Nowadays one of the most important lines of research is the study of the interaction between the wind and civil structures. However in some cases the wind is not the only source of dynamic excitation, for example on sailing boats and on floating off-shore wind turbines (FOWT) the contribution given by the sea cannot be neglected.

The experimental phase is fundamental to understand the aerodynamic and hydrodynamic phenomena that are involved: as a matter of fact by properly setting the control parameters of the inclination of the blades of wind turbines it's possible to have a better exploitation of the wind energy; instead as regards the sailboats it's important to understand which is the fluid-structure interaction during the change of direction.

Three kinds of experiment can be carried out: wind tunnel tests; tank tests; full scale simulations. Moreover, through these tests it is possible to validate CFD codes and to set all the parameters needed for the numerical simulations. As regards the wind tunnel tests, the real wind profile can be reproduced with a good approximation unlike the hydrodynamic effect of the sea waves. For this reason it's important to equip the wind tunnel of the Politecnico di Milano with a machine that takes as a input the forces and moments recorded by a 6-axes balance at the bottom of the scale models of FOWTs and boats, and on the basis of a mathematical model of the sea imposes a movement to the scale models [1].

In the beginning the hypothesis of using a serial robot was considered. Indeed these manipulators are definitely the most diffused commercially speaking and moreover don't suffer any problem in terms of workspace dimensions. However in serial manipulators the errors of each joint add up to those of the successive joints with the result of worsening the accuracy of the positioning of the end-effector. Due to their intrinsic embossed structure they have to bear with high bending loads and so the links that connect one actuated joint to the other become increasingly bulky starting from the end-effector towards the base joint. At this point the parallel kinematic manipulators (PKM) are taken into account. The performances of these machines strictly depend on their dimensions and so it becomes difficult to make a comparison between different architectures as stated by Merlet in [2]. However it's still possible to highlight their general peculiarities. For sure their working volume is lower than that of serial robots, but these machine are characterized by a high stiffness thanks to the fair distribution among the links of the loads acting on the end-effector. Furthermore in some PKMs the joints at the ends of the links are spherical or universal and so the loads the links are subject to are exclusively of axial nature. In parallel kinematic manipulators the interaction mechanisms between the errors of each joint and those of the end-effector are far too complex with respect to serial robots with the result of having a lower sensitivity of the positioning errors of the end-effector.

After these considerations two different architectures of hexapods were identified as suitable candidates to perform the hardware in the loop (HIL) simulations: (1) the Hexaslide which is made of a mobile platform connected to six linear guides by means of six links of fixed length, so that six independent kinematic chains can be identified belonging to the PUS family. A detailed description of 6-DoF PUS PKMs is provided by Bonev in [3]. The actuating system of the Hexaslide is fixed to the ground so that the resulting moving mass is represented just by the slender links and the thin platform. However this machine is characterized by a great footprint. (2) the second candidate is the well-known Gough-Stewart platform. This machine is made of six links of variable length that connect the mobile platform to six universal joints fixed to the ground. In this case the kinematic chains belong to the UPS family. Despite the lower dimensions with respect to the Hexaslide architecture, the actuation in this manipulator is provided directly by the links. Their length is changed by means of a ball-screw transmission unit which is characterized by a very high inertia. Furthermore the motors cannot be fixed to the ground but should be mounted on the links and so they contribute to increase the total moving mass.

The vertical available space for the machine in the wind tunnel is limited by the boundary layer in the zone near the ceiling. As a matter of fact here the wind profile is different with respect to the real one and so it's important that the scale models don't enter in this zone. For this reason the tool center point (TCP) of the manipulator should stay as low as possible. The two chosen architectures allow to have a very low TCP and in addition they avoid the risk of self collision between the machine elements which in general is a critical aspect of PKMs.

In order to respect the requirements in terms of working space and of constraints, the manipulator should undergo an optimization process. For both the Hexaslide and the Gough-Stewart a genetic algorithm was used to optimize the two machines.

The following sections are organized as follow: Section 2 describes the inverse kinematics of the Gough-Stewart platform and of the Hexaslide; Section 3 provides a description of the technical specifications and a detailed explanation of the cost function and of the constraints; numerical results and the comparison between the two architectures are presented in Section 4.

2 Inverse kinematics

In order to bring the robot mobile platform to the desired poses it is fundamental to establish the relationships between the TCP coordinates and the joint space ones. In general for PKMs it's easier to solve the inverse kinematics [4], that is the computation of the joint space coordinates once the TCP coordinates are known. The solution of the inverse kinematics requires to use two different reference systems, a global one denoted by unit vectors $\hat{\boldsymbol{x}}, \hat{\boldsymbol{y}}, \hat{\boldsymbol{z}}$, and a relative one attached to the mobile platform and denoted by unit vectors $\hat{\boldsymbol{x}}', \hat{\boldsymbol{y}}', \hat{\boldsymbol{z}}'$.

2.1 Gough-Stewart Platform

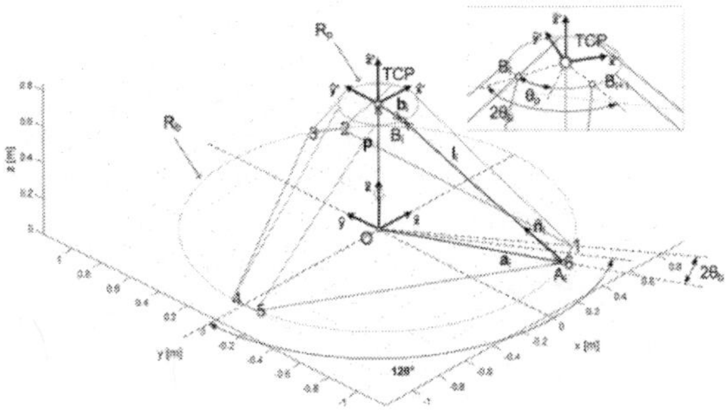

Figure 1. Gough-Stewart platform i-th kinematic chain.

The adopted strategy consists in closing a kinematic chain in order to determine the position of the platform joints with respect to the base joints. Once the coordinates of the i-th platform joint are known it is possible to compute the length of the i-th link. Looking at Figure 1 it is possible to write:

$$\boldsymbol{l}_i = \boldsymbol{p} + [R]\boldsymbol{b}'_i - \boldsymbol{a}_i \quad (1)$$

where $\boldsymbol{l}_i$ is the vector aligned with the i-th link, $\boldsymbol{p}$ is the vector containing the position and orientation of the mobile platform, [R] is the rotation matrix, $\boldsymbol{b}'_i$ expresses the position of the i-th platform joint with respect to the TCP in the relative frame and $\boldsymbol{a}_i$ is the is the vector that describes the position of the i-th base joint. By squaring equation (1) one gets:

$$l_i^2 = \boldsymbol{p}^T\boldsymbol{p} + \boldsymbol{b}'^T_i\boldsymbol{b}'_i + \boldsymbol{a}_i^T\boldsymbol{a}_i + 2\boldsymbol{p}^T[R]\boldsymbol{b}'_i - 2\boldsymbol{p}^T\boldsymbol{a}_i - 2\boldsymbol{a}_i^T[R]\boldsymbol{b}'_i \quad (2)$$

2.2 Hexaslide

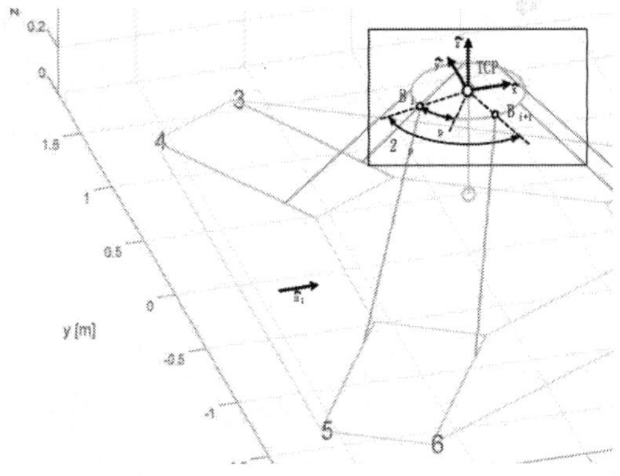

Figure 2. Hexaslide i-th kinematic chain.

For the Hexaslide a similar procedure can be used. Anyway in this case it's necessary to use two kinematic chains, the first one to find the position of the i-th platform joint with respect to the i-th guide and the latter to compute the position of the i-th slider on the relative guide.

So with reference to Figure 2 it is possible to write:

$$l_i = d_i + q_i \hat{u}_i \quad with \quad d_i = p + [R]b'_i - s_i \tag{3}$$

where $\hat{u}_i$ is a unitary vector aligned with the i-th guide axis, s_i is the position of the same guide with respect to the fixed frame, while q_i represents the position of the i-th slider along the corresponding guide axis.

After some simple mathematical passages the following expression is found:

$$q_i = d_i^T \hat{u}_i \pm \sqrt{d_i^T (\hat{u}_i \hat{u}_i^T - [I]) d_i + l_i^2} \tag{4}$$

2.3 Velocity Analysis

Once solved the inverse kinematics it is possible to find the relationship between the sliders/links velocity and the velocity of the TCP. For the Gough-Stewart it can be shown that for the i-th kinematic chain the following expression holds:

$$\dot{l}_i = [\hat{n}_i^T \ (b_i x \, \hat{n}_i)^T] W \tag{5}$$

where $W = [\dot{x}, \dot{y}, \dot{z}, \omega_x, \omega_y, \omega_z]^T$ is a vector containing both the translational and angular components of TCP velocity and $\hat{n}_i$ is the i-th link unitary vector.

For the Hexaslide a similar expression is obtained:

$$\hat{n}_i^T \hat{u}_i \dot{q}_i = [\hat{n}_i^T \ (b_i \, x \, \hat{n}_i)^T] W \tag{6}$$

Considering all the six links the previous expressions can be recast in a matrix form using the Jacobian matrix $[J]$:

$$W = [J] \dot{q} \tag{7}$$

2.4 Kinetostatic

The Jacobian matrix is fundamental also for the kinetostatic problem. As a matter of fact by applying the virtual work principle it's possible to compute the actuators forces τ_{act} needed to equilibrate the external forces and moments acting on the TCP $f_{e,TCP}$:

$$\delta X^T f_{e,TCP} + \delta q^T \tau_{act} = 0 \tag{8}$$

It is possible to recast the virtual work principle as follows:

$$\delta q^T \left([J]^T f_{e,TCP} + \tau_{act} \right) = 0 \tag{9}$$

Since this expression must hold for all virtual variation of the joint coordinates the following expression is found:

$$\tau_{act} = -[J]^T f_{e,TCP} \tag{10}$$

If the aim is to study how the forces acting on the TCP are transformed in those acting on the sliders along the guide axes it's sufficient to change the sign of the previous expression. In this sense the transposed Jacobian matrix assumes the meaning of a generalized transmission ratio. The vector $f_{e,TCP}$ is composed of three elements representing the external forces and three elements representing the external moments acting on the TCP. In order to normalize all the terms of the jacobian matrix and to consider the isotropy of the machine it's necessary to use a scale factor defined as characteristic length. The study of the isotropy of PKMs depending on their design parameters is a field in which many researchers are involved [5, 6].The characteristic length is defined in many ways in literature, in this paper the definition provided by Fassi and Legnani in [7] is used. This parameter allows to define a normalized Jacobian matrix which is used for the formulation of the kinetostatic constraint and is denoted as $[J_n]$.

3 Optimization

The optimization process requires to define a set of variables that constitute the quantities that can be varied in order to achieve a specific goal, in particular the minimization of a cost function. In this case as often happens in the optimization problems, the system to be optimized is subject to a set of constraints of various nature. The algorithm used to carry out the optimization is a genetic algorithm. Thanks to the symmetry of the two machines it is possible to use a single objective algorithm; on the contrary the optimization process of other PKMs may require a multi-objective algorithm. An example of a multi-objective genetic algorithm is presented in [8, 9]. In this paragraph the technical specifications, the cost function, the parameterization of the two manipulators and the constraints are described.

3.1. Technical Specifications

The researchers of the wind tunnel of the Politecnico di Milano studied the scaling of the aerodynamic models and the movements of the real FOWTs and boats. These studies allowed to define the workspace the TCP should be able to move in. The workspace is represented by a parallelepiped of dimensions 300 mm x 150 mm x 150 mm within which the TCP can assume any combination of the rotations described. Due to the problems related to the limited vertical space, the center of the workspace is set to be in the point of coordinates $[0,0,0.7]^T$.

3.2 Cost Function

The cost function for an individual characterized by a certain set of variables consists in the computation of the portion of volume of the desired workspace (WSd) that is not possible to cover. To this end the WSd is divided into sub-volumes using a three dimensional grid. Given the dimensions of the WSd and chosen a number of points N_{grid} on each edge, the WSd is divided into N_{grid}^3 sub-portions whose volume is equal to:

$$\Delta v = \frac{2\Delta x \ x \ 2\Delta y \ x \ 2\Delta z}{N_{grid}^3} \tag{11}$$

For a fixed orientation $\overline{\boldsymbol{\Theta}} = [\bar{\alpha}, \bar{\beta}, \bar{\gamma}]^T$ the not covered volume is computed. In other words it's necessary to assess how many points don't comply with all the imposed constraints:

$$v_{nc}(\overline{\boldsymbol{\theta}}) = \sum_i k_i \Delta v \qquad (12)$$

where k_i is equal to 1 if at least one constraints is violated and 0 otherwise. The procedure is repeated for each combination of the angular parameters α, β, γ and the total not covered volume is computed as:

$$V_{nc} = \sum_j v_{nc}(\overline{\boldsymbol{\theta}}_j) \qquad (13)$$

3.3 Constraints

It's important to guarantee that the kinematics of the manipulator is respected, to avoid the self-collision between the elements of the machine and that the force transmitted to the motors is not too high. For these reasons the following constraints are imposed:

I. *Kinematic constraints:* this kind of constraints guarantee to cover each point of the WSd:
   i. $\Delta_i \geq 0$ (Hexaslide only): for each link it's necessary to assess that $\Delta_i = \boldsymbol{d}_i^T(\hat{\boldsymbol{u}}_i\hat{\boldsymbol{u}}_i^T - [I])\boldsymbol{d}_i + l_i^2 \geq 0$. If it is not the case, it means that the distance between the platform joint and the corresponding base joint is greater than the link's length.
   ii. $q_{min} \leq q_i \leq q_{max}$: for the joint coordinates an upper-bound and a lower-bound are imposed. If one or more of the joint coordinates were out of this variation range, the manipulator would approach a singularity condition.
   iii. $\theta_{A,i} \leq \theta_{A,lim}$ *and* $\theta_{B,i} \leq \theta_{B,lim}$: each joint must respect the mobility range given by the constructor. The angular shift is computed with respect to the condition in which the TCP is coincident with the center of the WSd with null orientation.

II. *Kinetostatic constraint:* If for hypothesis a set of unitary forces and moments is applied to the TCP the maximum force transmitted to the motors for a prescribed pose is computed as:

$$\tau_{max}(\boldsymbol{X}) = \|J_n^T(\boldsymbol{X})\|_\infty \qquad (14)$$

The kinetostatic constraint is respected if the computed value is lower or equal to the imposed limit value $\tau_{max,lim}$.

III. *Geometric constraints:* The geometric constraints represent the respect of the minimum distance between the links and between each link and the mobile platform. The minimum distance between two elements is computed as the minimum distance between two segments in space. The platform is outlined as a 500 mm tall cylinder whose symmetry axis is orthogonal to the plane where the platform joint lie. For each pose it's necessary to assess that:

$$\begin{cases} d_{link,ij} \geq d_{link,min} & \forall i,j \ \ with \ \ i \neq j \\ d_{link-platform} \geq d_{link-platform,min} & \forall k \end{cases} \qquad (15)$$

3.4 Machines parameterization

Thanks to the symmetry that characterizes both the Gough-Stewart platform and the Hexaslide it is possible to use a four variable parametrization for the two manipulators.

The Gough-Stewart hexapod is made of a mobile platform on which three couples of spherical joints lie, each of them out of phase of 120° with respect to the others. On the fixed base the disposition of the universal joints is analogous to that of the platform joints. The chosen variables used to parametrize the GS-platform are: (1) R_b, radius of the circumference the base joints A_i lie on; (2) θ_b, semi angular displacement between the two segments that connect the origin with a couple of base joints; (3) R_p, radius of the circumference the platform joints B_i lie on; (4) θ_p , semi angular displacement between the two segments that connect the TCP with a couple of platform joints. Usually the couples of platform joints are organized as [(1,2) (3,4) (5,6)], while the couples of base joints are organized as [(1,6) (2,3) (4,5)].

In the Hexaslide architecture the length of the links is fixed and equal for all the links. On the base there are three couples of parallel guides, each of them out of phase of 120° with respect to the others about the z axis. For this architecture the parameterization variables are: (1) s, distance between two parallel guides; (2) l_{link}, length of the six links; (3) R_p, radius of the circumference the platform joints B_i lie on; (4) θ_p , semi angular displacement between the two segments that connect the TCP with a couple of platform joints.

4 Numerical Results

To optimize the machines an upper and a lower limits to the parameterization variables must be imposed. As told before one of the most critical aspects of PKMs is represented by the mobility range of the joints. Two kinds of joints are taken into account: (1) the spherical joints by Hephaist Seiko with a mobility range of ±25°; (2) common universal joints with a mobility range of ±40°. Even if the formers present lower backlash problem with respect to the universal ones, they prevent to have an acceptable coverage of the workspace.

The strategy used for the optimization process are: (1) gradual reduction of $\tau_{max,lim}$ starting from a high value and decreasing it in each new optimization as long as the coverage of the workspace is acceptable. (2) reduction of the upper-bound limit of the links length. During the optimization process one can notice that final value of the links length always converges to the highest limit. Since this parameter gives an idea of the total size of the manipulator, it's desirable to reduce it as much as possible. For this reason, once the minimum value of $\tau_{max,lim}$ is found, the next requirement is to reduce the upper limit of the links length as long as the workspace is entirely covered.

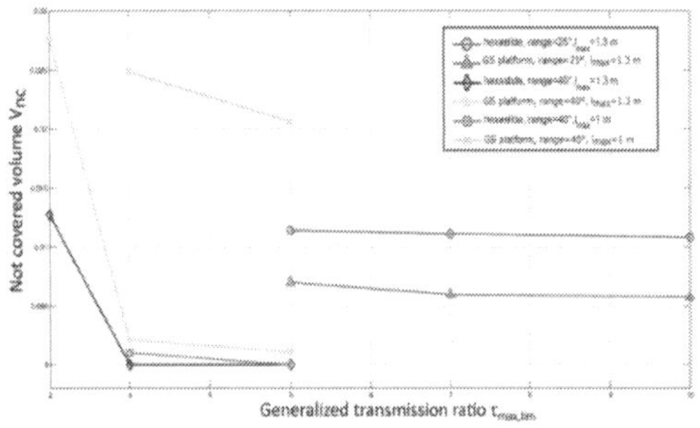

Figure 3. Not-covered volume for the two manipulators for various values of $\tau_{max,lim}$.

The comparison between the two machine can be seen Figure 3. It can be noticed that when using spherical joints the not covered volume is too high even if the value of $\tau_{max,lim}$ is set to 10. This means that in this case the limiting factor is represented by the joints mobility range. For this reason it's necessary to switch to the universal joints. If the parameter l_{max} is set to 1,3 m the Hexaslide allows to have a total coverage of the workspace with a $\tau_{max,lim}$ value that is lower than 3. However a 1,3 m long link is excessive and so, while still having a total coverage, the parameter l_{max} is lowered to 1 m accepting to have higher forces along the guides axes and consequently on the motors. On the other hand the Gough-Stewart is not able to guarantee a total coverage of the workspace even if the maximum link length is 1,3 m. After these considerations the chosen machine is the Hexaslide whose schematic design is reported in Fig 4.

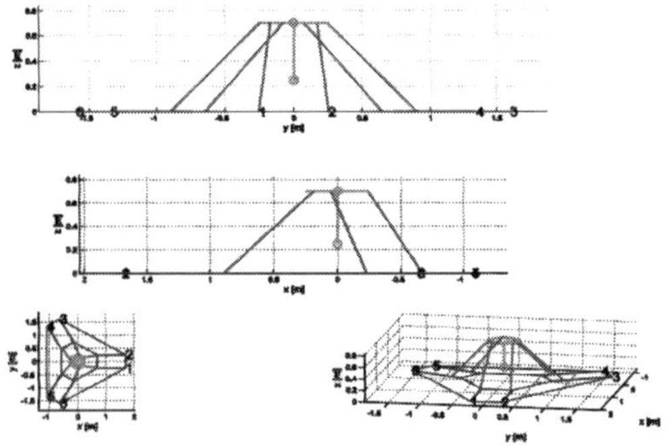

Figure 4. Hexaslide final structure.

5 Conclusions

This paper presents the optimization process of two parallel kinematic manipulators using a genetic algorithm given the technical specifications in terms of desired workspace for a particular application, that is the simulation of the sea waves through a mathematical model in wind tunnel tests. At the end of the optimization process it's clear that the only manipulator that is suitable for this specific task is the Hexaslide, a machine characterized by a high stiffness and a low inertia and that prevents the risk of collision between its elements. The optimization phase is the first and fundamental step of the machine synthesis process; as a matter of fact only once the dimensions and the characteristic parameters of the machine are well defined it will be possible to perform the following analyses.

For example it will be possible to built two dynamical models: (1) a rigid one used to size the drive system following the procedure described in [10, 11]; (2) a flexible one to perform a vibration analysis in order to map the system's first natural frequency all over the workspace as described in [12, 13].

References

1. Bayati I., Belloli M., Ferrari D., Fossati F. & Giberti H., "Design of a 6-DoF robotic platform for wind tunnel tests of floating wind turbines", *Energy Procedia*, pp. 313 (2014).
2. Merlet J. P., *Parallel Robots* , Springer, (2006).
3. Bonev I. A., *Mechanism and Machine Theory* , **36,** 1 (2001).
4. Merlet J. P., Gosselin C., *Handbook of Robotics* , (2008).
5. Legnani G., Tosi D., Fassi I., Giberti H., Cinquemani S., "The "point of isotropy" and other properties of serial and parallel manipulators", *Mechanism and Machine Theory*, vol. **45**, no. 10, pp. 1407-1423, (2010).
6. Legnani G., Fassi I., Giberti H., Cinquemani S., Tosi D., "A new isotropic and decoupled 6-DoF parallel manipulator", *Mechanism and Machine Theory*, vol. **58**, pp. 64-81 (2012).
7. Fassi I., Legnani G., Tosi D. , "Geometrical conditions for the design of partial or full isotropic hexapods", Journal of Robotic Systems, vol. 22, no. 10, pp. 507-518, (2005).
8. Cinquemani S., Ferrari D., "A genetic algorithm optimization for IMSC technique", 2014 *IEEE Conference on Control Applications*, CCA 2014, pp. 1942.
9. Giberti H., Ferrari D., "A novel hardware-in-the-loop device for floating offshore wind turbines and sailing boats", *Mechanism and Machine Theory*, vol. **85**, pp. 82-105, (2015).
10. Giberti H., Cinquemani S., Legnani G., "A practical approach to the selection of the motor-reducer unit in electric drive systems", *Mechanics Based Design of Structures and Machines*, vol. **39**, no. 3, pp. 303-319, (2011).
11. Giberti H., Cinquemani S., Legnani G., "Effects of transmission mechanical characteristics on the choice of a motor-reducer", *Mechatronics*, vol. **20**, no. 5, pp. 604-610, (2010).
12. Giberti H., Ferrari D., Fiore E., " Dynamics Modeling and Accuracy Evaluation of a 6-DoF Hexaslide Robot", Conference Proceedings of the Society for Experimental Mechanics Series, vol. **1**, pp. 473, (2015).
13. Tarabini M., Roure A., Pinhede C. Active control of noise on the source side of a partition to increase its sound isolation (2009) Journal of Sound and Vibration, 320 (4-5), pp. 726-743

MATEC Web of Conferences **45**, 04013 (2016)
DOI: 10.1051/matecconf/20164504013

Kinematic synthesis of a new 3D printing solution

Hermes Giberti[1], Enrico Fiore[1] and Luca Sbaglia[2]

[1]*Department of Mechanical Engineering, Politecnico di Milano, Campus Bovisa Sud - via La Masa 1, 20156 Milano, Italy*
[2]*Università degli studi di Parma, Dipartimento di ingegneria industriale Parco Area delle Scienze, 181/A, 43124 Parma,Italy*

Abstract. Low-cost production of metal parts is a challenge nowadays in the Additive Manufacturing world and new methods are being developed. The MIM technique is an innovative approach for 3D printing. This method requires a machine with suitable kinematics capable of generating the adequate movements.
The object of this article is the kinematic synthesis of a 5Dofs robot, based on two PKM machines, for additive manufacturing in order to compliant with the requirements of this new technology. Robot kinematics have been optimized by genetic algorithm in order to cover the required workspace and the design of the robot and outline of the control system are also given.

1 Introduction

Additive manufacturing, AM, is constantly growing due to its innovative design capacities. The forecast for this technology indicates that the market in 2017 will be about 5 billion dollars [1]. AM, also known as 3D printing, is based on several technologies based on the same idea: material deposition layer upon layer which task can be done in several ways and with different materials. Table 1 indicates some of the principal companies and technologies used in 3D printing[2],[3].

Tab. 1 Technologies for 3D printing

Company(foundation year)	3D printing technology
3D Systems(1986), USA	SLA,SLS
Statasys(1992), USA	FDM, Polyjet, WDNM
ExOne(2005), USA	Binder jetting
Arcam(1997), Sweden	EBM
Optomec(1998), USA	Aereoslojet, LENS

Among these technologies those related to the metal part production are based on the use of laser or electron beams [4]. Which are very costly. Stratasys is trying to provide an alternative means of production based on FDM [5] but at present there is still no commercial printer available. In order to reduce production costs of metal pieces, MIM technology (metal injection molding) could be a solution. In this way a feedstock made up of metal powder and a polymeric binder would be extruded on a platform layer by layer until completion of the object.

Thereafter the object will undergo heat treatment for the removal of the polymeric binder. Direct extrusion of the material greatly reduces the costs of process. This idea is rather simple but the main problem is related to the quality of the final product. Accordingly the accurate positioning of the worktable is of fundamental importance and this requires particular attention to the design of the mechanical structure.

Most industrial 3D printers have 3Dofs: one for the vertical translation of the platform and two for the movement in a plane of the extrusion head. According to 3D printing technology there can be some slight differences, but usually there are only 3Dofs (Optomec produces a 3D printer with 5Dofs). In general these machines are made up of serial robots based on a Cartesian layout where linear guides, driven by electrical motors, move the worktable and the head.

This article focuses on the kinematic optimization of a 5Dofs robot for a 3D printer based on MIM technology. In our solution the worktable has all the 5 dof while the extrusion head is fixed. This is why the extrusion system is powerful and heavy (about 25[kg]), made up of a feeder, a plasticizer and an injector connected to a final nozzle. It would be very difficult to move it quickly and accurately so it was necessary to develop a machine with a worktable completely moveable. This solution with the 5Dofs, 3 translations and 2 rotations, allows the possibility of new paths generation. This could help to solve typical problem of 3D printing like staircase effect, angle deposition [6] or reduce the use of support material [7]. For common 3D printers with only 3Dofs the only way to limit these problems is to reduce the thickness of each layer but this increases the production time and does not resolve the problem related to the use of the support material.

2 Robot kinematic

The robot requirements are the starting point of the kinematic synthesis, in particular: cube shape workspace

Article available at http://www.matec-conferences.org or http://dx.doi.org/10.1051/matecconf/20164504013

of side 100[mm], positioning precision under 0.1[mm] and robot 5 Dofs, translations x-y-z, rotations roll and pitch until 45°

In order to achieve these requirements, we decided to realize an hybrid structure characterized by two parallel robots in series. The first one carries out the translation x-y- z movement while he second one the rotation movements. This modular solution allows us to take advantage at a same time of parallel and serial robots characteristics, in particular, a good stiffness and precision from the first one and a rather wide workspace from the second. Note that with a parallel robot it would have been difficult to cover the required workspace mainly due to the large rotation angles required. Another important aspect is that the modular solution allows the use of two robots separately and we can optimize their design and workspace independently. This leads to the solution of two problems with smaller complexity compared to the study of a single 5Dofs robot. The chosen architecture for the first robot is called linear delta. Its purpose is to move the platform over which the second pkm is mounted. The name of this is agile eye and its purpose is to rotate the worktable where the material is deposited by the extruder. As follows, the kinematics of both the linear delta and agile eye is shown; Studies regarding these two robot architectures are presented in [8],[9]. Inverse and direct kinematic is necessary for the robot kinematic optimization and for the control system development.

2.1 Linear Delta

The machine which is responsible of the translational Dofs is a linear delta. This particular architecture is made up of three links of fixed length connecting the mobile platform with three different guides. The actuation is provided by three electrical motors connected to three linear transmission units, guides. Three independent PUS (Prismatic-Universal- Spherical) kinematic chains can be identified. The links are actually three parallelograms; in fact, this particular architecture ensures that the mobile platform is always parallel to the ground. In order to compute the kinematics of this machine two different reference systems are defined, the inertial frame placed on the ground and the Tool-Center-Point (TCP) which is fixed to the mobile platform. Since the links are parallelograms, the axes of the two frames are always parallel whatever the pose of the robot.

2.1.1 *Inverse Kinematics*

With reference to Fig. 1, for the i-th chain it is possible to follow the mathematical steps shown below in order to solve the inverse kinematics:

$$\boldsymbol{l}_i = \boldsymbol{d}_i - q_i\widehat{\mathbf{u}}_i \quad where \quad \boldsymbol{d}_i = \boldsymbol{p} + \boldsymbol{b}_i - \boldsymbol{s}_i \tag{1}$$

$$l_i^2 = \boldsymbol{l}_i^T\boldsymbol{l}_i = (\boldsymbol{d}_i - \boldsymbol{q}_i\hat{\mathrm{u}}_i)^T(\boldsymbol{d}_i - q_i\widehat{\mathbf{u}}_i) = \boldsymbol{d}_i\boldsymbol{d}_i^T - 2\boldsymbol{d}_i^T\widehat{\mathbf{u}}_i q_i + q_i^2 \tag{2}$$

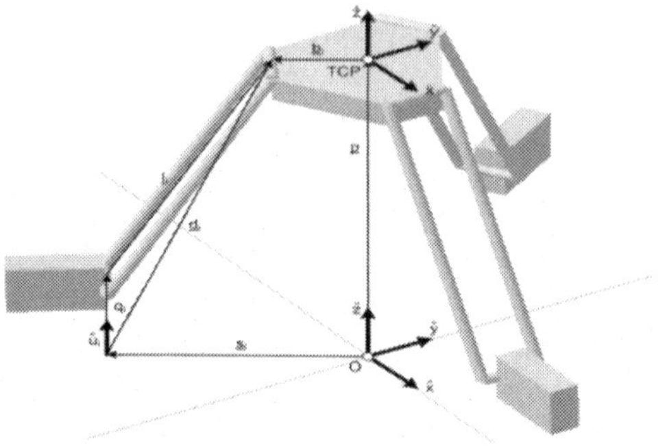

Fig. 1 Linear Delta

$$q_i = \boldsymbol{d}_i^T\widehat{\mathbf{u}}_i - \sqrt{\boldsymbol{d}_i^T(\widehat{\mathbf{u}}_i\widehat{\mathbf{u}}_i^T - [\mathrm{I}])\boldsymbol{d}_i + \boldsymbol{l}_i^2} \tag{3}$$

2.1.2 *Velocity analysis*

Deriving respect to time eq.2, it's possible to obtain the expression of the inverse Jacobian matrix:

$$2\boldsymbol{d}_i^T\dot{\boldsymbol{d}}_i - 2\boldsymbol{d}_i^T\widehat{\mathbf{u}}_i\dot{q}_i - 2q_i\widehat{\mathbf{u}}_i^T\dot{\boldsymbol{d}}_i + 2\dot{q}_i q_i = 0 \tag{4}$$

$$(d_i - \widehat{\mathbf{u}}_i q_i)^T(\dot{\boldsymbol{d}}_i - \widehat{\mathbf{u}}_i\dot{q}_i) = 0 \tag{5}$$

$$\boldsymbol{l}_i^T(\dot{\boldsymbol{d}}_i - \widehat{\mathbf{u}}_i\dot{q}_i) = 0 \tag{6}$$

$$\boldsymbol{l}_i^T(\widehat{\mathbf{u}}_i\dot{q}_i - \dot{\boldsymbol{p}}) = 0 \tag{7}$$

$$\widehat{\boldsymbol{n}}_i^T\widehat{\mathbf{u}}_i\dot{q}_i - \widehat{\boldsymbol{n}}_i^T\dot{\boldsymbol{p}} = 0 \tag{8}$$

Considering all the three kinematic chains it is possible to rearrange this expression in a matrix form and obtain the inverse Jacobian:

$$diag(\hat{n}_1^T\hat{u}_1, \dots, \hat{n}_3^T\hat{u}_3)\dot{q} - [\hat{n}_1^T, \hat{n}_2^T, \hat{n}_3^T]^T\dot{p} = 0 \tag{9}$$

$$[J]^{-1} = diag(\hat{n}_1^T\hat{u}_1, \dots, \hat{n}_3^T\hat{u}_3)[\hat{n}_1^T, \hat{n}_2^T, \hat{n}_3^T]^T \tag{10}$$

2.2 Agile eye

The worktable rotational Dofs are obtained through a spherical wrist named agile eye. This PKM is mounted on the mobile platform of the linear delta. In order to solve the kinematics of this manipulator the orientation evaluation of every its circular link is required.

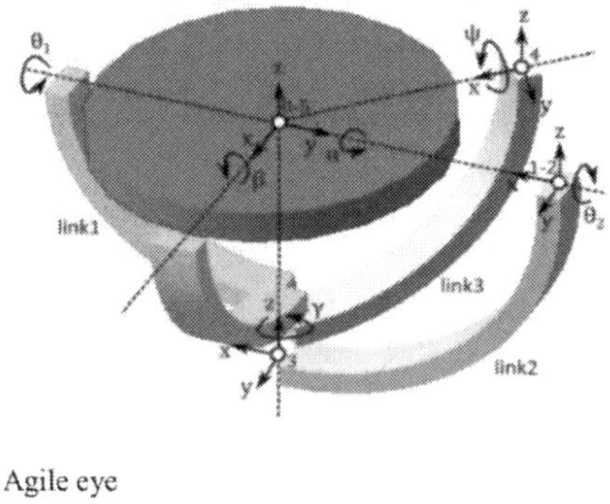

Fig. 2 Agile eye

Figure 2 shows the configuration chosen. The kinematics of this machine depend on the motor position and on the position joint of the links and platform. The expression that relates the rotations of the motors $(\theta_1; \theta_2)$ to the roll and pitch (α and β respectively) of the platform is:

$$\begin{cases} \theta_1 = \alpha \\ \theta_2 = \text{atan}\left(\frac{\sin\alpha - \cos\alpha\sin\beta}{\sin\alpha\sin\beta - \cos\alpha}\right) \end{cases} \tag{11}$$

2.2.1 *Velocity analysis*

Deriving Eq. 11 with respect to time, you get:

$$\dot{\theta} = \begin{bmatrix} 1 & 0 \\ \frac{d\theta_2}{d\alpha} & \frac{d\theta_2}{d\beta} \end{bmatrix} \begin{Bmatrix} \dot{\alpha} \\ \dot{\beta} \end{Bmatrix} = [J_\theta]^{-1} \begin{Bmatrix} \dot{\alpha} \\ \dot{\beta} \end{Bmatrix} \tag{12}$$

Where $\dot{\theta} = \{\dot{\theta_1} \quad \dot{\theta}_2\}^T$, and $[J_\theta]^{-1}$ represents the inverse Jacobian matrix associated to the rotational motions. The explicit equations have been omitted for the sake of simplicity, but it is rather easy to obtain them deriving the inverse kinematic expressions of $\theta_1. \theta_2$.

3 Kinematic optimization

Due to the modular solution of this project and to the fact that agile eye has no problem in reaching the desired rotation angles, it is possible to optimize only the linear delta kinematics provided that we take into account the behavior of the agile eye. The idea is to increase the working space of the linear delta in order to take into account the position reached by the piece in production caused by the worktable rotations. The result is an equivalent cube into which the center of the movable platform of the linear delta can go.

Linear delta for its own design has no limit for the movement in z axis direction, so we can study its kinematics in a xy plane. In order to optimize the system first we performed an adimensional study to understand the influence of the geometric parameters of the robot in the definition of its workspace, after a dimensional optimization has been performed to obtain the values of these parameters.

We decided to have the three linear guides of the delta robot at the same distance from the center of the machine. The boundary of the working area is the circle inscribed within the three guides. The robot is obliged not to overpass this circle even if it would be possible to do. Geometric parameters of the optimization are: D = diameter where linear guides are placed, d= mobile platform's diameter, l= legs length and α= angle between guides. The α angle defines the relative position of the guides on circle of D diameter. Considering a guide placed on the D circle the others two are placed on the same circle but $\pm\alpha$ degrees distant from the previous one. In this way the symmetry of the machine is preserved. To carry on this study a genetic algorithm with a single target has been used. Matlab function *ga* is used to this purpose.

3.1 Genetic algorithm

The matlab function *ga* works in this way: having chosen a set of n parameters, 4 in this case, and a specific range for each parameter, the function tries to minimize a desired function value. In our case we wanted to maximize an area so our function was the difference between the target area and the area covered by the robot, minimize this area means maximizing the workspace of the robot [10] [11]. The program works according to the genetic algorithm approach: making a random choice of the set values, inside the specified range, the objective function is evaluated. This process is carried on several times for a number of times, specified by the user, until the first family of the robot (with its own set of parameters) has been created. Later the program proceeds making a second family combining the values of the parameters obtained in the first family. In particular the program gives more importance to the parameters related to the lower values of the objective function found for the first family. The program continues creating several different families of robots and it can stop in different ways, for example because the maximum number of families, defined at the beginning of the optimisation, has been reached or after a situation of stall due to the impossibility to increase the minimization of the objective function.

In this project we had only 4 parameters, which is not a large number; so we made families of 300 members and we set the stop condition when the fiftieth family generation is reached. The objective function is an area evaluation. The robot workspace has been calculated in this way: from the center of the circle (the center of the machine) the platform is moved along a radial path until the position of the platform is reachable in accordance with the inverse kinematics of the machine; thereafter the platform is moved on a circle path for 360°, always checking that the position is reachable according to the kinematic equations. A set of points reached by the platform represents the workspace of the robot, which has to be subtracted from target area. The optimization has to take into account some limitations, in particular: joint limits and force and velocity transmission factors

The joints, used to connect the links (or robot legs) from one hand with the movable platform from the other hand and with the slider of the linear guides, have a limited mobility. We have to consider these limits using the genetic algorithms. With respect to the force and velocity transmission factors we set the value three as a limit in order to reduce the force level that the actuation system has to bear. Lower values means that it can bear more weight on the platform, reach higher velocities and using smaller actuators. To evaluate force and velocity transmission factors, an infinite norm of the Jacobian matrix has been used.

$$\tau_{force} = \|[J]^T\|_\infty < 3$$
$$\tau_{velocity} = \|[J]^{-1}\|_\infty < 3$$

Notice that the Jacobian, and τ_{force} and $\tau_{velocity}$, depend only on the unitary vectors which defines the space directions of each leg.

3.2 Adimensional study

In this first part of the optimization we wanted to understand the influence of the 4 parameters on the robot workspace using an adimensional study. Each parameter was divided by *D*, so the robot diameter was fixed to one and the optimization parameters became three. The legs could change their length from 0.4 to 2 and could vary from 0.2 to 1. Angle α was free to change between 10 to 170 degrees. Table 3 shows that *d* has to be as low as possible while α more than 90°, and the length of the legs equal to *D*.

Tab. 2 Results adimensional study

Parameter	Value
l	1
d	0.2
α	95°

3.3 Dimensional optimization

Using the results achieved by the adimensional study the kinematics optimization needed to reach the required dimensional workspace has to be performed. The space to cover is a 100[mm] cube but in order to compensate the movements of the agile eye the dimension of this cube has to be increased. Considering rotations of the worktable up to 45° and evaluating the displacement of the platform center, it is relatively simple to evaluate that the cube has to be increased from 100x100[mm] to at least 220x220[mm]. Actually in order to be sure of covering the required workspace at the end of the optimization task, it is a good idea to increase further the estimated value to 300x300 [mm].

We have to consider that to cover a workspace bigger than what is strictly necessary means to have joints with bigger rotational angle or have a bigger machine with a large dimension *D*. We want to have a compact robot in order to cover the 300x300[mm] workspace. We set the parameters *d* to 400[mm], that is because from the adimensional study we saw how *d* should be as short as possible and this is the minimum value we need in order to have an adequate space to install the agile eye. In Tab.3 we see the ranges of the 3 parameters and the results we obtained from the optimization.

Looking to the results we notice how there is no more equality between length legs and diameter D, which is due to the different shape of the workspace we have to cover: a square compared to circle one. As for the angle, we see that it changes between 90° and 125°. The solutions to the mathematical problem were more than one so the genetic algorithm couldn't converge always to the same solution. Notice that 120° allows us to have a machine with three symmetry planes, which means a better isotropy inside the workspace.

Moreover we wanted to use the guides not only to move the platform but also as a frame for the robot, as can be seen in figure 4 we decided to fix the angle α equal to 120° and make the last optimization considering only 2 parameters variable: *l* and *D*. At the end of the optimization task, the length of the legs is set to 600[mm] and the diameter D to 913[mm]. In Fig.3 it is shown the workspace reached with the values found and the actual limits of the revolute joints. We see how we can cover the target square space 220x220[mm] but not the 300x300[mm], which is because of the limits in the revolute joints. Anyway the use of a bigger workspace has led to reach a margin on the 220x220[mm] as we expected.

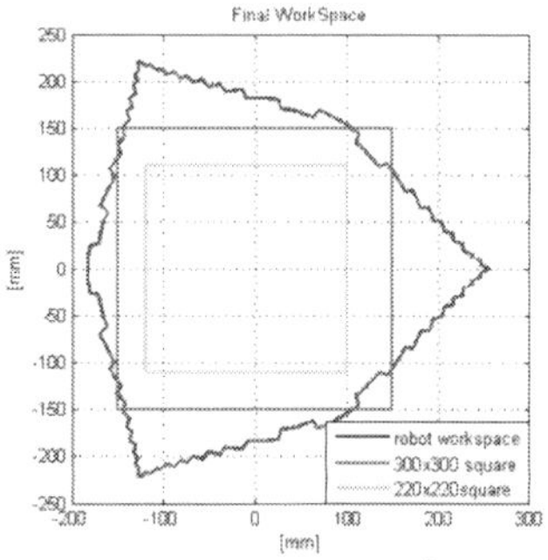

Fig. 3 Linear Delta workspace

Tab. 3 Results dimensional study

Parameters	range	Results 1	Results 2	Results 3
l[mm]	400-1000	660	610	580
D[mm]	600-1000	0.95	0.92	0.89
α	90°-170°	93°	124°	117°

4 Design and control system

The results obtained in the optimization were used to make the design of the machine. The robot is structured by a mobile platform connected to three parallelograms which are the three legs of the robot and which are connected to the three sliders of the linear guides. The constraints between legs and platform and legs and sliders are rotational joints which allows 2 rotational Dofs. In Fig.4 is shown the complete robot. It's possible to see the three guides where at the bottom there are the motors with their respective gear motors. At the center of the machine there is the mobile platform where the agile eye is mounted. We can see how there are two plates, one at the bottom of the machine and the other on the upper part. These two plates are needed two guarantee the correct mounting of the guides, furthermore the upper plate is necessary to bear the extruder that is fixed on the top of the machine. To increase the all stiffness of the guides and limit their vibration tie beams between each guide will be built.

The all robots will be controlled by means of an industrial control system. This system is principally based on: PLC, Motion control unit until 16 axis and Servo drives fully closed loop integrated. With this kind of system it is possible to control and coordinate all the 7 machine axis: 3 for the Linear Delta, 2 for the Agile Eye and 2 for the extrusion system. The control loops of the Linear Delta and Agile eye can be based either on the encoders of the respective motors or on external sensors.

For instance it is possible to close the position loop for the Linear Delta on the position of the three sliders using magnetic bands along the guides. Direct measure of the sliders position allows one to avoid the imprecision due to the backslash between motors and gear motors, and we don't have to consider the elasticity of the belts on which the linear guides are based.

Fig. 4 Machine Design

Fig. 5 Machine Protipe

5 Conclusion

In this paper we presented a kinematic synthesis of a 5Dofs robot based on a modular solution given by a linear delta and an agile eye.

According to the results obtained we can see how this new machine concept can give possibilities of movement which are not feasible with classical 3D printer and so it will be possible to develop an innovative system of production based on MIM technology. The robot kinematic allows the generation of paths for the platform which can solve the problems of staircase effect, angle deposition and use of support material. The design of the robot has been made with the aim of covering a specific workspace. It is rather easy following the same procedure shown to cover any desired workspace and so to extend this study to many other situations and different necessities. Simply increasing mobility of revolute joints, which is mainly a question of materials and costs, it will be possible to compensate even bigger rotations of the agile eye, which are already possible, and so develop new paths generation for 3D printing. The simple design and the commercial control system can simplify the use of this technology and so permit a cheaper production of complex shape metal parts with additive manufacturing which is a growing demand nowadays.

References

1. B. N. Turner, R. Strong and S. A. Gold , *A review of melt extrusion additive manufacturing processes: I. Process design and modeling*, Rapid Prototyping Journal, Vol. 20 Iss 3 pp. 192 – 204, (2014)
2. Douglas R. Gress, Ronald V. Kalafsky, *Geographies of production in 3D: Theoretical and research implications stemming from additive manufacturing*, Geoforum,Vol.60, March 2015, Pages 43-52, (2014)
3. K.P. Karunakaran Alain Bernard S. Suryakumar Lucas Dembinski Georges Taillandier, *Rapid manufacturing of metallic objects*, Rapid Prototyping Journal, Vol. 18 Iss 4 pp. 264 - 280, (2012)
4. N. Guo, M. C. Leu, *Additive manufacturing: technology, applications and research needs*, Front Mech Eng, 2013, 8(3): 215-243, (2013)
5. Patent publication number US2009/0314391 A1,assignee Stratasys Inc., *system and method for building three-dimensional object with metal based Alloys,* (2009)
6. A. Boschetto,L. Bottini, *Accuracy prediction in fused deposition modeling*, The international journal of advanced manufacturing technology, Volume 73, Issue 5-8, pp 913-928, (2014)
7. W.C. Lee, C.c. Wei, S.C. Chung, *Development of a hybrid rapid prototyping system using low-cost fused deposition modeling and five-axis machining,* Journal of Materials Processing Technology 214 (2014)
8. M. J. Uddin, S. Refaat, S. Nahavandi, H. Trinh, *kinematic and dynamic modeling of a robotic head with linear motors,* International Conference on Manufacturing and Management (2004), Queensland, pp 879-884
9. G. Palmieri,M. Callegari, L. Carbonari, M.C. Palpacelli, *Design and testing of a spherical parallel mini manipulator*, 10th IEEE/ASME International Conference on Mechatronic and Embedded Systems and Applications. IEEE MESA, (2014), pp 1-6
10. D'Antona, G., Davoudi, M., Ferrero, R., Giberti, H. *A model predictive protection system for actuators placed in hostile environments*, 2010 IEEE International Instrumentation and Measurement Technology Conference, I2MTC 2010 - Proceedings, art. no. 5488111, pp. 1602-1606.
11. Giberti, H., Ferrari, D. *A novel hardware-in-the-loop device for floating offshore wind turbines and sailing boats* (2015) Mechanism and Machine Theory, 85, pp. 82-105.

MATEC Web of Conferences **45**, 05001 (2016)
DOI: 10.1051/matecconf/20164505001

High Efficiency On-Board Hyperspectral Image Classification with Zynq SoC

Ning Ma [a], Shaojun Wang, Syed Mohsin Ali, Xiuhai Cui and Yu Peng

Department of Automatic Test and Control, Harbin Institute of Technology, Harbin, China

Abstract. Because of the downlink bandwidth bottleneck and power limitation on satellite, the demands for low power cost high performance on-board payload data processing which can reduce the volume of communication data are growing as well. This paper propos es a high efficiency architecture for on-board hyperspectral image classification in a Zynq Soc to achieve real-time performance. The Hamming-distance based Support vector machine (SVM) is adopted to get a high accuracy and low energy consumption for multi-class classification. The sequential control and the computing data path are realized in ARM processor and Programmable logic respectively. By the pipelined computing data path, a satisfying speedup is reached and thus lowers the energy consumption. The experiments on real hyperspectral image datasets demonstrate that our architecture can achieve 97.8% overall accuracy, 2.5~330x speed up and 11~835x energy saving compared with different state-of-art embedded platforms. For the AVIRIS spectrometer in real NASA application, it can realize real-time image classification.

1 Introduction

The spatial, spectral and temporal resolutions of hyperspectral image (HSI) in remote sensing keep increasing in recent years. However, the limited downlink bandwidth between satellites and ground station cannot catch up with the drastically increased data volume from high resolution spectroscopy instruments [1]. One approach to resolve this downlink bottleneck is to process the image data onboard and in real-time to decrease the data size to be transferred by downlink [2]. This requires the onboard processing system not only to achieve high performance to process the online continuous image data, but also to meet the harsh limitation of satellite, such as power consumption, size and weight.

In recent years, significant efforts have been made to cope with above challenges. In [3], Graphics Processing Units (GPUs) was used to process hyperspectral images (HSI) and to gain a superb performance while it is difficult to employ it for onboard satellite applications because of its high power consumption. Due to the superiority of FPGA devices for onboard application, Bernabe et al. [4] developed an automatic target generation process system for HSI using FPGAs.

Classification is one of main tasks for remote sensing image processing. Support vector machine (SVM) algorithm which can perform equal or better accuracy than other classifiers is widely used for hyperspectral classification [5]. It has a number of advantages over its counterparts, such as asymmetry of computation in training and classification, generalization capacity and good performance on small training data set [6]. HSI classification is multi classification problem, and different strategies to build a multi-classifier by SVM binary classifier has been developed in [7]. Hamming-distance SVM is the most potential choice by overcoming deficiencies of other methods [8].

A lot of researchers have already worked on SVM implementation utilizing different hardware. A. H. M. Jallad et al. [9] proposed a binary SVM classifier on SRAM based FPGA which was designed to identify the Cloud and Non-Cloud pixels. However, in most applications, multi classifier is required. J. Manikandan et al. [10] designed a SVM multi-classifier with System-on-programmable-chip (SOPC) technology on a Cyclone II FPGA. The accurate and high efficiency onboard HSI multi-classifier satisfying hardware resources and power consumption requirements are still ongoing research concerns.

In this work, a high efficiency multi-classifier using Least Square SVM (LS-SVM) and Hamming-distance judging strategy is proposed and its implementation on a state-of-art Xilinx hybrid chip, ZYNQ, is presented. ZYNQ is a better option for onboard application owing to its low-power consumption and advanced ARM processor with abundant logic resources. To achieve an optimum design solution, an iterative procedure is applied where variant implementation techniques are utilized to produce best possible efficiency. By simultaneously utilizing the control intensive module in processor and computing intensive module in logic cells, a high performance and

[a] Corresponding author: maning @hit.edu.cn

low-power consumption computing architecture is designed. The main contributions are:

- Proposing a parallelism multi-classifier based on Hamming-distance judging strategy by combining hardware and software co-design for high performance and energy efficiency to achieve real-time hyperspectral image processing,
- Full experiments performed with real HSI dataset on a ZYNQ platform which is designed for satellite data processing.

2 Mulit-Classifier algorithm

According to above discussed features of SVM, it is quite suitable for HSI classification. While SVM is a binary classifier, a multi-classifier is necessary in real processing task which can be implemented by using several binary classifiers in a certain architecture. In order to reduce the requirement of logic resource, LS-SVM is proposed which is the optimum option for onboard application [11].

2.1 Multi-classifier by Hamming-distance decision

To realize a multi-classifier, a *Hamming-distance decision* approach is used to combine the results of several binary classifiers. The result of every binary classifier represents one bit of a code. The bit width of the code is equal to $k(k-1)/2$. For a binary classifier, the outputs are assigned to 1 or 0 to represent different labels. Therefore after the processing of all binary classifiers for a test data, a new result code will be generated. Likewise, for every class label, an identifying code is created when training the sample dataset. The Hamming distance between result code and every identifying code is computed by counting the number of different bits. The class label with minimum distance is then assigned to test data. So for a new result code, Hamming distance (H) between all of the identifying code should be calculated, however not all of the bits in result code are useful for different class labels, a pre-mask should be used for selecting the bits. The Hamming distance is calculated as the following formula:

$$H = \sum_{l=0}^{n} b_l \tag{1}$$

Where H is the Hamming distance, n is the width of result code and is equal to $k(k-1)/2$, b_l means the lth bit value of T_res which is calculated in (2).

$$T\_res = (R\_Code \,\&\, mask) \oplus I\_code \tag{2}$$

Where R_Code is the new result code, mask is the one of the pre-mask which is different for different class label. I_code is the identifying code. The value of identifying code and pre-mask code is dependent on the number of class and the class label of every binary classifier.

In our research we will focus on this approach owing to its higher accuracy than previously described approaches [8, 12].

2.2 SVM binary classifier

An improved version of SVM, LS-SVM is adopted as the binary classifier, because it has assured computing process and needs less logic resources [11].

A LS-SVM classifier should be trained first. The training process can be done in the ground before satellite launching. The on-board processing only involves classification which is described as following:

$$f(x) = sign\left[\sum_{i=1}^{l} \alpha_i K(x, x_i) + b)\right] \tag{3}$$

where x_i is the training data, $K(x, x_i)$ is kernel function, α are Lagrange multipliers. b is a real constant. α and b can be calculated from training process.

The RBF kernel is the most used kernel function in hyperspectral classification. It is described as:

$$K(x,y) = exp\{-\|x-y\|_2^2/\sigma^2\} \tag{4}$$

where σ is the width parameter, and the optimum value of σ in Equation (4) is ascertained by Cross Validation (CV) method together with α and b in training processing.

3 Implementation

To implement such complicated algorithm in heterogeneous platform, several challenges should be overcome. Firstly, a special hardware platform should be designed with considering the computing architecture (including SOPC), logic resource, data cache and data communication interface with the satellite. Secondly, when designing the algorithm, a fully pipeline design should be carefully implemented to get high performance and energy efficiency of the algorithm, the balance point of cycle and resource should be find. The logic part in logic resource and the software part in processor must be working synchronously.

As comprising of processors and special computing elements, a new type of heterogeneous chip, ZYNQ, its lower-power costing and light weight, is used in this research work for onboard classification. The Zynq chip contains two ARM processors (PS partition) and logic resources(PL partition).

Parallelism is implemented to get high performance, and it should be reasonable considering the limitation of hardware resource, data cache size, bus bandwidth, speed of data source and the power supply.

3.1 Implementation of a binary classifier

In this paper, each multi-classifier is made up with a binary LS-SVM classifier and Hamming-distance judging module.

For each binary classifier, the formulas are as described in (3) and (4), Fig.3 shows the architecture of a binary classifier and the architecture of the whole algorithm. For every test pixel which contain several (in this paper, it is 9) different band data, the binary classifier will run *k(k-1)/2* times to get every bit of result code.

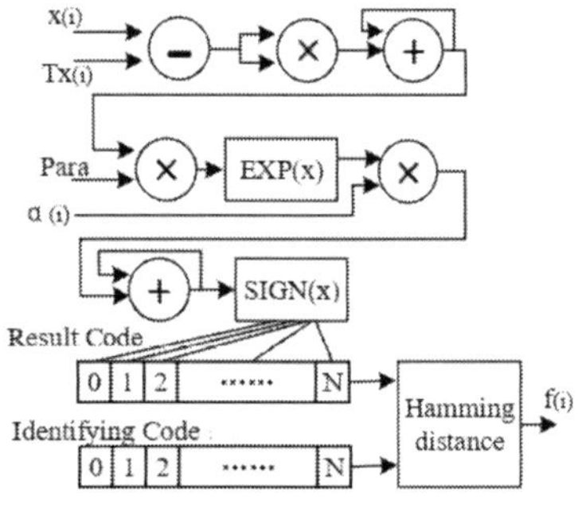

Fig.1. Architecture of. Multi-classifier

According to section II, the function of a 1-vs-1 SVM binary classifier is completed in four steps. In Fig.4, "Bd" stands for spectral band numbers and the "Dim" stands for the training dimension. In the "Two Norm Value" step shown in Fig.4, the value of $\|x-y\|_2^2$ as in Equation (4) is computed, and we need the pixel data and the training pixel data. For one pixel data, it should be processed together with all the training data in above formula, every spectral data should participate the processing which is shown in Fig.3. In the end of this step, there is an iteration which is a challenge for a full pipeline design, all the operations should be finished before starting the second step. Considering that this loop is the inner loop, unrolling can reduce a huge number of processing cycles but not too much extra resources, this step will be implemented in parallel.

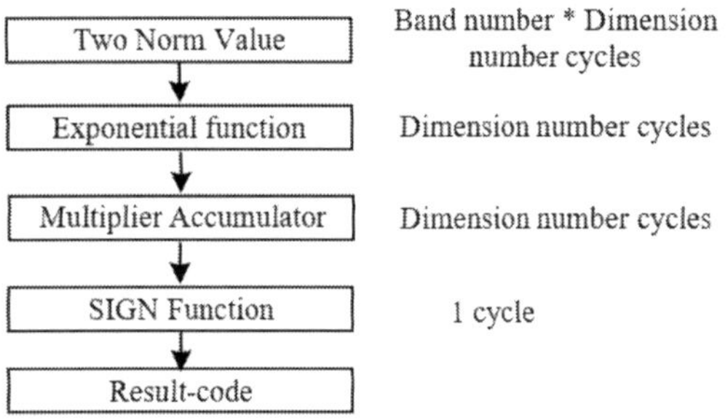

Fig.2. Computing flow of binary classifier

For the step of "Exponential function", σ is required to calculate the value of Equation (4), and is implemented by using the Exponential IP of Xilinx by using DSP48E and logic resource.

For "Multiplier Accumulator" step, a and b is required. The results of "Exponential function" step are multiplied by a, then accumulating together, this iteration is an outer, unrolling will take a great number of logic resource, and not reduce the processing cycle sharply, this step will be implemented in loop.

Those step can be executed faster by promoting the parallelism at every step but the resource consumption and data communication bandwidth will be sharply increased. So the balance point should be calculated between processing cycles and such factors. For example, in first step, using 8 couples of subtractor and multiplier, instead of one couple to process 8 training pixels data simultaneously. The cycles of the first step will be reduced to 1/8 compared with raw design. After experiments, we found that a float-point adder required 2 DSP48Es which is the most limited and useful resource in a computing intensive application design, and for subtractor, multiplier and exponential function module the requiring number are 2, 3 and 26 respectively. Considering the total number of DSP48E in our platform is 220, so, when class number is over 4, the device cannot support so much DSP48E for the *k(k-1)/2* counts of binary classifiers. To realize a scalable design for different number of classes, each multi classifier uses only one unparalleled binary classifier, six multi classifier is designed in our application. For every binary unparalleled classifier, the theoretical total cycle requirement is described in (3). Higher parallelism can be carried out by unrolling the computing loop.

$$Cyc\_one = \left(\left(Bd + 2 \right) \times Dim + 1 \right) \tag{3}$$

The architecture of the multi classifier by using Hamming-distance is shown in Fig.5.

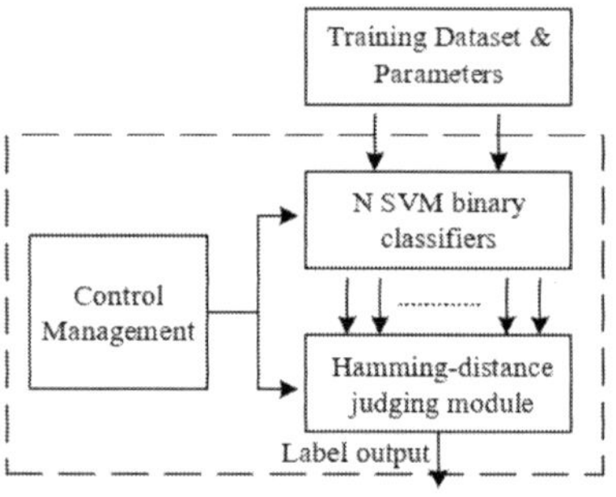

Fig.3. Architecture of Hamming-distance classifier.

It contains three modules: Control Management, SVM binary classifiers and Hamming-distance decision module. Control Management module manages the flow of computing including data input and output, and selects the training data and parameters which decides the final bit location in result code for computing Hamming-distance, those parameters and training data are saved in Training Dataset & Parameters module. There are two approaches to implement SVM binary classifier module, using *k(k-1)/2* classifiers in parallel to get the results at the same time, this method consumes *k(k-1)/2* times resources than using only one which required more cycles for processing. Considering the resource limitation of onboard application, so in this paper, one binary classifier is used in each multi classifier.

To find the nearest class, Hamming-distance between result code and identifying code should be computed. By the advantage of logic gates, the result code will be processed in exclusive OR operation with identifying code of every class, then, the result of above operation will be processed in AND operation with a pre-mask code of every class. The number of 1 in every result will be accounted, the minimum one have the most similar degree with the class label, and set as the data label. Table 1. shows the identifying and pre-mask codes for class number upto six

Table 1. The identifying and Pre-mask codes

Class Number	Identifying Code	Pre-mask Code
Class 1	111110000000000	111110000000000
Class 2	000001111000000	100001111000000
Class 3	000000000111000	010001000111000
Class 4	000000000000110	001000100100110
Class 5	000000000000001	000100010010101
Class 6	000000000000000	000010001001011

3.2 A scalable control program in ARM processor

By just uploading the training dataset and parameters, another new multi-classifier can be launched based on the old classifiers under the control of program. The program which is run on ARM processor controls the flow of computing. It reads HSI dataset and configures the classifier. Program flow is shown in Fig.6.

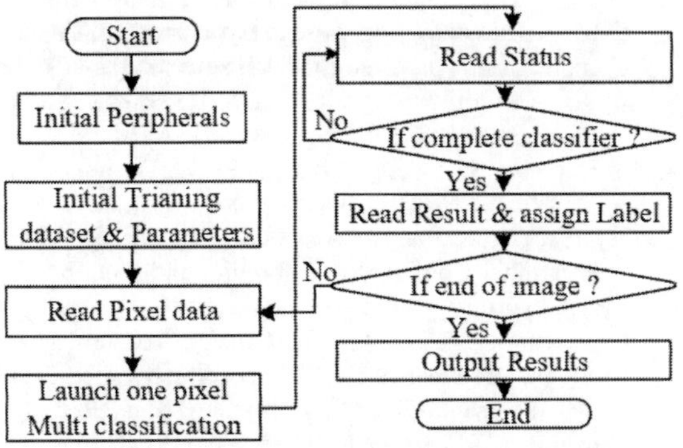

Fig.4. The flow of program in ARM processor.

The program can assure and control the classifier by reading and writing the registers in classifier through AXI Lite interface. In "Initial Peripherals", the peripherals and the program are prepared for computing, then in "Initial Training dataset & Parameters", the training dataset and parameter α is sent to Storage module by DMA controller through AXI HP ports, and this can get a high transfer speed and realize a heavy work on processor. In the "Read Pixel data", the image pixel are transmitted to the binary classifier, the status of the classifier is monitored, after computing the Hamming-distance, multi classifier send the class label of input data to ARM processor. Different classifiers are launched by the program works in parallel. After the processing, results can output through CPCI interface in the demo system. In our experiment, the time and the classifying accuracy will be shown through UART ports on PC.

4 Experiments &Result

To compare efficiency with different platforms, the experiment is carried out on two well-known real HSI datasets. These are produced by Airborne Visible Infra-Red Imaging Spectrometer (AVIRIS) [13], are engaged for testing. AVIRIS can capture 224 bands data for every pixel, from its instrument features, its scan rate is 12Hz, and in every scan, 677 pixels will be produced, the sampling rate is approximately 123.1μs/pixel.

To prove our design can realized online real time processing for AVIRIS, the processing time of every pixel is evaluated from the experiments. These two datasets are both provided by University of the Basque Country [14]. The first image contains 145 × 145 pixels, and it comes from a mixed agriculture/forestry photo in Northwestern Indiana on June 1992. This image is gathered over the Indian Pines Test Site and shown in Fig.7. The second image is collected from Salinas Valley, California, and its spatial resolution is 3.7meter/pixels comprising of 512 × 217 pixels.

In our experiments, considering water absorption bands and information redundance, only 9 spectral bands and 6 classes are used for training and identifying in both datasets. For each class, the number of training and testing pixels is 50 and 100 respectively.

In order to evaluate the performance of our proposed design, four reference designs in different hardware platform are developed in the experiment II. The first one is implemented on HP XW8600 workstation with the configuration of Intel Xeon X5482 having 8 cores at 3.2GHz frequency and 64GB of memory. The same algorithm is implemented in C language on visual studio 2010 development environment. The second reference design is implemented on an embedded system comprising of an ARM cortex-A9 processor at 666.7MHz, Vector Floating Point Unit and 32KB Cache is employed to speed up the computing. The third one is on a state-of-the-art Texas Instruments DSP of TMS320C6778 at 1000 MHz which contain 8 cores to accelerate the processing.The last reference design is on Power PC440 processor, which runs at 400MHz and a FPU in it. The power consumption of these platform are also measured. In order to measure the power consumption more accurately on workstation, the difference of power consumption in running and idle state is used to calculate the power consumption of algorithm. The same 600 pixels data is tested on all the platforms with in the same training dataset, parameters and data precision.

For the Indian Pines Test Site dataset, the time consumption and speedup are shown in Table 2.

Table 2. The speed and power consumption comparing on ZYNQ, ARM, DSP, PowerPC and HP workstation

Platform	ZYNQ	PC	ARM	DSP	PPC
T (μs/pixel)	25.8	216.3	1321.2	65.7	8516
Power (W)	3.9	103.8	3.3	16.0	9.8
E(mJ/pixel)	0.1	22.4	4.3	1.1	83.5
Speedup	1.0	8.3	51.2	2.54	330
OA (%)	98.3	98.3	98.3	98.3	98.3

In above table, OA means overall accuracy, E stand for energy consumption. From this table, our Hammning-distance multi-classifier on ZYNQ gets 8.3x speedup with about 224x energy saving compared to HP workstation. Compared with ARM, DSP and PPC, our heterogeneous design gain 51.2x, 2.54x, and 330x speedup, and 43x, 11x, and 835x energy saving respectively. By the high frequency clock, embedded FPU, and fast cache, ARM platform gain higher computing speed than DSP and PPC platform. As the lower frequency and the bandwidth limitation between FPU and CPU core, the PPC shows slowest process than other platforms.

Our design gains about 98.3% overall accuracy. With the same training and test data, especially the same classification algorithm, the overall accuracies are the same on different embedded platforms and PC. Under the onboard resources and power limitation, our heterogeneous platform and algorithm architecture is efficient for HSI classification applications.

Comparing with other research on the same dataset, the comparison of overall accuracy is shown in table 3.

Table 3. Comparison with other researches

Methods	OA(%)
Proposed Approach in this paper	98.3
ANN based Adaboost[15]	98.02
Wavelet Networks[16]	82.0
MLRsub[17]	92.5
HA-PSO-SVM[18]	98.2
PGNMF[19]	93.36

From above table, the proposed approach in this paper gain higher accuracy than other research.

For the Salinas Valley dataset, the comparison on power consumption and speedup are shown in Table 4.

Table 4. The speed and power consumption comparison on Zynq, ARM, DSP and HP workstation

Platform	ZYNQ	PC	ARM	DSP	PPC
T (μs/pixel)	26.5	220.3	1367.2	66.8	8487.4
Power (W)	3.7	106.8	3.1	16.5	9.2
E(mJ/pixel)	0.1	23.5	4.2	1.1	78.1
Speedup	1.0	8.3	51.6	2.6	320.3
OA (%)	97.8	97.8	97.8	97.8	97.8

From above table, for Salinas Valley dataset, the proposed approach can reach 97.8% overall accuracy, and compared to the embedded platform of ARM, DSP and PowerPC, it gains speedup and energy efficiency ranging from 11x ~ 781x.

For online application of AVIRIS, the sampling rate of spectrometer is 123.1μs/pixel [20], and our design can realize 27μs/pixel classification. So it can fully fill the real-time processing requirements.

5 Conclusions

In this paper, we propose a novel Hamming distance judging strategy based multi-LS-SVM-classifier for HSI classification. By employing parallel logic architecture and flexibility of software in hybrid ZYNQ SOC, we realized the proposed multi classifier with high performance and power efficiency for satellite onboard application. The experiments results on two datasets from AVIRIS demonstrate that the proposed multi-classifier reaches up to 2.5x ~ 330x speed up with 11x ~ 835x energy saving compared to different embedded platforms. At the same time, it gains over 97.8% overall accuracy. So it can realize high overall accuracy and low power consumption real time hyperspectral image classification.

References

1. B. B. Marschallinger, D. Sabel, W. Wagner, "Optimisation of global grids for high-resolution remote sensing data," Computers & Geosciences, vol. 72, pp.84-93, November 2014.
2. K. L. Wagstaff, D. R. Thompson and W. K. Abbey, "texture-sensitive instrument classification for in situ rock and layer analysis," Geophysical Research Letters, vol. 40, no. 16, pp. 4188-4193, August 2013.
3. S. Lucana, M. Enrico, V. Raffaele, "Highly-Parallel GPU Architecture for Lossy Hyperspectral Image Compression," IEEE Journal of Selected Topics in Applied Earth Observations and Remote Sensing (J-STARS), vol. 6, no. 2, pp. 670-681, April 2013.
4. S. Bernabe, S. Lopez, A. Plaza, "FPGA Design of an Automatic Target Generation Process for Hyperspectral Image Analysis," in IEEE 17th International Conference on Parallel and Distributed Systems, 2011, pp. 1010-1015.
5. G. Camps Valls, L. Bruzzone, "Kernel-based methods for hyperspectral image classification," in IEEE Transaction on Geoscience and Remote Sensing, vol. 43, no. 6, pp.1351-1362, 2005.
6. H. Guo, W. Wang, "An active learning-based SVM multi-class classification model," in Pattern Recognition, vol.48, no.5, pp. 1577-1597, May 2015.
7. W. H. Chih, J. L. Chih, "A comparison of methods for multiclass support vector machines," in IEEE Transactions on Neural Networks, vol. 13, no. 2, pp. 415-425, March 2002.
8. S. Hari, F. Agrafioti and D. Hatzinakos, "Design of a Hamming-distance Classifier for ECG-Biometrics," in Acoustics, Speech and Signal Processing (ICASSP), 2013, pp. 3009-3012.
9. A. H. M. Jallad and L. B. Mohammed, "Hardware Support Vector Machine (SVM) for satellite on-board applications," in Adaptive Hardware and Systems (AHS), 2014, pp. 256-261.
10. J. Manikandan, B. Venkataramani, "System-on-programmable -chip implementation of diminishing learning based pattern recognition system," in International Journal of Machine Learning and Cybernetics, vol. 4, no. 4, pp. 347-363, Augst 2013.

11. J. Suykens and J. Vandewalle, "Least square support vector machine classifiers," in Neural Processing Letters, vol.9, no.3, pp.293-300, March 1999.
12. A. Vempaty, L. R.Varshney, P. K. Varshney, "Reliable Crowdsourcing for Multi-Class Labeling Using Coding Theory," in IEEE Transactions on Signal Processing, vol.8, no.4, pp.667-679, AUGUST 2014.
13. R. Green, M. Eastwood, M. Sarture, "Imaging spectroscopy and the Airborne Visible/Infrared Imaging Spectrometer (AVIRIS)," in Remote Sensing of Environment, vol.6, no.3, pp. 227-248, September, 1998.
14. Computational Intelligence Group, Hyperspectral Remote Sensing Scenes datasets, 2015 [Online]. Available: http://alweb.ehu.es/ccwintco/index.php?title=Hyperspectral_Remote_Sensing_Scenes.
15. Q. Sami ul Haq, L. Tao, S. Yang, "Neural Network based Adaboosting Approach for Hyperspectral Data Classification," in International Conference on Computer Science and Network Technology, 2011, pp.241-245.
16. P. H. Hsu, H. H. Yang, "Hyperspectral Image Classification Using Wavelet Networks," in International Conference on Geoscience and Remote Sensing Symposium, 2007, pp.1767-1770.
17. M. Khodadadzadeh, J. Li, A. Plaza, P. Gamba, "A New Framework for Hyperspectral Image Classification Using Multiple Spectral and Spatial Features," in International Conference on Geoscience and Remote Sensing Symposium, 2014, pp.4628-4631.
18. Z. H. Xue, P. J. Du, H. J. Su, "Harmonic Analysis for Hyperspectral Image Classification Integrated With PSO Optimized SVM," in International Conference on Computer Science and Network Technology, 2011, pp.241-245.
19. Q. Sami ul Haq, L. Tao, S. Yang, "Neural Network based Adaboosting Approach for Hyperspectral Data Classification," in IEEE Journal of Selected Topics in Applied Earth Observations and Remote Sensing, vol. 7, no. 6, pp. 2131-2146, 2014.
20. Jet Propulsion Laboratory, AVIRIS Instrument, 2015 [Online].Available: http://aviris.jpl.nasa.gov/aviris/instrument.html.

MATEC Web of Conferences 45, 05002 (2016)
DOI: 10.1051/matecconf/20164505002

Design and Implementation of 3 Axis CNC Router for Computer Aided Manufacturing Courses

Mehmet Emin Aktan[1], Nihat Akkuş[2], Abdurrahman Yılmaz[3] and Erhan Akdoğan[1]

[1]Yıldız Technical University, Department of Mechatronics Engineering, 34349 Istanbul, Turkey
[2]Marmara University, Department of Mechatronics Engineering, 34722 Istanbul, Turkey
[3]Istanbul Technical University, Department of Control and Automation Engineering, 34469 Istanbul, Turkey

Abstract. In this paper, it is intended to make the mechanical design of 3 axis Computer Numerical Control (CNC) router with linear joints, production of electronic control interface cards and drivers and manufacturing of CNC router system which is a combination of mechanics and electronics. At the same time, interface program has been prepared to control router via USB. The router was developed for educational purpose. In some vocational schools and universities, Computer Aided Manufacturing (CAM) courses are though rather theoretical. This situation cause ineffective and temporary learning. Moreover, students at schools which have the opportunity to apply for these systems can face with various dangerous accidents. Because of this situation, these students start to get knowledge about this system for the first time. For the first steps of CNC education, using smaller and less dangerous systems will be easier. A new concept CNC machine and its user interface suitable and profitable for education have been completely designed and realized during this study. To test the validity of the hypothesis which the benefits that may exist on the educational life, enhanced traditional education method with the contribution of the designed machine has been practiced on CAM course students for a semester. At the end of the semester, the new method applied students were more successful in the rate of 27.36 percent both in terms of verbal comprehension and exam grades.

1 Introduction

In some vocational schools and universities, Computer Aided Manufacturing (CAM) courses are though rather theoretical. Therefore, students face with ineffective and temporary learning. Besides, students at schools that have the opportunity to applying real Computer Numerical Control (CNC) machines, can face with various accidents. To eliminate these problems, various simulation systems are used. Before being applied to actual CNC machine, the corresponding meaning of the code written in simulation environment can be first observed on the virtual CNC machine. As a result, without being confronted with adverse accident which they may face while trying their own codes, students can make the necessary arrangements in a virtual environment. Thus, by applying a sort of debugging process to written code instead of making directly real-life testing of applications that actually confronted and observed, training more efficient and effective results with CNC machines are becoming possible. Starting from this situation, to accustom students to workbench in the first step of CNC training, creating a compact and relatively less dangerous system will have great help in education. There are several studies on this subject. Aktan, in his thesis, designed and manufactured low cost, three axis, multi-purpose cartesian robot system for education [1]. Lee and Chang, in their study, have developed CAD/CAE/CAM system. They also stated that this system could be used as supplementary training tool in courses [2]. Köbeloğlu, in his thesis, designed and manufactured low cost, two axis, CNC lathe for training in vocational and technical schools [3]. Tseng and Kolluri, in their study, are designed microcontroller controlled 3 axis CNC router [4]. Their aim is was try out new technologies and use them in educational applications, especially. In this study, a compact, USB controlled 3 axes CNC router is designed and manufactured. An interface program is prepared with Microsoft Visual C#. Unlike other studies, the produced system has been tested on students. Mechanical design of the system is described in Section 2. In Section 3, electronics design is explained. User interface and control method are shown within the Section 4. Methodology and testing procedure are also described in Section 5. The results and conclusions are shown and discussed in Section 6 and 7, respectively.

2 Mechanical Design

The system has 3 linear axes. In these axes, transmission of motion is provided with the help of linear bearings. Radial movement which is produced by stepper motors, is transformed to linear movement by linear bearings. Stepper motors are linked to linear bearings via couplings. Axes are able to work independent from each other, at the same time interpolation process can be done with the concurrent movement of multiple axes. Technical properties of the router are shown in Table 1 and mechanical design of the router is shown in Figure 1.

Table 1. Technical properties of the router.

Actuator Type	Stepper Motors
X,Y,Z Limits	500 mm, 400 mm, 180 mm
Cutting Depth	2 mm
Cutting Velocity	1 m/min
Free Velocity	3 m/min

Figure 1. Mechanical structure of the router [1]

Necessary amount of motor power is calculated by regarding the general structure of the system and axes weight [5].

$$P = (2\pi N_m T_t) / 60 \qquad (1)$$

$$P_a = (2\pi N_m / 60)^2 (J_L / t_a) \qquad (2)$$

Where;
P = Motor power while cutting **(W)**
T_t = Total moment while cutting **(Nm)**
N_m = Motor speed **(rpm)**
P_a = Motor power while acceleration **(W)**
J_L = Total inertia **(kg-m^2)**
t_a = Acceleration time **(sec)**

As a result of calculations, findings about motor power requirements are pointed in Table 2.

Table 2. Motor power requirements.

	X Axis	**Y Axis**	**Z Axis**
During Acceleration (W)	16,03	13,47	5,77
During Cutting (W)	30,16	15,07	39,57

3 Electronics Design

The router has 3 stepper motors and for each stepper motor, drivers shown in Figure 2 were designed and produced. These drivers are combination of L297 and L6203. L297 is integrated circuit of stepper motor control. From step, direction and enable signals, which are given to input, it produces step motor phase signals at output. It can work at full step and half step modes. It is necessary to use adjunct integrated circuits like L298 and L6203 to drive motor with L297. L6203 is integrated circuit of H-bridge driver. There is one bridge inside. For this reason, to drive one stepper motor, it is necessary to use two L6203. It heats less and has ability to give current more in comparison to L298. It can provide approximately up to 5 A output current.

Figure 2. Motor drivers [1]

3.1. Power Supply

Full wave rectifier circuit is prepared as power source of motors. 300W 24 Volt transformer is used at circuit input. Rectification is managed with 1000V 50 ampere bridge diode at circuit output. 100V 1000uf parallel capacitor is connected to output. Whole power supply circuit and its schematic drawing shown in Figure 3 and 4, respectively.

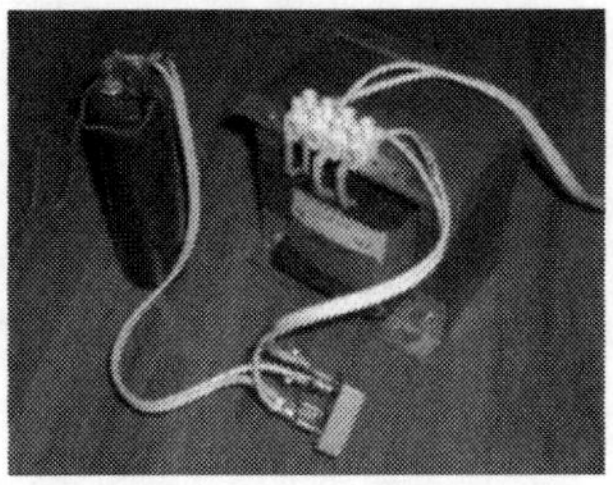

Figure 3. Power supply [1]

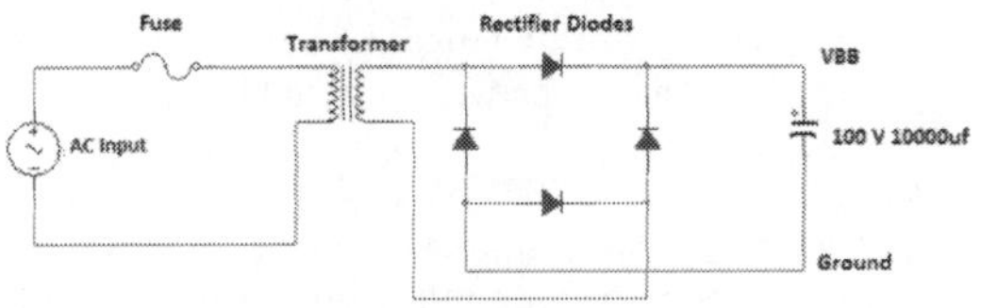

Figure 4. Circuit of power supply

3.2 Control Box

The control box (Fig. 5), which is consisting of electronic parts of the system, gives opportunity for user to make control and adjustment of processes. There are power source, driver and other electronic equipment inside the box. The front panel includes on-off switch, power indicator, axes movement indicators, X-Y-Z manual movement buttons, spindle switch and emergency stop button. The back panel has motor and limit switch connections.

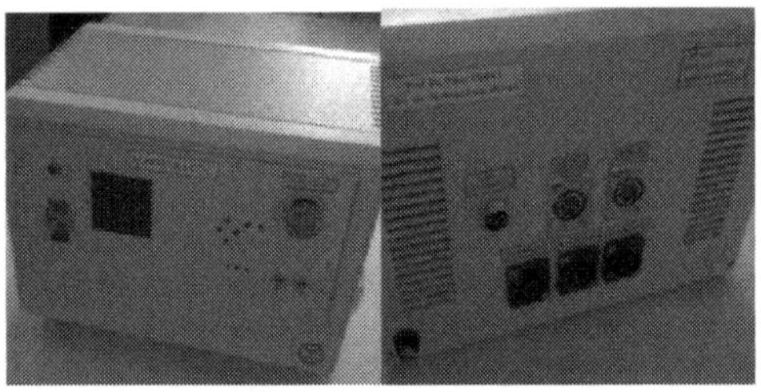

Figure 5. Control box [1]

4 Interfaces and Control

The router system's control type is open loop. By making essential calculations with microcontroller, number of steps are calculated and sent to L297 stepper motor driver. According to incoming signals, pulses are produced and sent to L6203 MOSFET driver. Necessary movements are carried out through driving motor with the help of driver. The block diagram of the system is shown in Figure 6.

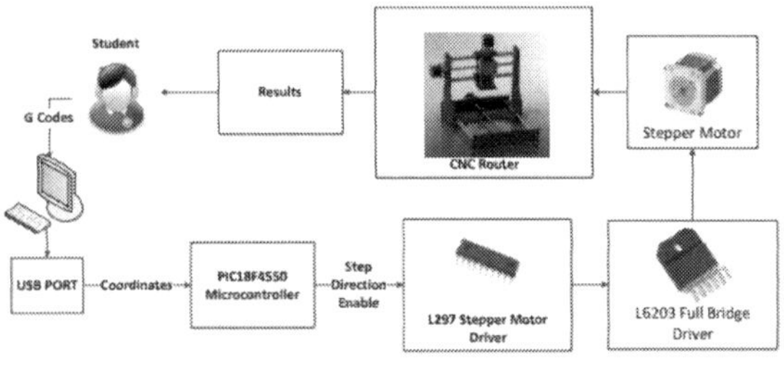

Figure 6. Block diagram

So far mechanic, electronic and program interface parts of system were mentioned. Now, it will be explained how the system works. Data transmission is carried out via USB. So, the developed system is available to work on every computer that has USB port. The first step is the data analysis of G and M codes which are entered by user.

Our sample code is: "G01X97Y85"

In this code, G letter indicates working mode. What is to be done is determined according to the ensuring value. In our example, the value is 01. This value signifies that router will start cutting process. Velocities of router at the time of cutting and going idle must be different. Because, while going idle, there is no force against and machine must be send to the place of cutting rapidly for time saving. Here, because of 01 value, axes should move slowly. Ensuring letter and the following number demonstrate which axes should move and the amount of movement in millimeters. According to the example, if the line had finished here, it would have been understood that we should go 97 mm in X axes at slow velocity. But it is seen that Y value is present next to X. It is the sign of linear interpolation to be done. After the line extraction in this process, velocity, axis and distance data are appointed to variables and send via USB channel to the microcontroller. From these incoming data, according to distance value which motor needs to take how many steps is calculated and pulses are sent to the driver. Through these pulses, motor rotates as required and a linear movement is performed via ball bearings. After end of the movement, microcontroller sends a signal to computer that the process is finished. Then, the computer sends values of next line to microcontroller. In this way, process finishes after all lines were processed.

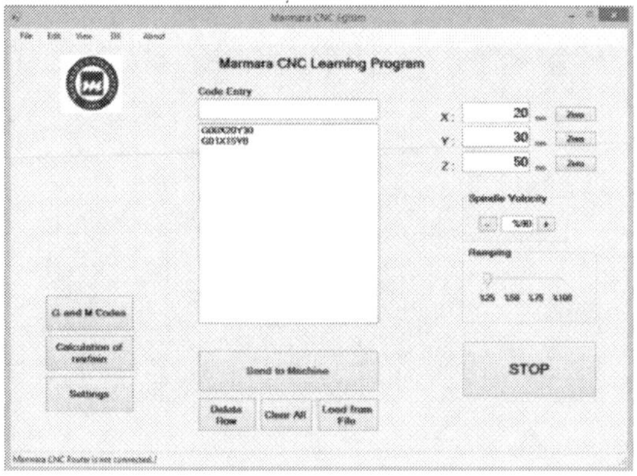

Figure 7. User interface [1]

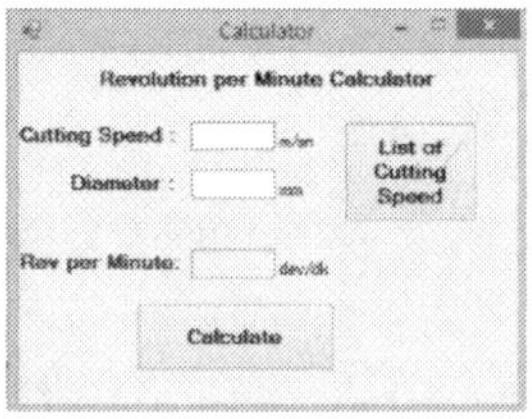

Figure 8. Revolution per minute calculator

Figure 9. CNC router system [1]

The user interface which is written with Microsoft Visual C# is shown in Figure 7. Here, G and M codes are entered from Code Entry section. The user writes the code and presses enter. The code passes the lower box. After all lines are written, the codes send to router by pressing "Send to Machine" button. By "Load from File" option, the code can also be downloaded from any text file. With indicators on the right side of the interface, can be seen the axis positions and by pressing "Zero" button, they can send to home positions. With the "Spindle Velocity" button, speed of the spindle can be adjusted. Required spindle speed can be calculated by pressing the "Calculation of rev/min" button on the left side of the interface. The calculator showed in Figure 8. From the ramping section, acceleration and deceleration ramping rate of the stepper motors can be set. In Figure 9, it is tried to visualize the entire mechanical design of the systemfor using CAM courses.

5 Methodology of Testing

After completing design and implementation of desired system, testing process the last step of the study has been started. During this step, the designed system has been integrated to CAM course taken by 96 undergraduate students and the aimed students are randomly divided into two groups. Addition to this, one of the groups was selected randomly and classic education methods applied during semester. The other group also had been educated via new, system integrated training scheme. In order to eliminate the effects of lecturer mentioned in the paper of Kane and et al [6], the influences of lecturing class size explained in Angrist and et al [7] study and also the possible impacts of lecturing at different times of a day, for both of the groups have been educated by the same lecturer at the same class and the same hours of the day. For comparing education methodologies during the semester two written and one oral examination including same kind of questions were applied to both groups. By averaging exam grades of the students, their semester-end scores were acquired.

The average semester - end grades of the students taken given CAM course can be seen as a bar graph in Figure 10. When the distribution of the grades between 0 and 100 investigated statistically, normal (Gaussian) distribution approximately has been achieved. The graph of average grades is clearly shaped like a Gaussian distribution. Due to the fact that any Gaussian distribution is a symmetrical distribution [8], skewness in other words coefficient of asymmetry should be equal to zero [9, 10]. Kurtosis which means convexity is one of the characteristics for comparing any random distribution with normally distributed ones [9, 10]. Kurtosis coefficient is equal to 3 for Gaussian distributions [9, 10]. As can be seen from Table 3, the parameters were calculated for obtained grades distribution as 3.1301 and 0.0648 respectively. Consequently, the distribution was roughly accepted in the form of Gaussian distribution.

On the other hand, so as to make the Gaussian distribution more meaningful and eliminate individual success or failure of the students, 95 percent was selected as confidence interval [11]. If the given Table 3 is examined, it has been also presented the whole class, end-average grade and standard deviation of the grades. By putting the average and standard deviation information to the given Equation 3 [11], the upper and lower boundaries of the confidence interval were derived as shown on Table 3. The boundaries of confidence interval and the eliminated students, emphasized by red color, with respect to them are demonstrated on Figure 10. The number of students eliminated due by confidence interval lower boundary and upper boundary, is 3 and 4 respectively. At this point, the students within the confidence interval were determined and listed in their groups. With final selection, average score calculation process for selected students was applied again in terms of their groups. Additionally, at the end of the semester a question concerning whether novel educational model are more efficient was directed by conducting a survey on related students.

$$(Confidence\ Interval)_{95\%} = \mu \pm 1.96\,\sigma \qquad (3)$$

μ: Average of distribution

σ: Standard deviation of distribution

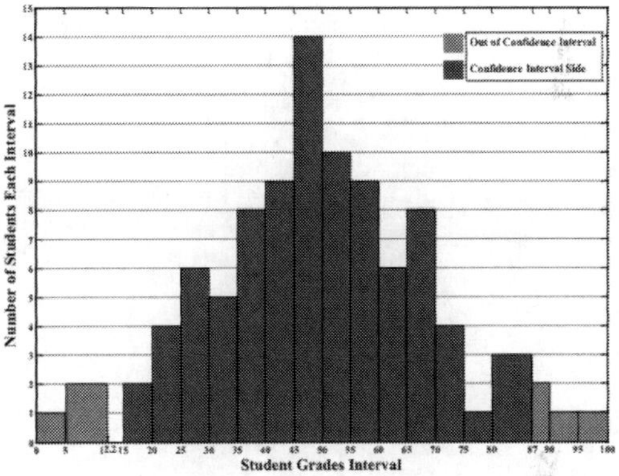

Figure 10. Semester - end grade averages of all students

Table 3. Main specifications of semester - end grades distribution.

Number of Total Students	96
Average Grades of All Class	49.6017
Standard Deviation of Grades	19.0719
Kurtosis of the Distribution	3.1301
Skewness of the Distribution	0.0648
Upper Boundary of Confidence Interval	86.9826
Lower Boundary of Confidence Interval	12.2207
Average Grades *(New Method Applied Students)*	55.2209
Average Grades *(Conventional Method Applied Students)*	43.3581

6 Results

Targets were performed in educational 3 axis CNC router system and it was observed that system worked properly. Students trained by new method of learning had the opportunity of testing their experiences on real system during semester. Thanks to the recently designed CNC machine, without any accidents or injury the students reached to the end of the semester. Working with smaller and less detrimental system has encouraged the students to be self-confident and all students trained by new methodology has had the experience of conduct an experiment on real system at least once.

It is obvious that statistically obtained results are master indications of the study success. The evaluations at the end of the semester in terms of both exam grades and verbal dominance clearly show that the students trained by new method are 27.36% more accomplished than other students. Moreover the questionnaire made on modified learning process applied students results support the previous statistical outputs. As can be seen in Figure 11, 64.44 percent of the participants declared that new methods for learning are more efficient. The 24.4 and 11.12 percent of the participants also claimed respectively that no difference between new or classical methods, and backing up classical methods of education. When the obtained statements evaluated, it is deduced the new designed system used CAM courses will be more efficient and successful.

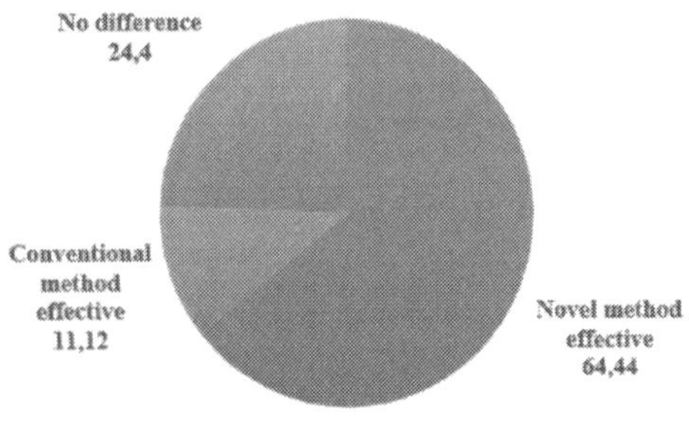

Figure 11. Questionnaire results on students

7 Conclusions

The students have difficulty to design or use real systems caused by the theoretical experiences in CAM courses. In order to overcome this problem, a novel methodology including simple and less dangerous CNC machine which is designed through this study was prepared and applied to CAM courses. A student group taken CAM course has been randomly divided into two groups. One of the groups was educated by traditional theoretical methods and the other one by recently designed method. The effects of the education methods have been tested by one verbal and two written examinations. The results show that new proposed and applied method to CAM courses is more efficient and successful in the rate of 27.36 percent. Also, the conducted survey outputs emphasize that new education methodology is widely acclaimed.

In order to improve proposed methodology and get suitable for any CAM course without any hesitation, several adding and regulations will be performed in further studies. Axis dimensions can be enlarged. Instead of step motors, more precision and powerful servo motors can be used. With the addition of divisor like structures, axis number can be increased and more varied geometrical shaped products can be processed. Open loop control can be converted to close loop control with the usage of encoders.

References

1. M. E. Aktan, Çok amaçlı 3 eksen Kartezyen robot sisteminin tasarımı ve imalatı, *Master of Science Thesis*, Marmara University, (2012)
2. H. S. Lee, S. L. Chang, Dev. of a CAD/CAE/CAM sys. for a robot man., *Journal of Materials Processing Technology*, 100-104 (2003)
3. A. Köbeloğlu, Eği. amaçlı masaüstü CNC torna tezgahı tasarımı ve prototipi, *Master of Science Thesis*, Karabük University, 1-6, (2011)
4. A.A. Tseng, S. P. Kolluri, A CNC machining system for education, *Journal of Manufacturing Systems*, 207-214, (1989)
5. M. Kutlu, "Üç eksenli masa tipi CNC freze tezgahı tasarım ve imalatı", *Master of Science Thesis*, Afyonkarahisar Kocatepe University, 39-42, (2006)
6. T. J. Kane, and O. S. Douglas, Estimating teacher impacts on student achievement: An experimental evaluation, *National Bureau of Economic Research*, (2008)
7. J. D. Angrist, V. Lavy, Using Maimonides' rule to est. the effect of class size on student achievement, *National Bureau of Economic Research*, (1997)
8. L.B. Castaneda, V. Arunachalam, D. Selvamuthu. *Introduction to Probability and Stochastic Processes with Applications*, Wiley, 151-161, (2012)
9. D. Doric, E. N. Doric, V. Jevremovic, J. Malisic, On meas. skewness and kurtosis. *Quality and Quantity*, 43(3), 481 – 493, (2009)
10. L. Miodrag, *International encyclopedia of statistical science*, Springer-Verlag, 1012 – 1013, (2013)
11. W. C. Navidi, *Statistics for engineers and scientists*, McGraw-Hill Higher Education, 344 - 354, (2008)

MATEC Web of Conferences **45**, 05003 (2016)
DOI: 10.1051/matecconf/20164505003

Mobile Phone Based Falling Detection Sensor and Computer-Aided Algorithm for Elderly People

Jong-Ha Lee , Hee-Jun Park , and Seon-Chil Kim

Department of Biomedical Engineering, School of Medicine, Keimyung University, Daegu, South Korea

Abstract. Falls are dangerous for the elderly population; therefore many fall detection systems have been developed. However, previous methods are bulky for elderly people or only use a single sensor to isolate falls from daily living activities, which makes a fall difficult to distinguish. In this paper, we present a cost-effective and easy-to-use portable fall-detection sensor and algorithm. Specifically, to detect human falls, we used a three-axis accelerator and a three-axis gyroscope in a mobile phone. We used the Fourier descriptor-based frequency analysis method to classify both normal and falling status. From the experimental results, the proposed method detects falling status with 96.14% accuracy.

1 Introduction

Falling, which means losing one's balance or tripping over something, occurs in all age groups, but especially in the elderly. In Asia, senior citizens over the age of 65 experience over half of the body injuries caused by falling [1]. In spite of extensive preventive efforts, falls remain the leading cause of accidental deaths among the elderly. Research estimates that over a third of the people aged over 65 suffer from a fall each year, and the number of falls increases progressively with age in both genders and in all racial and ethnic groups [2]. Because of vision and other problems, the falls typically occur in certain environmental conditions. Falls cause serious damage, including fractures, brain damage, and anxiety toward exercise. Quickly dealing with a fall injury such as a broken bone will reduce the time the bone fragment remains in the body. Moreover, some fractured bone regions never completely heal if they do not receive proper care within four hours [3]. If a person falls and bumps his or her head, he or she can black out, and any resulting brain damage should be cared for in a short amount of time to reduce side effects such as paralysis. Reducing the time until treatment can also help prevent a second injury and pain. For rapid treatment, the most important challenge is fall detection, which simultaneously provides communication of the fall. Therefore, fall detection is critical. A mobile phone with a three-axis accelerometer and gyroscope is a device that can easily carried by humans. In this paper, a mobile phone is used to classify human falls from normal walking motion. The three-axis accelerometer is used to measure three direction activity data. A gyroscope sensor is used to detect angular momentum; it is also useful in determining the user's rotation when he or she falls. To verify the accuracy of the proposed method, the system and algorithm were validated with a real human data set.

2 Materials and Method

2.1 Data acquisition

Subjects either walked or fell down while holding a mobile phone in their hand. A three-axis accelerometer was used to detect users' activity such as walking, falling, sitting, etc. When a person falls, one's body rotates on the body's center. A gyroscope that measures angular momentum was used to detect this rotation. The sampling frequency of each data was 20 Hz. The device used for the data acquisition was the Galaxy Note 1 (Samsung Electronics, Seoul, South Korea). The total data set of walking and falling was 212 and 202, respectively. The mean age of the subjects was 23.2; the age range was from 21 to 24.

2.2 Data analysis

For analysis and synthesis of the plane closed curves, the Fourier descriptor (FD) is used. The normalized Fourier transformed coefficients are called the Fourier descriptors [4]. The discrete Fourier transformation of the signature function is given by the following equation:

$$a_n = \frac{1}{N}\sum_{t=0}^{N-1} r(t) exp(\frac{-j2\pi nt}{N}), n = 0,1,\cdots,N-1 \quad (1)$$

The data was produced using a one-point fast Fourier transformation. The FD was extracted from the x-accelerometer, y-accelerometer, z-accelerometer, x-gyroscope, y-gyroscope, and z-gyroscope signals: a total of 96 features were considered. Figure 1 shows the x-axis accelerometer signal while walking (up) and falling (down) and Figure 2 presents the x-axis gyroscope signal while walking (up) and falling (down). From the results, we clearly see the difference between the accelerometer signal and the gyroscope signal while walking and falling in the time domain.

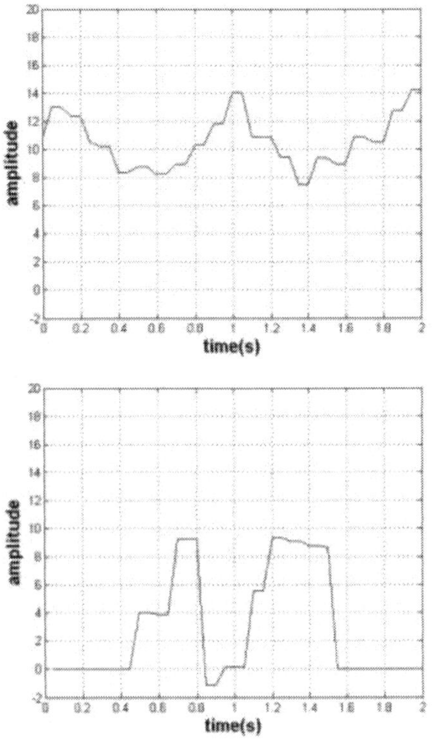

Figure 1. X-axis accelerometer signal while walking (up) and falling (down).

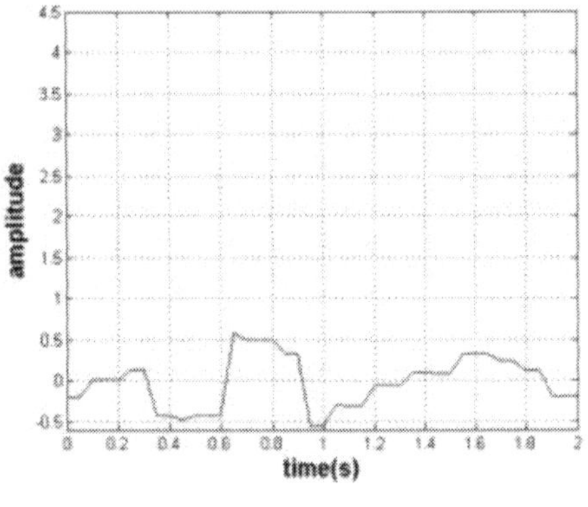

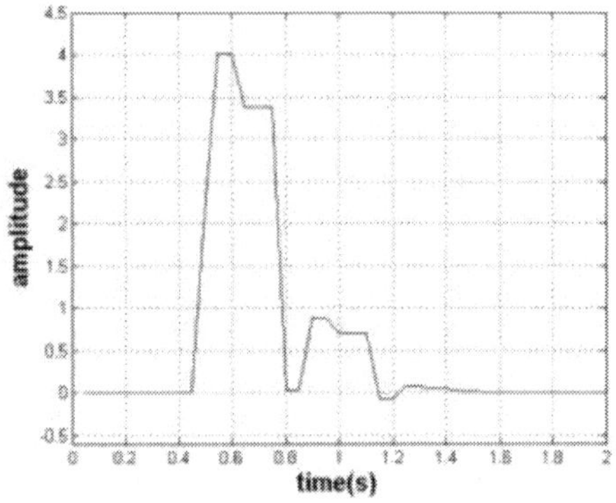

Figure 2. X-axis gyroscope signal while walking (up) and falling (down).

2.3 Fourier Descriptor

For analysis and synthesis of the plane-closed curves, the Fourier descriptor (FD) is used. FD is obtained by applying Fourier transform on a shape signature derived from shape boundary coordinates in Section III-A and III-B.

The normalized Fourier transformed coefficients are called the Fourier descriptors. The discrete Fourier transform of signature function is given by

$$a_n = \frac{1}{N}\sum_{t=0}^{N=1} r(t)\exp\left(\frac{-j2\pi nt}{N}\right), n = 0,1,\dots,N = 1 \qquad (2)$$

Since the CDF and CSCF are invariant to translation and rotation, the Fourier coefficients have to be further to make them scale invariant and start point independent shape descriptors. The relation between Fourier coefficients of original shape and transformed through scaling and change of start point is given by

$$a_n = \exp(jn\varphi) \cdot s \cdot a_n^o \qquad (3)$$

where φ is the angles incurred by the change of start po-int and and s is the scale factor. Then normalized Fourier coefficients b_n of the transformed shape is obtained by

$$b_n = \frac{a_n}{a_1} = \frac{a_n^o}{a_1^o}\exp[j(n-1)\varphi] = b_n^o\exp[j(n-1)\varphi] \qquad (4)$$

where b_n^o is the normalized Fourier coefficients of the original shape. From Eq. (4), it is clear that, if we ignore phase of the coefficients, then magnitudes $|b_n|$ and $|b_n^o|$ are the same. In other words, $|b_n|$ is invariant to translation, rotation, scaling and change of start point. The values of $|b_n|$ are the shape features that we used for the breast tumor classification. For the improvement in the classification of ultrasound images, we have combined both Fourier descriptors from CDF and CSCF through staked vector approach (SVA) technique, where both Fourier descriptors are concatenated to form a long feature vector.

2.4 Support vector machine (SVM) classifier

The support vector machine (SVM) is a widely used method for classification and regression tasks. For a two-class problem with training samples $\{x_i, y_i\}$,i = 1,2,⋯, N in d-dimensional feature space, $x_i \in R^d$, with associated targets $y_i \in \{-1,1\}$, the discriminate function of the separating hyperplane for linearly separable classes is given as follows:

$$f(x) = w \cdot \Phi(x) + b \qquad (5)$$

where w $\in R^d$ is the vector normal to the hyperplane and b ∈ R is the bias. In addition, x_i is the featured vectors extracted from the gyroscope and accelerometer sensors and y_i is the falling and normal status.

The SVM fits a hyperplane to the training samples of two classes in the feature space by minimizing the cost function, which consists of two criteria: margin maximization and error minimization [5]. SVM used real number vector. In this paper, a two-category walk or fall can be showed as (1,0) and (0,1).

2.5 Evaluation model

The model's accuracy is estimated using a cross-validation method. The data is divided into K to K-fold cross-validation [6]. The subsample left validation data. The other (K-1) data use training data. Validation data is tested using the classifier trained on the remaining training data. In addition, K is repeated, changing the subsample. Each instance of the whole training set is predicted once so the cross-validation accuracy is the percentage of data, which are correctly classified [7]. After that, the results take an average. Y_i is the real response values and X_p is the vector covariates.

$$\frac{\sum_i (Y_i - a - b_1 X_{1i} - \cdots - b_p X_{pi})^2}{n}, \ \mathrm{n} = \mathrm{K}, \mathrm{i} = 1,2, \cdots \mathrm{n}, \ \mathrm{p} = 1,2, \cdots \mathrm{n} \qquad (6)$$

TP and TN are the correctly classified occurrences of either walking or falling. On other hand, walking and falling are incorrectly classified as FP and FN. An evaluation was completed using a tenfold cross-validation.

3 Experimental Results

To calculate the classification accuracy, the K-fold cross-validation method was used and K was set as 10. The performance indices were evaluated by five performance indices, including classification accuracy (TP+TN)/(TP+TN+FP+FN), sensitivity (TP/[TP+FN]), specificity (TN/[TN+FP]), positive predictive value (TP/[TP+FP]), and negative predictive value (TN/[TN+FN]), where TP is the number of true positive findings correctly classified as positive; TN = true negative; FP = false positive; and FN = false negative [8]. From the results, we noticed that the proposed method detects 384 data (falling or walking) out of 414 original data (falling or walking). The statistical analysis shows that the resulting accuracy was 96.14%, sensitivity was 100%, specificity was 92.08%, positive predictive value was 92.98%, and negative predictive value was 100%.

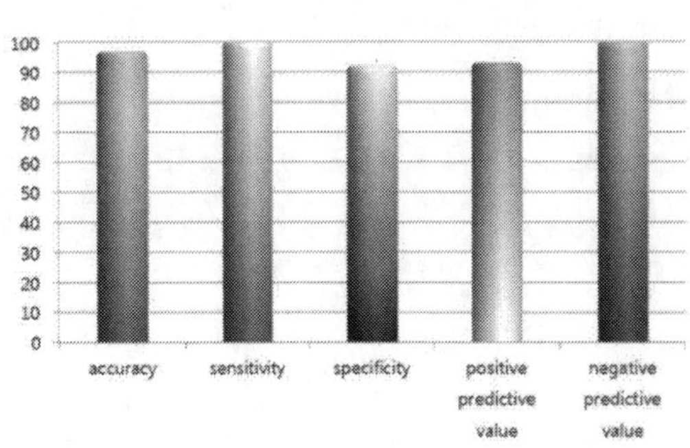

Figure 3. Statistical analysis results for proposed method verification.

4 Discussion

In this paper, we assumed that the elderly person always carries the proposed platform due to the fact that the ordinal mobile phone motion sensors are what detect the falling status. The current platform can be interrupted with events such as an incoming call, or if the motion sensors are occupied with other applications, the motion-detecting algorithm might be hold. Further, we assumed that the small and portable sensor attached to the elderly person will be carried by the same elderly person. Finally, the deliberate fall of young people is quite different from the fall of elderly people, and their walking patterns might also be quite different. In future work, the clinical trials of real data sets from both old and young people will be obtained and analyzed to optimize the fall detection algorithm for ordinary people.

5 Conclusion

In this paper, we propose a mobile phone-based falling detection sensor and algorithm. We used a three-axis accelerator and three-axis gyroscope to determine the direction and rotation of the subjects' bodies. The support vector machine classifier was used to analyze the sensor's signal. From the experimental results, the proposed method detects falling status with 96.14% accuracy, which indicates that human activity can be classified using a mobile phone. These detection methods can warn others of a user's fall.

Acknowledgements

The research partly was supported by the Ministry of Science, ICT and Future Planning Through the Development for IT·SW industrial convergence original technology (ID: R0101-15-0147) and Basic Science Research Program through the National Research Foundation of Korea (NRF) funded by the Ministry of

Education (2014R1A1A2056420) and supported by the Technological Innovation R&D Program (S2173477) funded by the Small and Medium Business Administration (SMBA, Korea).

References

1. M.C. Hornbrook, V.J. Stevens, D.J. Wingfield, J.F. Hollis, M.R. Green lick, and M.G. Ory, Preventing falls among community-dwelling older persons: Results from a randomized trial, The Gerontologist 34, 16-23 (1994).
2. Y. Li, G. Chen, Y. Shge, Y. Zhu, and Z. Cheng, Accelerometer-based fall detection sensor system for the elderly, IEEE CCIS (2012).
3. T.S. Sorensen, H. Colding, C. Schroeder, and V.T. Rosdahl, The penetration of cefazolin, erythromycin and methicillin into human bone tissue 49, 549-553 (1978).
4. C.T. Zhan and R.Z. Roskies, Fourier descriptors for plane closed curves, IEEE Trans Computer 21, 269-281 (1972).
5. J.-H. Lee and Y. K. Seong, Fourier-based shape feature extraction technique for computer-aided b-mode ultrasound diagnosis of breast tumor, 34th Annual Internal IEEE EMBS Conference (2012).
6. R.M. Simon, J. Subramanian, M.-C. Li, and S. Menezes, Using cross-validation to evaluate predictive accuracy of survival risk classifiers based on high-dimensional data, Oxford Journal 12, 203-214 (2011),.
7. C.-W. Hsu, C.C. Chang, and C.-J. Lin, A practical guide to support vector classification, https://www.cs.sfu.ca /people/Faculty/teaching/726/spring11/svmguide.pdf (2003).
8. R.M. Rangan, *Biomedical image analysis*, CRC Press (2005).

MATEC Web of Conferences 45, 05004 (2016)
DOI: 10.1051/matecconf/20164505004

Interference Cancellation in Aircraft Cockpit by Adaptive Filters

Arun.C

Assistant Professor, Department of ECE, College of Engineering Perumon, Kollam, Kerala, India

Abstract. This paper investigates on the development and implementation of adaptive noise cancellation (ANC) algorithm meant for mitigating the high level engine noise in the cockpit of an aircraft, which makes the speech signal unintelligible. Adaptive filters configured as interference canceller have the potential application in mitigating the above interference. A comparative study of Gradient based adaptive Infinite Impulse Response (IIR) algorithm and its modified version is performed using MATLAB simulator in terms of converging speed. From the simulation result the best IIR algorithm is used for implementation in Performance Optimized with Enhanced RISC PC (Power PC) 7448.

1 Introduction

High level of engine noise is present in the cockpit of an aircraft. Consider the communication of a pilot by radio from the cockpit to an air traffic controller, the engine noise would be interfered with the speech signal because the noise occupy the same frequency band as speech .The microphone into which pilot speaks also picks the engine noise, which makes the speech signal unintelligible. A conventional filter cannot be used to remove the engine noise because the intensity and frequency of noise components change with engine speed, load and pilot's head. If a second microphone is placed at a convenient location in the cockpit, it is possible to obtain the ambient noise field without pilot's speech. This noise could be filtered and removed from the transmission with the help of adaptive filters configured as noise canceller. There are some situations where a speech signal is interfered by periodic engine noise and no noise field free of pilot's speech is available. In such situations adaptive Line Enhancer (ALE) can be utilized as noise canceller. This paper focused on ALE as noise canceller

An adaptive filter is a digital filter that can adjust its coefficients to give the best match to a given desired signal [1]. There are different adaptive algorithms (eg. Least Mean Square (LMS), Normalized Least Mean Square (NLMS) etc) that can be used in time domain. Some widely used methods with Adaptive FIR filter have been explained by S. Haykins [2] & B.Widrow [3]. Fir filters are more stable than IIR filters. Adaptive IIR filters are attractive: Many fewer coefficients may be needed to achieve the desired performance in some applications [4]. However, it is more difficult to develop stable IIR algorithms, they can converge slowly. Adaptive IIR algorithms are used in some applications (such as low frequency noise cancellation) where the need for IIR-type responses is great. In some cases, the exact algorithm used by a company is a tightly guarded trade secret. However a gradient-based Adaptive IIR algorithm, with some additional features that enable it to adapt more quickly is explained by Don R.Hush [5]. A modified version of the above algorithm gives even faster convergence. So the modified algorithm can be used for implementation.

2 Adaptive signal processing

The term Adaptive can be understood by considering a system which is trying to adjust itself so as to respond to some phenomenon that is taking place in its surroundings. This is what adaptation means. Moreover there is a need to have a set of steps or certain procedure called algorithm by which this process of adaptation is carried out [6].

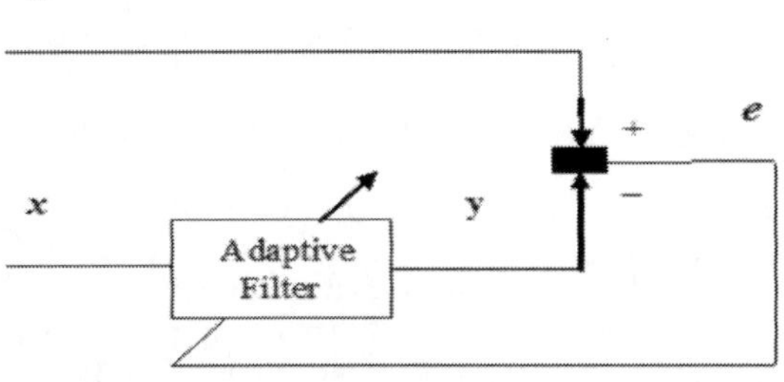

Figure 1. Basic adaptive filter structure

The adaptive filter structure shown in the Figure 1 in which the filter's output y is compared with a desired signal *d* to produce an error signal *e*, which is fed back to the adaptive filter. The error signal is given as input to the

Article available at http://www.matec-conferences.org or http://dx.doi.org/10.1051/matecconf/20164505004

adaptive algorithm, which adjusts the variable filter to satisfy some predetermined rules. In stochastic frame work (based on Wiener filter theory), the optimum coefficients of a linear filter are obtained by minimization of its mean-square error (MSE). All the adaptive algorithms take the output error of the filter, correlate that with the samples of filter input in some way, and use the result in a recursive equation to adjust the filter coefficients iteratively.

3 Adaptive noise cancellation (ANC)

The ANC method uses a "primary" input containing the corrupted signal and a "reference" input containing noise correlated in some unknown way with the primary noise. The reference input is adaptively filtered and subtracted from the primary input to obtain the signal estimate [1].

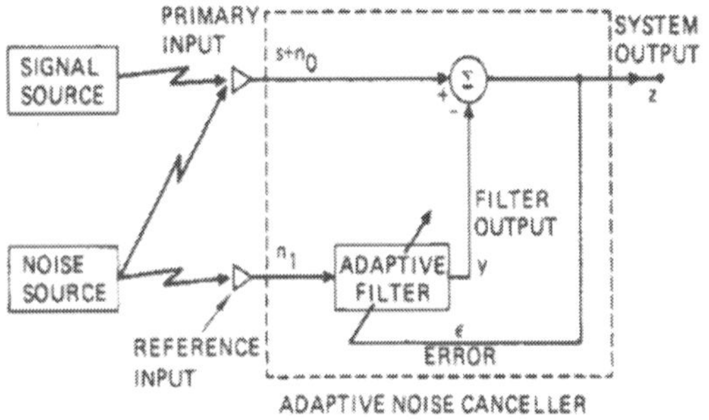

Figure 2. Adaptive noise cancelling concept

A signal s is transmitted over a channel to a sensor that also receives a noise n_0 uncorrelated with the signal. The combined signal and noise $s+n_0$ form the primary input to the canceller. A second sensor receives a noise n_1 uncorrelated with the signal but correlated in some unknown way with the noise n_0. This sensor provides the reference input to the canceller. The noise n_1 is filtered to produce an output y that is as close a replica as possible of n_0. This output is subtracted from the primary input $s+n_0$ to produce the system output $z=s+n_0-y$. Hence the output of the system contains the signal alone [1]- [7].

4 Adaptive line enhancer IIR structure

The Adaptive Line Enhancer (ALE) enhances the sinusoidal component of the reference input so that output has high signal to noise ratio SNR. From the ALE we also obtain an estimate of the sinusoidal frequency. This ALE can be utilizes as noise canceller. Several forms of the ALE are available. The most popular is the FIR ALE. It has the advantage of being inherently stable and easy to adapt. However, it often requires a large number of filter weights to provide adequate enhancement of narrow-band signals. In an effort to reduce the number of weights, several forms of adaptive recursive filters (IIR) can be used.

The purpose of Δ is to decorrelate the input noise and its delayed version present at the filter input. This causes the adaptive process to respond only to the sinusoid. As such, H (z) forms a band pass filter around the sinusoid, applying at the same time a phase shift to compensate for the delay so that the sinusoidal component of x_k is removed at the summer. At the output of the adaptive filter x'_k, the enhanced sinusoid is obtained. To simplify the analysis assume n_k is white noise so that $\Delta = 1$ is sufficient for decorrelation.

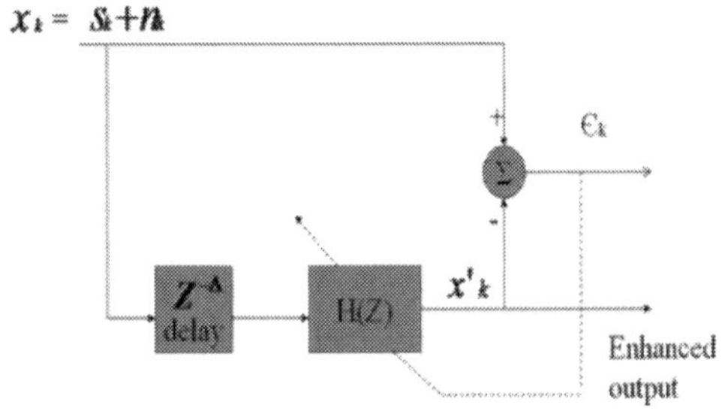

Figure 3. Adaptive line enhancer [2]

4.1 Gradient based adaptive algorithm for IIR ALE

The technique used here is a gradient-based algorithm with some additional features that enable it to adapt more quickly. The filter output is given by

$$x'_k = \{\frac{1-r^2}{1+r^2}\} w_k x_{k-1} - (1-r^2)x_{k-2} + w_k x'_{k-1} - r^2 x'_{k-2} \qquad (1)$$

Where $0 << r < 1$

$-2r < w < 2r$, r is the radius of the circle just inside the unit circle.

Also from Figure 3, the error output is defined to be

$$\varepsilon_k = x_k - x'_k \qquad (2)$$

The coefficient update for the adaptive algorithm can be expressed as

$$w_{k+1} = w_k + \rho \varepsilon_k \alpha_k \qquad (3)$$

Where ρ is a parameter which controls the rate of convergence, ε_k is the error, and α_k is the partial derivative

$$\alpha_k = \frac{\partial x'_k}{\partial w_k} \qquad (4)$$

From (1) and the definition of (3), recursive relationship for

α_k $$\alpha_k = w_k \alpha_{k-1} - r^2 \alpha_{k-2} + (\frac{1-r^2}{1+r^2}) x_{k-1} + x'_{k-1} \qquad (5)$$

The coefficient update is then defined by (2),(3) and (5).The normalization factor is incorporated by reexpressing the coefficient update in (3) as

$$w_{k+1} = w_k + \mu_1 \frac{\varepsilon_k \alpha_k}{\psi_k} \qquad (6)$$

Where ε_k and α_k are as defined above and

$$\psi_k = v\psi_{k-1} + (1-v)\alpha_k^2 \quad (7)$$

The "forgetting factor "v is in the range 0<<v<1

The performance of this algorithm can be enhanced even further by using an approximate partial derivative in place of α_k in (5). The modified derivative, denoted by α_{mk}, is obtained from (5) by suppressing the recursive terms, i.e.,

$$\alpha_{mk} = (\frac{1-r^2}{1+r^2})x_{k-1} + x'_{k-1} \quad (8)$$

4.2 Simulation results

After performing the adaptive process, at the output of the filter we get the enhanced version of the sinusoid. At the output of the summer, (error signal) sinusoidal component of input is removed. The simulation result shows the Power Spectral Density (PSD) of pure signal(Figure4), noise corrupted signal(Figure5), comparisons of PSD of error signals (Figure6)and enhanced signals(Figure7) using both the algorithms. Mean Square Error (MSE) of two algorithms is also shown (Figure8). From the PSD of error signal it is clear that a very large reduction of power occur at the output of the summer. The modified algorithm provides more reduction in power.

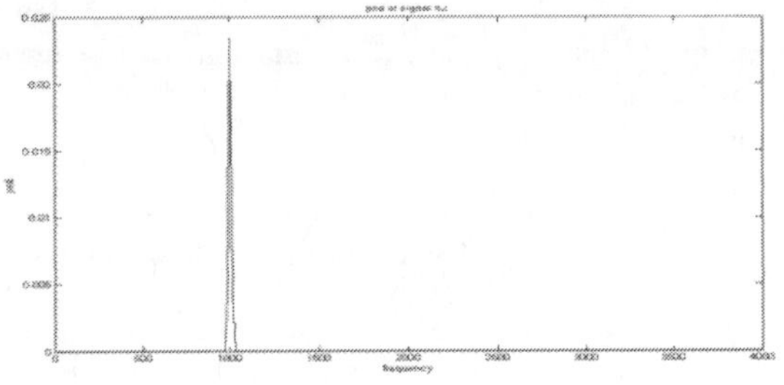

Figure 4. PSD of pure signal

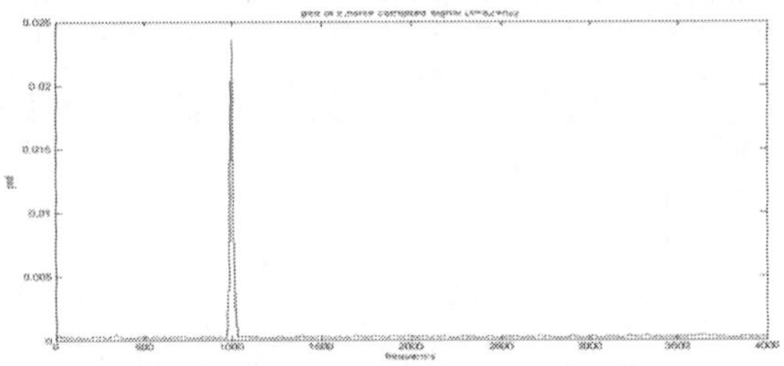

Figure 5. PSD of noise corrupted signal

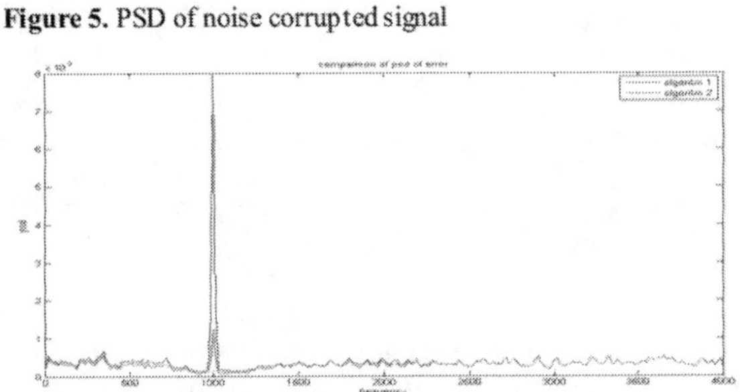

Figure 6. Comparison of PSD of error

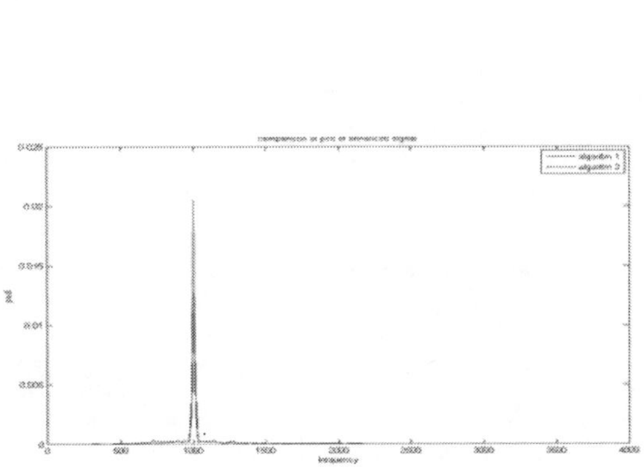

Figure 7. Comparison of PSD of enhanced signal

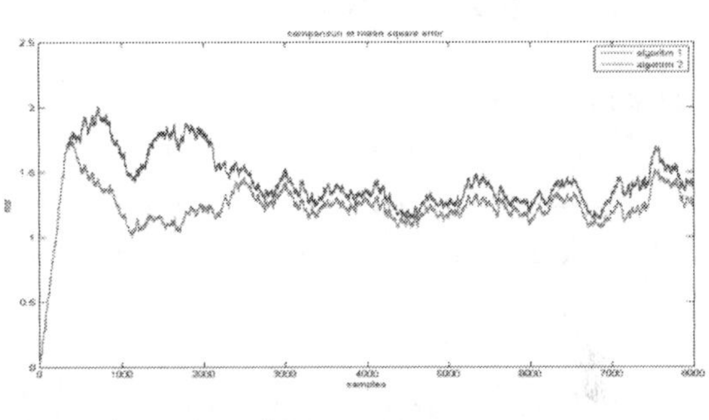

Figure 8. Comparison of mean square error

The Figure9 shows the convergence time of both the algorithms. Modified algorithm converges faster

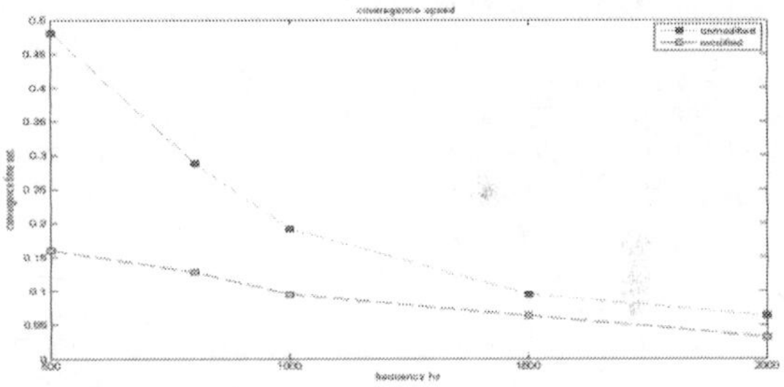

Figure 9. Comparison of convergence time

5 Adaptive IIR noise cancellation

ALE can be utilized as noise canceller with experimental data. ALE is used as the basic structure for noise cancellation. Here the input signal X contains the speech signal and engine noise. The purpose of Δ is to decorrelate the input speech and its delayed version present at the filter input. This causes the adaptive process to respond only to the noise As such, H (z) forms a band pass filter around the noise, applying at the same time a phase shift to compensate for the delay so that the noise component of **X** is removed at the summer. By taking the autocorrelation of speech signal one can find out suitable value for Δ. So at the output of the summer (error signal) contains the speech alone. i.e., the engine noise is adaptively cancelled out. Let the signal to be interfered with noise such that the SNR is controllable and verify the performance of IIR structure in highly noise environment. The system is provided with the permission to input a particular SNR (user's choice), a noise with corresponding amplitude is generated and added to signal. This noise corrupted signal will be processed.

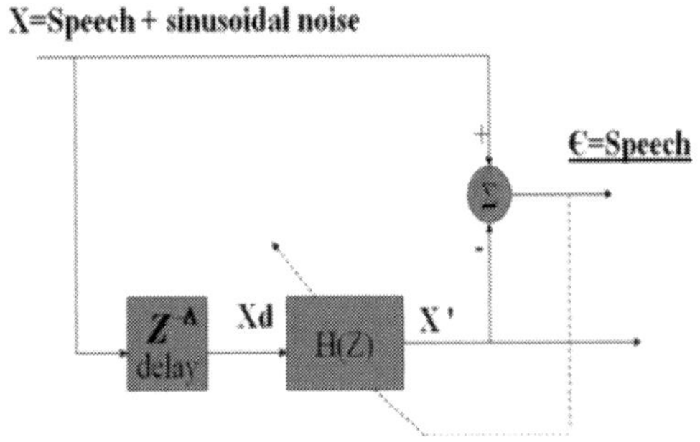

Figure 10. IIR noise canceller

Table 1. SNR and power levels

SNRin dB	Power level between signal and noise
0 dB	Signal power = noise power
-3 dB	Signal power =(1/2) noise power
-5 dB	Signal power << noise power

5.1 Simulation results

Verify the performance of algorithm in highly noise environment. Produce the simulation result with SNR = -5 dB, the speech power is too low compared to engine noise.

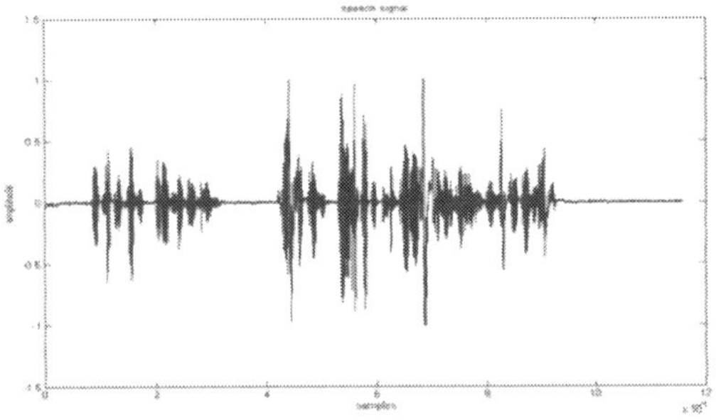

Figure 11. Speech signal

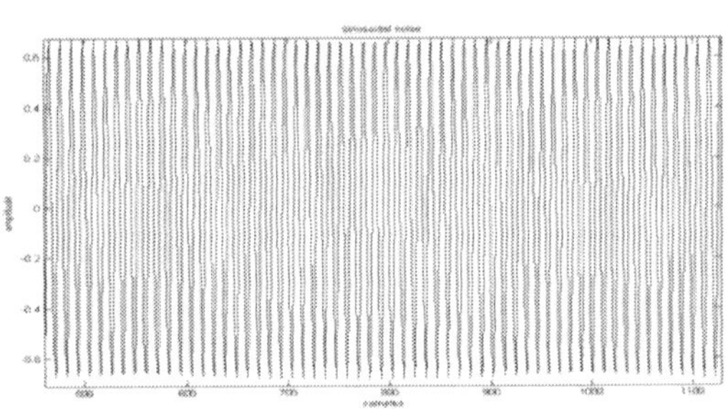

Figure 12. Engine noise

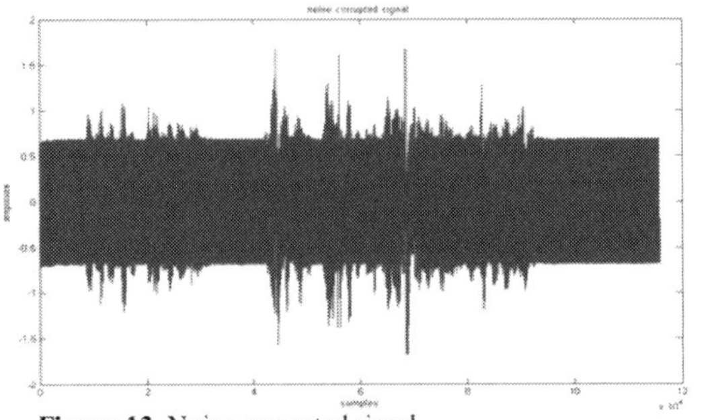

Figure 13. Noise corrupted signal

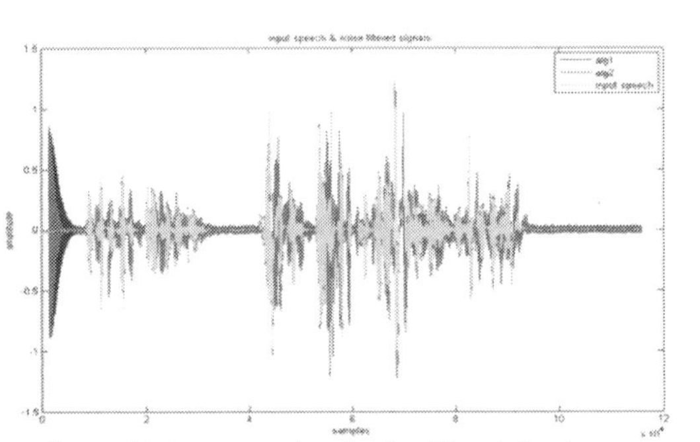

Figure 14. Input speech and noise filtered signals

The Figure11 shows the pure speech and the Figure12 shows the engine noise. Figure13 shows the noise corrupted signal. As in ALE here also two adaptive algorithms for noise cancellation were performed, a gradient based algorithm and a modified version of the same.

The Figure14 shows the comparison among the pure speech and noise filtered signals obtained from two algorithms. From the simulated result it is clear that even if the signal is interfered by a noise whose power is very high compared to signal power, adaptive IIR filter is able to pick out the noise. The blue colored signal is obtained from normal algorithm. The red colored signal is produced by modified algorithm. From the figure below it is clear that the modified algorithm is converging faster than the normal one. Hence modified version of the gradient based algorithm is used for implementation on PowerPC.

6 Implementation

Interference cancellation using adaptive IIR algorithm is implemented in PowerPC 7448 SBC(Single Board Computer).

6.1 Implementation process flow

The input signal contains the speech signal interfered by engine noise. The purpose of delay is to decorrelate the input speech and its delayed version present at the filter input. This causes the adaptive process to respond only to the noise. So at the output of the summer (error signal) contains the speech alone. i.e., the engine noise is adaptively cancelled out.

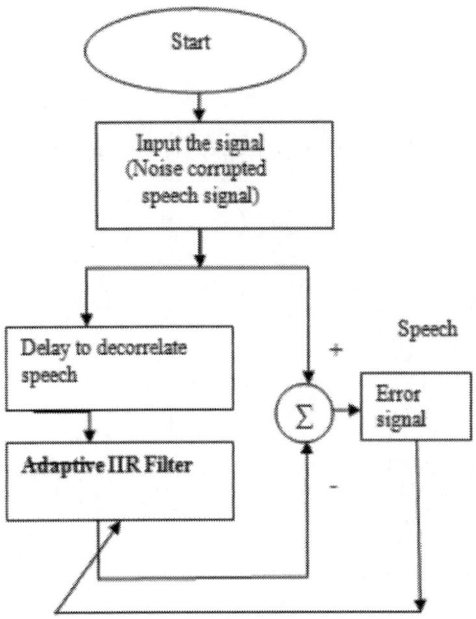

Figure 15. Implementation flow chart

6.2 Results and discussion

The interference cancellation using IIR filter has been implemented successfully on PowerPC 7448 with Modified Gradient Based Adaptive algorithm and verified the output. Here the signal applied to the input is speech corrupted by engine noise. After adaptive filtering process speech alone is produced at the output.

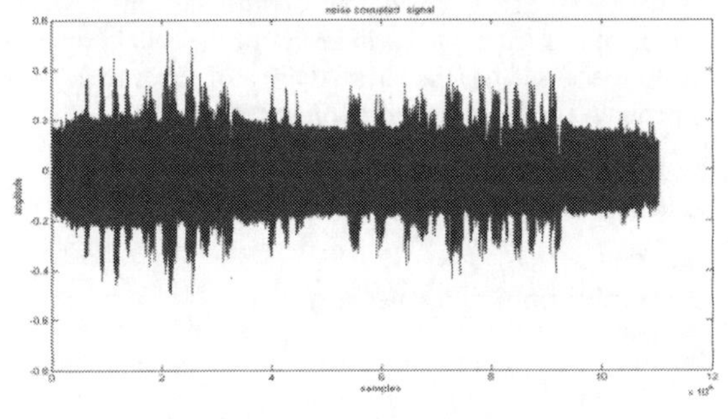

Figure 16. Noise corrupted speech signal

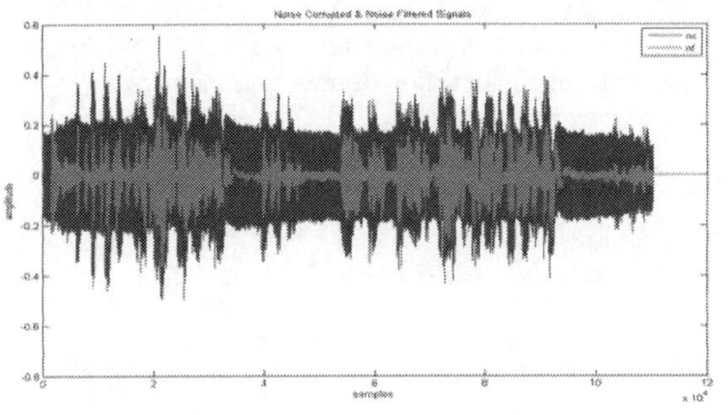

Figure 17. Noise corrupted and noise filtered speech signals

Figure 17 show the noise corrupted signal and noise filtered signal together. The Adaptive filter filtered the engine noise and produced the speech alone at the output.

7 Conclusion and future scope

A detailed survey on ANC is carried out and observed that many fewer coefficients may be needed to achieve the desired performance for IIR filter when compared to FIR. From simulation under different noise environment, it is concluded that the modified gradient based algorithm is more suitable for Adaptive IIR noise cancellation. Even if the signal is interfered by a noise whose power is very high compared to signal power, adaptive IIR filter is able to pick out the noise. i.e., even in the highly noise environment adaptive IIR filter is powerful.

Future work includes the implementation of the pipelined version of the filter for improved speed, and to optimize the programmable processing element at circuit level for efficient ASIC implementation of the reconfigurable fabric.

References

1. B. Widrow *et al.*, "Adaptive noise cancelling: Principles and applications," *Proc. IEEE*, vol. **63**, *pp.* 1692-1716, Dec. (1975).
2. S. Haykin: Adaptive Filter Theory, Englewood Cliffs, N.J.: Prentice-Hall, Inc., 4th Edition (2001).
3. B. Widrow and S. Stearns: Adaptive Signal Processing. Prentice Hall, p. 301(1985).
4. Bijan Sayyarrodsari et al.,"An Estimation-Based Approach to the design of Adaptive IIR Filters", Proceedings of the American Control Conference Philadelphia, Pennsylvania l June (1998).
5. Don R. Hush *et al.*, "An Adaptive IIR Structure for sinusoidal Enhancement, Frequency Estimation and Detection", IEEE Trans. Acoust., Speech, Signal Processing, vol.ASSP-**34**, No.**6**, pp.1380-1390, Dec(1986).
6. J.J. Shynk. Adaptive IIR filtering. IEEE ASSP Magazine, Vol. **6**, No. **2**, pp. 4-21, April (1989).
7. M. Harteneck, R.W Stewart, "Adaptive Digital Signal Processing Java Teaching Too*l"* Submitted to IEEE Transactions on Education - Special CDROM Issue, November (1999).
8. Farhang-Boroujeny,B. Adaptive filters: theory and applications, John Wiley& sons Ltd,(1998).
9. Freescale semiconductor, "MPC 7448 RISC Microprocessor Hardware Specifications" a 60 page PDF manual, (2005).
10. O.L Frost III ,"An algorithm for linearly constrained adaptive array processing"Proc.IEEE,vol.**60**.pp.926-935 Aug(1972).
11. Adaptive Filter Theory by Simen Haykin: 3rd edition,Pearson Education Asia.LPE

MATEC Web of Conferences **45**, 05005 (2016)
DOI: 10.1051/matecconf/20164505005

Performance prediction of a horizontal axis wind turbine using BEM and CFD methods

Younes El khchine and Mohammed Sriti

Sciences engineering laboratory, University of Sidi Mohamed Ben Abdellah, Polydisciplinary faculty of Taza, B.P.1223, Morocco

Abstract. In this study, a combination of CFD (Computational Fluid Dynamics) and BEM (Blade Element Momentum Method) methods are used to simulate the flow field around a wind turbine rotor with horizontal axis. The main objectives are to predict the aerodynamic performances such as forces and torque imposed on the rotor blades which are essential to its structure or design. This approach requires much less computing time and memory than three-dimensional simulation flow around the wind turbine rotor with simple CFD method. The flow is assumed unsteady, incompressible and fully turbulent. This work consists of two parts:

-The Calculation of aerodynamic coefficients by the CFD method, using Ansys / Fluent software.

-The Simulation of 3D flow field through the rotor of the wind turbine using the BEM method.

The obtained results are in good agreement with the experimental measurements.

1 Introduction

The study of aerodynamic performance of horizontal axis wind turbines has attracted many researchers in recent years because of potential applications in the production of electricity based on wind kinetic energy that falls into the category of clean renewable energy to the environment.

This study is based on CFD simulation (Computation Fluid Dynamic) using the software ANSYS/Fluent [1] that solves the equations of fluid mechanics in 2D and 3D flows, incorporating highly advanced turbulence models. This software uses the finite element method, appropriate digital techniques and features a mesh adapted to very complex configurations.

Many research studies have been conducted to determine the fine structure of the flow through the blades of horizontal axis wind turbines. As examples, we cite some work done in recent years, such as that of Chandrala et al. ([2], 2012), which deals with the aerodynamic analysis of a wind rotor of the NACA 4402 type by comparing the calculation results with CFD measurements for different angles of attack, the Working Yilei He et al. ([3] 2014) which focuses on optimizing the S809 airfoil, using a multi-objective algorithm MOGA, and the work of Bekhti et al. ([4], 2010) worked on the simulation of the airflow around a NREL S809 airfoil to predict the dynamic stall phenomenon and understand the physical behavior of the flow around a rotor wind. This work is based on solving the Navier-Stokes equations averaged RANS using advanced turbulence model.

In the present work, our study was focused on the numerical simulation using Ansys/Fluent software to understand and analyze the fine structure of the flow around wind turbine S809, and also determine the characteristics for aerodynamic optimization and the improvement of the net power. In addition, CFD simulation is used to calculate the aerodynamic forces and to identify the necessary correlations that will be incorporated into a global model for the calculation of aerodynamic performance equivalent wind turbines. The choice of a three-bladed wind rotor of the S809 type was made because of the availability of geometric and functional data provide a benchmark study.

2. CFD method

2.1 mathematical modelling

In this study, we are interested in the numerical simulation of the flow around airfoil wind turbine S809 exposed to the wind speed at a different angle of incidence α (Figure 1). The flow is assumed axisymmetric two-dimensional, unsteady and fully turbulent. The flow is therefore governed by the equations of conservation of mass and momentum, which are written respectively:

$$\frac{\partial \rho}{\partial t} + \vec{\nabla}(\rho \vec{V}) = 0 \qquad (1)$$

$$\frac{\partial(\rho\vec{V})}{\partial t} + \vec{\nabla}(\rho\vec{V}.\vec{V}) = -\vec{\nabla}p + \vec{\nabla}\bar{\bar{\tau}} + \rho\vec{f} \quad (2)$$

Where t represents the time (in s), ρ is the density of the fluid, V is the free stream velocity of a fluid particle, p is the static pressure, τ is the viscous stress tensor, f is the resultant of the inertial forces acting on a fluid particle. The equation of state for a perfect gas is also added to express the density function of the static pressure.

The SST k-ω turbulence model is used in turbulent zones due to good prediction in separated flow simulation. The model uses the standard k-ω model near the wall, but switches to a k-ε model away from the wall.

For the flow analysis of the wind turbine blade mesh is created in the Ansys/workbench software. Figure 2 shows the meshing drawing of the airfoil below.

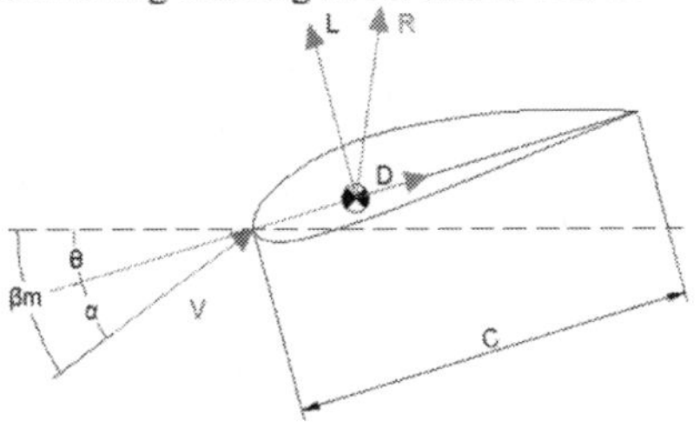

Figure 1. Geometric model

2.2. Mesh topology and boundary conditions

The mesh quality and domain size of the 2D CFD model do affect the computation accuracy and convergence time. A good mesh should be large enough to avoid boundary effects. There are many mesh topologies for airfoil CFD modeling. The most popular mesh topology is the C mesh which is designed to have a C-type topology around the airfoil as shown in figure 2.The mesh used here is the C-topology structured mesh contains 181200 quadric-elements. A large number of grids around the airfoil surface are used to capture the pressure gradient accurately at the boundary layer. In the far-field, the mesh resolution can become progressively coarser since the flow gradients approach zero. The dimensions of the domain of the following calculations: A height of 12 chord length, a length of 18 including 6 upstream of the trailing edge 12 and downstream.

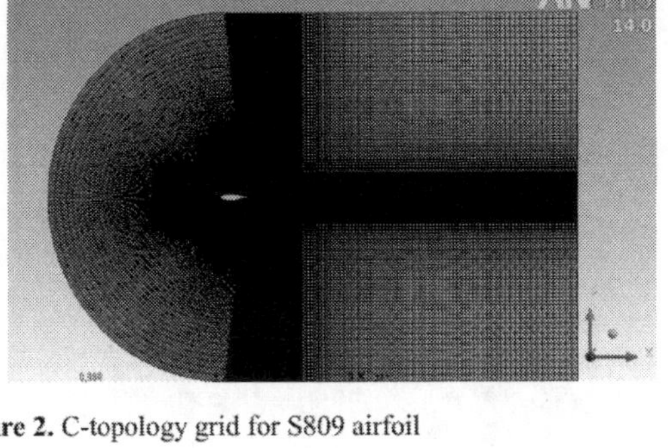

Figure 2. C-topology grid for S809 airfoil

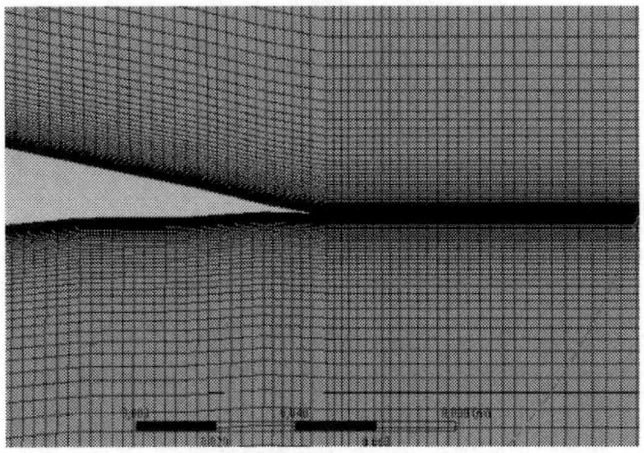

Figure 3. Grid mesh around trailing edge

For the boundary conditions, at the inlet, the velocity components are calculated for each case of angle of attack as follows. The x-component and y component of velocity is calculated respectively by $V_x = V\cos\alpha$ and $V_y = V\sin\alpha$, where α is the angle of attack in degrees. The velocity inlet (V) is 14.6 m/s corresponds to a Reynolds number of one million. The density of the air is ρ=1.225 kg.m^{-3} and the dynamic viscosity is μ=1.7894 10^{-5} kg/m.s, the fluid is assumed incompressible unsteady. Outlet pressure is considered as atmospheric pressure P_{atm}=101325 Pa. No slip boundary condition at the airfoil wall has been considered. The turbulent intensity is 2.84% and operating conditions for temperature and pressure are assumed respectively 288.16 K and 101325 Pa. FLUENT it set little parameters which are:

Table 1: Fluent parameters and boundary conditions

Turbulence model	k-ω SST
Fluid	Air, incompressible, unsteady
Density (ρ)	1.225 kg/m^3
Dynamic viscosity (μ)	1.7894*10^{-5} kg/m.s
Turbulent intensity	2.84 %
Inlet velocity (V)	14.6 m/s
Atmospheric pressure (P_{atm})	101325 Pa
Chord length (c)	1m
Solved	Pressure based unsteady
CFD algorithm	Coupled
Scheme of discretization	Pressure (Second order upwind) Momentum (Second order upwind)

2.3. Results analysis

In first part, we calculated the coefficients of lift and drag of S809 airfoil wind turbine. The calculation results are presented in Figures 4 and 5. In this simulation, the incidence angle α of the flow varies from 0 to 20°.

In Figure 4, we see that the curve is a straight line until α=9°, then undergoes a shift beyond this value. This inflection point is expressed by a detachment of the flow from the upper surface at the trailing edge airfoil.

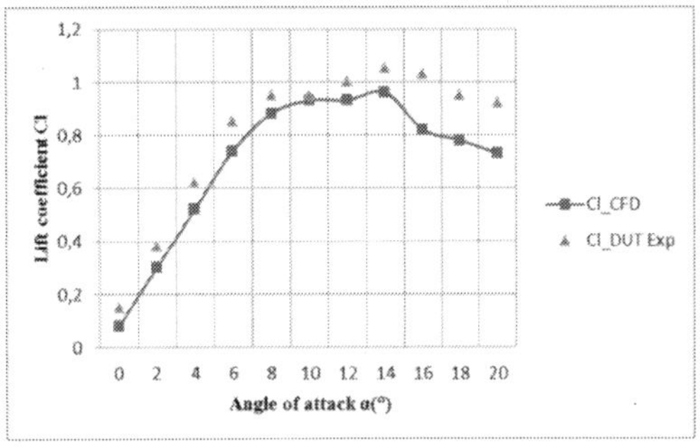

Figure 4. Lift coefficient versus the attack angle

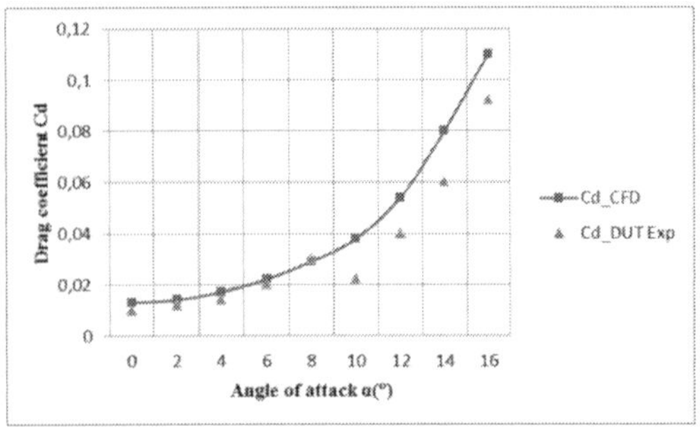

Figure 5. Drag coefficient as a function of attack angle

Figures 4 and 5 show the variations of the lift and drag coefficients as a function of incidence angle for S809 airfoil, these results are obtained for optimal mesh. We see that lift coefficient increases with incidence up to 8°. In this angle range, the curve is similar to a straight, after the lift increasing with incidence slowly, and then it undergoes a sharp fall: the static stall.

Figure 5 show that the drag coefficient sudden increase in the stall angle. These results correlated well with the experimental data.

3. BEM method

BEM theory is a compilation of both momentum theory and blade element theory. Momentum theory, which is for the determination of forces acting on the rotor to produce the motion of the fluid. This theory has no connection to the geometry of the blade, thus is not able to provide optimal blade parameters. Blade element theory determines the forces on the blade as a result of the motion of the fluid in terms of the blade geometry. By combining the two theories, BEM theory, also known as strip theory, relates rotor performance to rotor geometry.

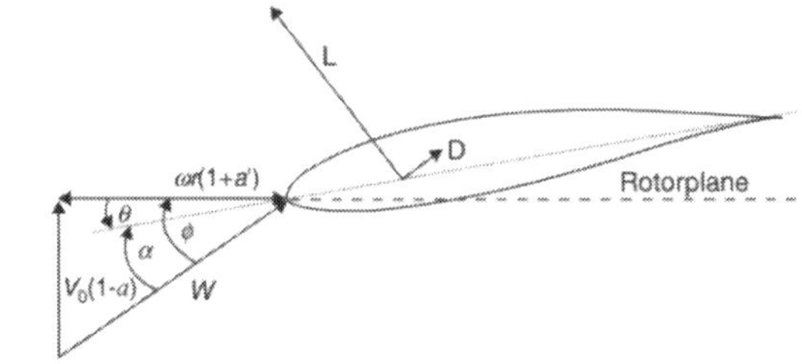

Figure 6: Blade section for a given radius r

The mathematical code in this work for wind turbine design is based on BEM method. By applying axial and angular momentum conservation, the incremental axial force dT and angular torque dM on blade sectors can be obtained:

$$dT = \rho V_0^2 4a(1-a)\pi r dr \tag{3}$$

$$dM = \rho\omega V_0 4a'(1-a)\pi r^3 dr \tag{4}$$

Where ρ is the air density, V is the mean air flow velocity, r is the local radius, ω is the blade angular rotational speed, a and a' are the axial and the tangential induction factors, respectively.

According to the blade element theory and by referring to Figure 7:

$$dT = \rho\sigma V_0^2 \frac{(1-a)^2}{\sin^2\phi}(C_l \cos\phi + C_d \sin\phi)\pi r dr \tag{5}$$

$$dM = \rho\sigma V^2 \frac{(1-a)^2}{\sin^2\phi}(C_l \sin\phi - C_d \cos\phi)\pi r^2 dr \tag{6}$$

Where C_l and C_d are the lift and drag coefficients respectively, c is the local chord length. $\sigma = {}^{Bc}/_{2\pi r}$, is the local solidity, here B is the number of blades and ϕ is the angle of relative wind.

Thus, the total torque (M) is the summation of dM for all the blades elements, and the wind turbine power is given by P=M.ω. Eventually, the wind turbine power prediction lies in solving the axial induction factor a and the tangential induction factor a'.

By combining Equations 3, 4, 5, and 6, one can solve the induction factors a and a' as given below:

$$a = \frac{\sigma(C_l \cos\phi + C_d \sin\phi)}{\sigma(C_l \cos\phi + C_d \sin\phi) + 4\sin^2\phi} \tag{7}$$

$$a' = \frac{V_0(C_l \sin\phi - C_d \cos\phi)}{\omega r(C_l \cos\phi + C_d \sin\phi) + 8\pi r \sin^2\phi} \tag{8}$$

The axial thrust coefficient C_T is expressed as:

$$C_T = 4a(1-a) \tag{9}$$

By substituting the two induction factors into Equation 4, the wind turbine performance can be assessed.

3.1 Corrections for BEM method

The BEM method was developed by Glauert [5], it was not to give reliable results, and many authors have introduced corrections on the method.

These corrections relate to tip and Losses factor, to correct the detached from tip and hub effects of tourbillions to model the velocity distribution induced at the rotor.

An approximate method of estimating the effect of tip losses has been given by Prandtl, given by equation (10):

$$F = \frac{\pi}{2}\cos^{-1}(e^{-f}) \tag{10}$$

Where $f = \dfrac{B(R-r)}{2r\sin\phi}$ is tip loss factor, R is the radius of blade.

The application of this equation for the losses at the blade tips is to provide an approximate correction to the system of equations for predicting rotor performance and blade design, we find expressions of thrust force and torque:

$$dT = F\rho V_0^{\,2}\, 4a(1-a)\pi r dr \tag{11}$$

$$dM = F\rho\omega V_0 4a'(1-a)\pi r^3 dr \tag{12}$$

The new formulas of axial and tangential induction factors and thrust coefficient becomes respectively:

$$a = \frac{\sigma(C_l\cos\phi + C_d\sin\phi)}{\sigma(C_l\cos\phi + C_d\sin\phi) + 4F\sin^2\phi} \tag{13}$$

$$a' = \frac{V_0(C_l\sin\phi - C_d\cos\phi)}{\omega r(C_l\cos\phi + C_d\sin\phi) + 8F\pi r\sin^2\phi} \tag{14}$$

$$\begin{cases} C_T = 4aF(1-a) & \text{if } a \le a_c \\ C_T = 4F\left[a_c^2 + (1-2a_c)a\right] & \text{if } a > a_c \end{cases}$$

a_c is approximately 0.2.

3.2. Analysis results

BEM method is programming in MATLAB, the results of the optimal distribution of the chord length and the twist angle are summarized in the graphs below. Figure 7 shows the optimum distribution of chord length airfoil S809 as a function of the local radius.

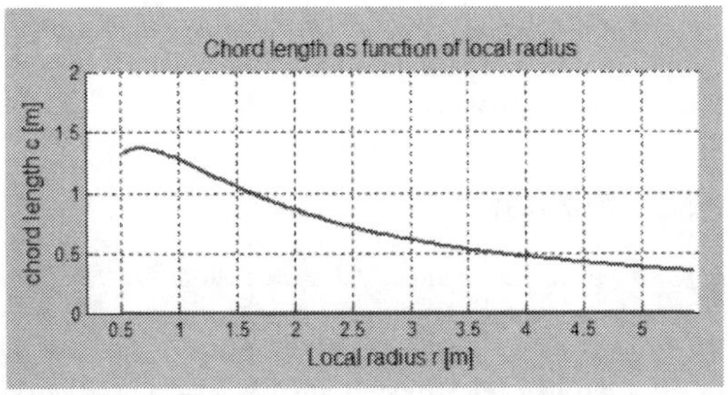

Figure 7. Chord length distribution versus local radius

The chord line decrease with increase of local radius r, and reach to maximum value at r=0.65 m, then decrease steeply.

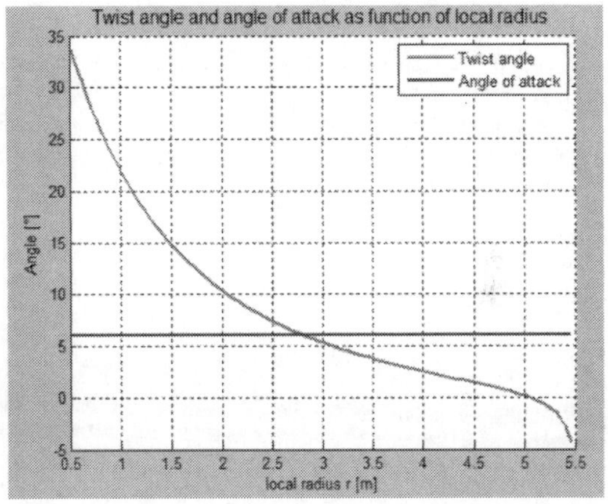

Figure 8. Twist angle distributions and optimal attack angle versus a local radius

Figure 8 represent the relationship between twist angle θ and local radius r, the curve shows the twist angle decrease toward blade's tip and there are large twists in regions of close to blade's root.

In figure 7 and 8, the distribution calculated for the two parameters: the chord length and twist angle are not linear. In practice, these distributions are not respected because of manufacturing difficulties. The compromise would be to use linear distribution close to those calculated without sacrificing too much energy performance.

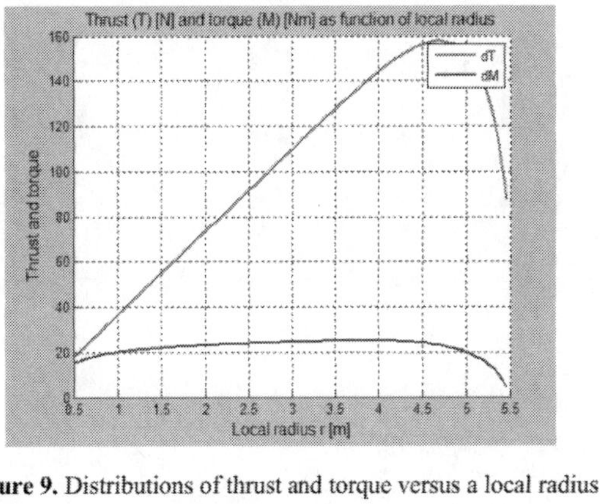

Figure 9. Distributions of thrust and torque versus a local radius

In figure 9, we see that the distribution of the thrust force is an increasing linear function until r = 5.7, but the torque remains constant along the blade.

4. Conclusion

The S809 airfoil with different attack angles throughout the length is analyzed to find the maximum L/D ratio. This is done at the angle ranging from 0° to 20° for the velocity 14.6 m/s. The maximum C_L/C_D ratio is achieved at 6° of the attack angle. The airfoil optimal design, for a given site, must be based on the wind speed corresponding to the maximum average power of wind.

The BEM theory is very successful in horizontal axis wind turbine blade design. The modification factor and model were also combined into the BEM theory to predict the blade performance and there are good results of torque and thrust in each section between the improved BEM theory and CFD simulation.

The most difficult issues for the BEM theory are the mathematical representation of the correct lift and drag coefficient values, and the evaluation of axial and tangential induction factors.

References

1. Ansys Fluent Release 14
2. M. Chandrala, A. Choubey and B. Gupta, Aerodynamic Analysis Of Horizontal Axis Wind Turbine Blade, IJERA, V. 2, **6**, pp. 1244-1248, (2012)
3. Yilei He and Ramesh K. Agarwal, Shape Optimization of NREL S809 Airfoil for Wind Turbine Blades Using a Multiobjective Genetic Algorithm, IJAE, Vol 2014, ID. 864210, 13 pages, (2014)
4. A. Bekhti and O. Guerri, Simulation de l'écoulement autour d'un profil de pale d'éolienne, Revue des Energies Renouvelables SMEE, pp. 49-56, (2010)
5. Glauert, H: Airplane propellers. In: Durand, WF (ed.) Aerodynamic Theory, 4th EDN, pp. 169–360. Springer, Berlin (1935)

AUTHOR INDEX